Two Manuals Combined into One Book

AF886469

Cavalry Troop

And

Military First Aid

Department of Defense

Pentagon
Publishing

***FM 17-97**

Field Manual
No 17-97

HEADQUARTERS
DEPARTMENT OF THE ARMY
Washington, DC 3 October 1995

CAVALRY TROOP

TABLE OF CONTENTS

DISTRIBUTION RESTRICTION: Approved for public release; distribution is unlimited.

*This publication supersedes FM 17-97, September 1988.

Preface

This publication serves as a doctrinal guide for commanders and subordinate leaders of both light (HMMWV equipped) and heavy (M1 and M3 equipped) cavalry troops. Heavy cavalry troops are found in Armored Cavalry Regiments (ACR), in Armor/Mechanized Divisions, and in Separate Armor/Mechanized Brigades. Light cavalry troops are found in Light Armored Cavalry Regiments (LACR), in Light Infantry Divisions, and in Separate Light Infantry Brigades.

Although FM 17-97 does not specifically address the TOE of the Light Division Cavalry Troop or the troops of a Separate Heavy or Light Brigade, the tactics, techniques, and procedures outlined in this publication still apply.

This publication lays out the organization, command and control, tactical employment, and service support of the cavalry troop in combat. It establishes the responsibilities and duties of key personnel in the troop during combat.

FM 17-97 is one of three publications in the cavalry troop family. It provides troop leaders with a doctrinal reference to link training and fighting. ARTEP 17-487-30-MTP, *Mission Training Plan for the Regimental Armored Cavalry Troop*, outlines how to train the troop. FKSM 17-97-3, *Cavalry Troop Common SOP*, describes troop operating procedures. This family of publications supports FM 17-15, *Tank Platoon*; FM 7-91, *Tactical Employment of Antiarmor Platoons, Companies, and Battalions*; and FM 17-98, *Scout Platoon*.

The information in this manual is based on the objective table of organization and equipment (TOE) for the ACR and LACR troops. Differences between actual unit organizations and equipment and those described in this manual are caused by differences in the modified table of equipment and organization (MTOE).

The proponent for this publication is HQ TRADOC. Submit changes for improving this publication on DA Form 2028 (Recommended Changes to Publications and Blank Forms), and forward it to Commandant, USAARMS, ATTN: ATSB-SBA-F, Fort Knox, KY 40121-5211.

Although this manual does not implement any particular international agreement, the material presented herein is in accordance with related international agreements. A list of related international agreements and other references can be found in the References section. The paragraph in Chapter 2, Section I, entitled Issue a Warning Order is in compliance with STANAG 2014. Chapter 7, Section III, Relief in Place, and Section IV, Battle Handover and Passage of Lines, are in compliance with STANAG 2082. Tables 7-2 and 7-4 are in compliance with STANAG 2083.

Unless this publication states otherwise, masculine nouns and pronouns do not refer exclusively to men.

Article I. Chapter 1

Section 1.01 Introduction

The role of the cavalry troop in Army operations remains unchanged from the traditional role of cavalry throughout the history of warfare. The troop is organized, equipped, and trained to protect and preserve the fighting ability of other combined arms forces. While its primary missions are reconnaissance and security, the cavalry troop may be called upon to execute attack, defend, and delay missions as part of squadron and regimental missions. *The troop accomplishes its missions by communicating, moving, and shooting in that order.*

The purpose of this chapter is—

- To depict organizations of the cavalry troop in—

  Heavy Troop:

  - The Armored Cavalry Regiment (ACR) (M1 and M3 equipped).
  - The Heavy Division Cavalry Squadron (M1 and M3 equipped).

  Light Troop:

  - The Light Armored Cavalry Regiment (LACR) (HMMWV equipped).

Contents

- To outline missions each troop performs and to highlight each troop's capabilities and limitations.

- To establish responsibilities of key personnel in combat.

(a) Section I. Organizations

(I) THE HEAVY CAVALRY TROOP

The heavy cavalry troop consists of 6 officers and 126 enlisted soldiers. The troop is organized into a headquarters section, two scout platoons, two tank platoons, a mortar section, and a maintenance section (see Figure 1-1).

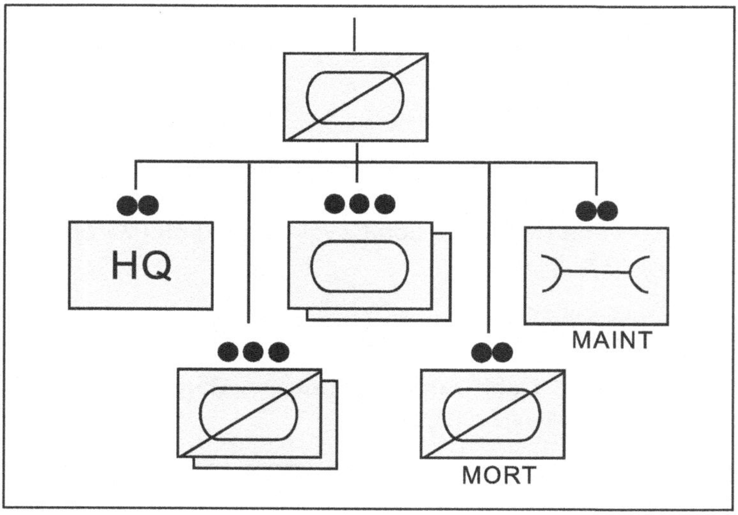

a) Figure 1-1. Heavy cavalry troop organization.

2) The Heavy Troop Scout Platoon

The scout platoon is organized and equipped to conduct reconnaissance and screening in support of its parent troop. However, when the parent unit is performing missions within an economy-of-force role, the scout platoon may conduct offensive, defensive, and retrograde operations in support of the troop mission. The platoon consists of 1 officer and 29 enlisted soldiers, and is equipped with six M3 cavalry fighting vehicles (CFV) organized into three scout sections (see Figure 1-2).

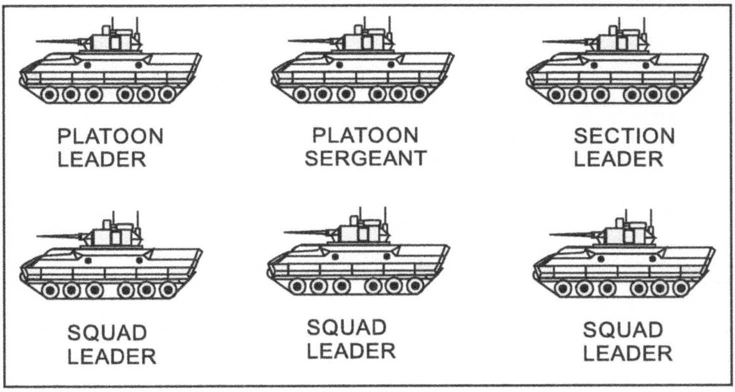

3)

a) Figure 1-2. Scout platoon organization.

4)

5) The Heavy Troop Tank Platoon

The tank platoon is organized and equipped to perform its three primary missions—attack, defend, and move. The platoon consists of 1 officer and 15 enlisted soldiers manning four M1A1 Abrams main battle tanks organized into two sections (see Figure 1-3).

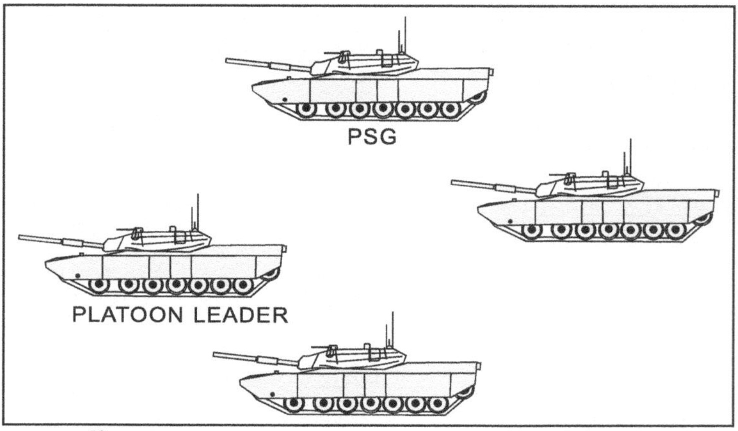

6)

a) Figure 1-3. Tank platoon organization.

7) The Heavy Troop Mortar Section

The mortar section is organized and equipped to provide immediate indirect fires in support of troop operations. Such supporting fires are usually suppression, screening, obscuration, or illumination. The section consists of nine enlisted soldiers. It is equipped with two 107-mm mortars mounted in two self-propelled mortar carriers (see Figure 1-4).

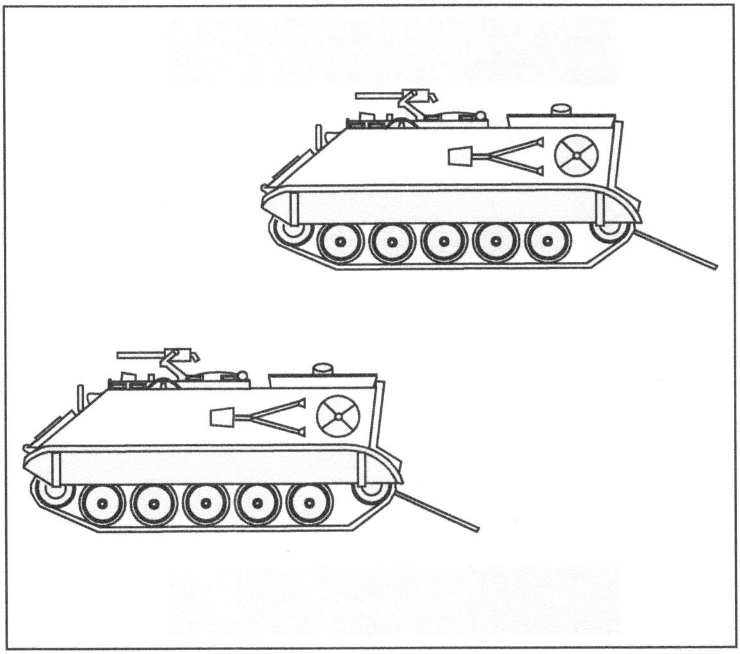

Figure 1-4. Mortar section organization.

8) The Heavy Troop Maintenance Section

This section is organized and equipped to diagnose and repair most equipment faults at troop level. It has the capability to recover all troop vehicles and to maintain the troop's equipment records. The section consists of 18 enlisted soldiers equipped with one armored personnel carrier (APC), one heavy recovery vehicle, one utility truck with cargo trailer, and two cargo trucks with cargo trailers (see Figure 1-5).

a) Figure 1-5. Maintenance section organization.

9) The Heavy Troop Headquarters Section

a)

The troop headquarters section is organized and equipped to perform command and control and logistical support functions for the troop. The section consists of 2 officers and 11 enlisted soldiers. It is equipped with one main battle tank, one command post (CP) carrier, one APC, one M3 CFV, one cargo truck with a 400-gallon water trailer, and two utility trucks with one cargo trailer (see Figure 1-6).

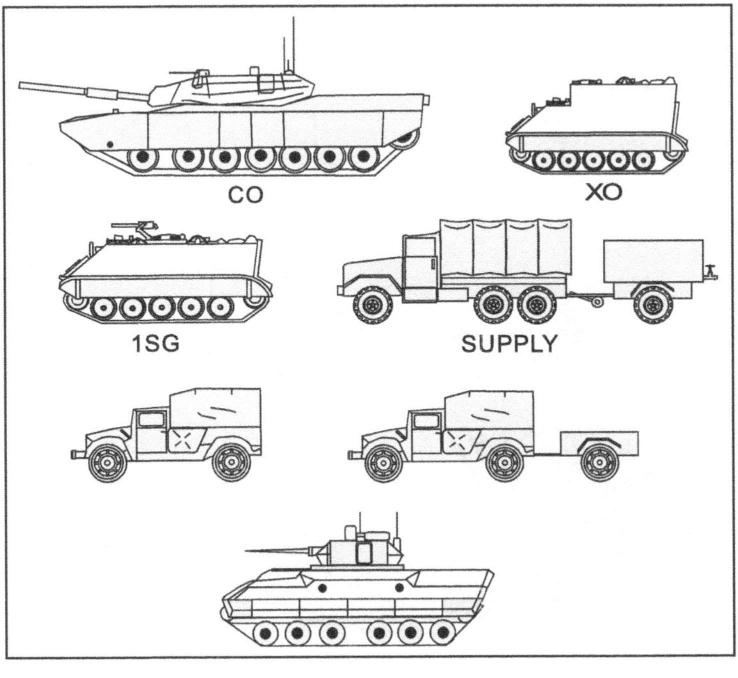

b) Figure 1-6. Headquarters section organization.

10) The Fire Support Team (FIST)

The FIST comes from the squadron howitzer battery in the heavy cavalry regiment and the division artillery in the division cavalry squadron. The FIST consists of one fire support officer (FSO), one NCO (team chief), one enlisted fire support specialist, and one enlisted radio operator. The FIST is responsible for coordinating indirect fires for the troop. In the heavy troop the team is mounted on an M981 FIST vehicle (see Figure 1-7).

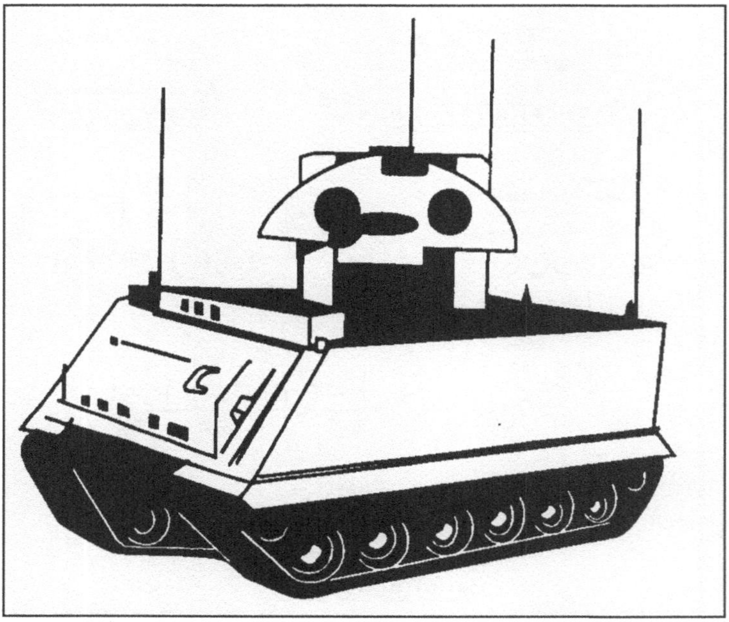

11)

a) Figure 1-7. FIST organization.

(II) THE LIGHT CAVALRY TROOP

The light cavalry troop consists of 6 officers and 107 enlisted soldiers. The troop is organized into a headquarters section, two scout platoons, two antitank (AT) platoons, a mortar section, and a maintenance section (see Figure 1-8).

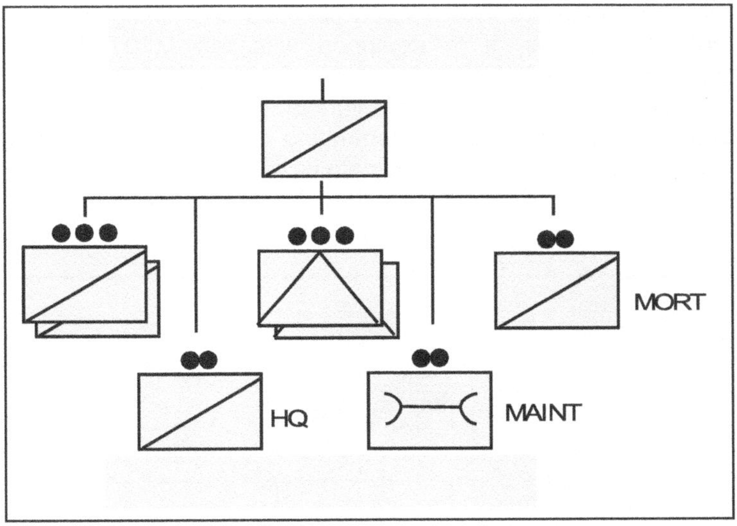

a) Figure 1-8. Light cavalry troop organization.

2) The Light Troop Scout Platoon

The scout platoon is organized and equipped to conduct reconnaissance and screening in support of its parent troop. However, when the parent unit is performing missions within an economy-of-force role, the scout platoon may conduct offensive, defensive, and retrograde operations in support of the troop mission. The platoon consists of 1 officer and 29 enlisted soldiers. It is equipped with ten M1025/M1026 HMMWVs (five MK-19 equipped and five cal .50 equipped). The platoon can organize into various configurations, usually from two to five sections, depending on factors of METT-T (see Figure 1-9). FM 17-98 details the tactics, techniques, and procedures for employing the HMMWV scout platoon.

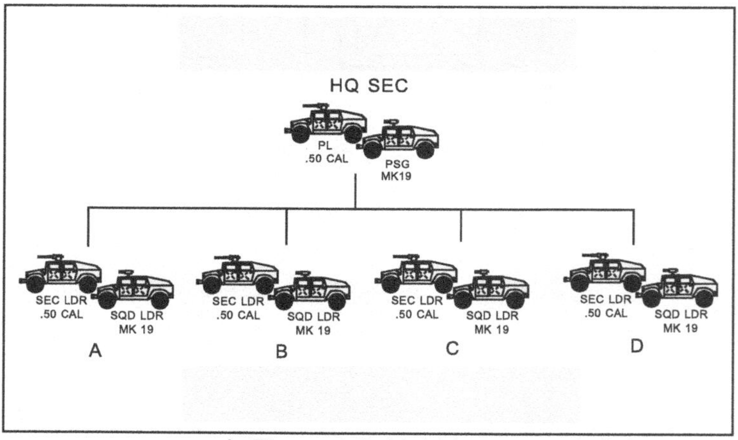

a) Figure 1-9. Light troop scout platoon organization.

3) The Light Troop Antitank Platoon

The AT platoon is organized and equipped to perform its three primary missions—attack, defend, and move. The platoon consists of 1 officer and 11 enlisted soldiers manning four M996 HMMWVs (TOW carriers) organized into two sections (see Figure 1-10).

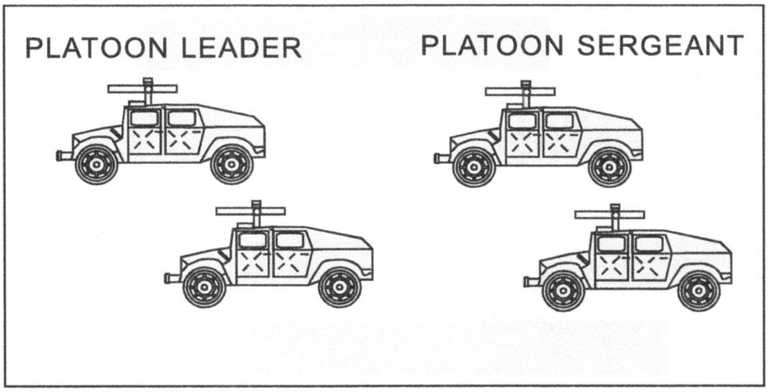

a) Figure 1-10. Light troop AT platoon organization.

4) The Light Troop Mortar Section

The mortar section is organized and equipped to provide immediate indirect fires to support troop operations. Such supporting fires are usually suppression, screening, obscuration, or illumination. The section consists of nine enlisted soldiers. It is equipped with two 120-mm mortars trailered by three M1025/M1026 HMMWVs (see Figure 1-11).

a) Figure 1-11. Light troop mortar section organization.

5) The Light Troop Maintenance Section

This section is organized and equipped to diagnose and repair most equipment faults at troop level. It has the capability to recover all troop vehicles and to maintain the troop's equipment records. The section consists of eight enlisted soldiers equipped with one M998 cargo/troop carrier HMMWV, one cargo truck with cargo trailer, and one 5-ton wrecker (see Figure 1-12).

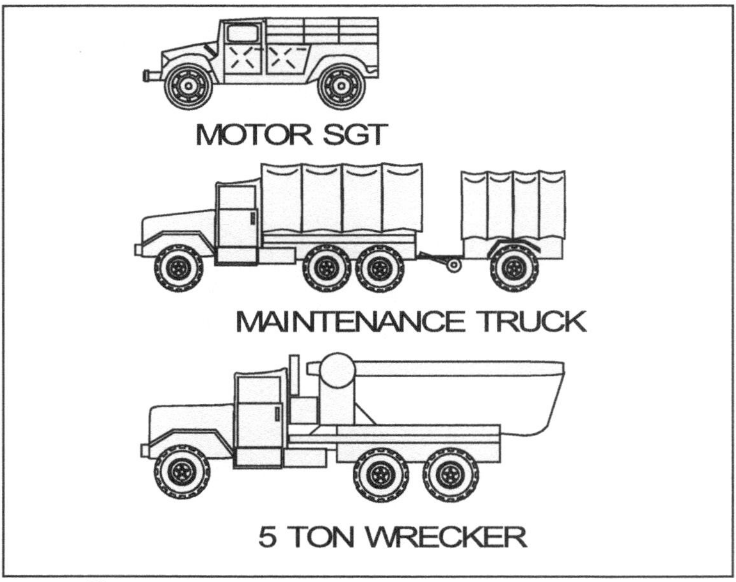

6)

a) Figure 1-12. Light troop maintenance section organization.

7) The Light Troop Headquarters Section

The troop headquarters section is organized and equipped to perform command and control and logistical support functions for the troop. The section consists of two officers and ten enlisted soldiers. It is equipped with one M1025 HMMWV, one M1037 HMMWV shelter carrier with S250 shelter, one M998 HMMWV with cargo trailer, and one 2 1/2-ton cargo truck with 400-gallon water trailer (see Figure 1-13).

a) Figure 1-13. Light troop headquarters section organization.

8) The Fire Support Team (FIST)

The FIST comes from the squadron howitzer battery in the light cavalry regiment. The FIST consists of one officer (FSO), one NCO (team chief), one enlisted fire support specialist, and one enlisted radio operator. The FIST is responsible for coordinating indirect fires for the troop. In the light troop the team is mounted on a HMMWV (see Figure 1-14).

(b)

a) Figure 1-14. Light troop FIST organization.

(c) Section II. Missions, Capabilities, and Limitations

(I) MISSIONS

Heavy and light cavalry troops perform reconnaissance and security missions to protect and preserve the fighting ability of the units to which they are assigned or attached. Both troops also conduct offensive, defensive, and retrograde operations in an economy-of-force role. Cavalry troops in general have limitations and capabilities associated with their TOEs and METT-T that must be considered when employing them in a specific mission role (see Figure 1-15).

	Hvy Trp	Lt Trp
Reconnaissance Missions		
Route Recon	✓	✓
Zone Recon	✓	✓
Area Recon	✓	✓
Security Missions		
Screen	✓	✓
Area Security	✸	✸
Convoy Security	x	x
Economy-of-Force Role		
Offensive Missions		
Hasty Attack	✓	✸
Attack	✓	✸
Movement to Contact	✓	✸
Defensive Missions		
Defend a Battle Position	✓	✸
Defend in Sector	✓	✸
Retrograde Missions		
Delay	✓	✸

✓ = fully capable
x = capable when reinforced
✸ = capable under permissive METT-T

a) Figure 1-15. Cavalry troop mission profiles.

(II) CAPABILITIES

a) The heavy cavalry troop is a unit that can be deployed by rail, sea, or both into a theater of operations. It can perform its missions under all visibility conditions and in any terrain that supports heavy armor movement. The integrated thermal sights on the M1A1 tank and M3 CFV along with ancillary passive night sights and night vision devices authorized to the troop provide an outstanding nighttime/reduced visibility acquisition and fighting capability. Additionally, the firepower and survivability organic to the heavy cavalry troop allow it to execute missions aggressively across the spectrum of warfare.

The light cavalry troop is a unit that can be rapidly deployed by air, rail, or sea into a theater of operations. It can perform its missions under all visibility conditions and in any terrain that supports wheeled vehicle movement. The ancillary thermal and night vision devices authorized to the troop provide an excellent reduced visibility acquisition capability. However, the light cavalry troop's ability to fight under reduced visibility conditions is limited because of its lack of integrated passive and thermal sights on stabilized weapons platforms.

(III)

(IV) LIMITATIONS

a) Given suitable terrain conditions, the ability of heavy and light cavalry troops to accomplish their assigned missions is mainly limited by the size and strength of threat forces encountered. Both troops are limited in close terrain, especially urban environments, due to their limited number of dismounted scouts. Also, the troops are dependent on their parent squadrons and regiments for additional combat support (CS) and combat service support (CSS) assets.

(d)
(e)

(f) Section III. Responsibilities

(I) THE TROOP COMMANDER

The troop commander is responsible to the squadron commander (SCO) for the discipline, combat readiness, and training of the troop, and for the maintenance of its equipment. He must be proficient in the tactical employment of the troop and its assigned and attached CS elements. He must also know the capabilities and limitations of the troop's personnel and equipment as well as those of CS elements attached to him.

a) The troop commander's responsibility in combat is twofold. He will—

- Accomplish all missions assigned to the troop in accordance with the SCO's intent.
- Preserve the fighting capability of the troop.

(II) THE EXECUTIVE OFFICER

In combat, the troop executive officer (XO) is second in command. He supervises the troop tactical operations center (TOC), where he stays abreast of the tactical situation in the troop's area of operations (AO). He manages the flow of combat information between the troop and the squadron from the troop TOC. With the assistance of the troop first sergeant (1SG), he plans and coordinates CSS for the troop.

(III) THE FIRST SERGEANT

The primary responsibility of the troop 1SG is sustaining the troop's ability to fight. He supervises the procurement and distribution of fuel, ammunition, food, water, clothing, equipment, replacements, and repair parts. He receives incoming personnel and assigns them to subordinate elements as needed. He is responsible for the medical evacuation of sick, injured, and wounded soldiers to the supporting medical treatment facility. He is also responsible for the evacuation of soldiers killed in action to the supporting graves registration collection point. He is also responsible for the recovery and evacuation of damaged combat equipment.

(IV) THE TROOP FIRE SUPPORT OFFICER

The troop FSO is responsible for the planning and coordination of the troop fire support plan. He relays troop requests for indirect fire and advises the commander on the employment of indirect-fire weapons in support of the troop's maneuver. The troop FSO may control and position the mortars during combat operations.

(V) THE PLATOON LEADER

The platoon leader is responsible to the troop commander for the discipline, combat readiness, and training of the platoon, and for the maintenance of its equipment. He must be proficient in the tactical employment of the platoon and know the capabilities and limitations of the platoon's personnel and equipment.

The platoon leader's responsibility in combat is twofold. He must—

- Accomplish all missions assigned to the platoon in accordance with the troop commander's intent.

- Preserve the fighting capability of the platoon.

(VI) THE PLATOON SERGEANT

The platoon sergeant (PSG) leads elements of the platoon as directed by the platoon leader, and assumes command of the platoon in his absence. The PSG assists the platoon leader in maintaining discipline, conducting training, and exercising control. He supervises platoon CSS, which includes supply and equipment maintenance.

(VII) THE MORTAR SECTION SERGEANT

The mortar section sergeant is responsible for providing responsive indirect fires to support the troop commander's concept of the operation.

(VIII) THE SUPPLY SERGEANT

The supply sergeant picks up, transports, and issues supplies and equipment to the troop. He works closely with the 1SG to accomplish these tasks. He also evacuates enemy prisoners of war and assists in the evacuation of soldiers who are killed in action to the graves registration collection point.

(IX) THE MAINTENANCE SERGEANT

The maintenance sergeant is responsible for the prompt battlefield repair or recovery of damaged or inoperable equipment. He works closely with the 1SG to accomplish these tasks.

(X) THE COMMUNICATIONS SERGEANT

The communications sergeant ensures the troop TOC and its crew are prepared for combat operations, and assists the XO in the TOC during combat operations. Within his capability, he repairs the communications equipment of subordinate elements.

(XI) THE NUCLEAR, BIOLOGICAL, CHEMICAL (NBC) NCO

The troop NBC NCO is responsible for troop NBC defense activities. He supervises radiological monitoring, chemical detection, and decontamination operations. He assists in maintaining NBC equipment and training NBC equipment operators and decontamination teams.

Chapter 2

Battle Command

Battle command is the art of battle decision making, leading and motivating soldiers and their organizations into action to accomplish missions. Battle command includes visualizing the current state and the future state, then formulating concepts of operations to get from one state to the other at least cost. Other functions of battle command include assigning missions; prioritizing and allocating resources; selecting the critical time and place to act; and knowing how and when to make adjustments during the fight.

Instantaneous response to orders is the hallmark of effective battle command in any cavalry operation. Leaders must quickly analyze information, make tactical decisions, and turn those decisions into successful battlefield actions.

Battle command of cavalry units is typically decentralized due to the size of the area of operations, vagueness of the enemy situation, and terrain unknowns. This places the burden of sound, timely decision making at the lowest levels. Leaders must develop a keen sense of situational awareness and constantly track the actions of subordinate units as well as those to the front, flank, and rear.

Contents

Effective battle command begins in the planning phase and continues through the execution phase of each mission. This chapter outlines the tools and techniques a troop commander needs to effectively command and control a cavalry troop in combat.

Section I. Troop Leading Procedures

The nine troop leading procedures are—

- Receive and analyze the mission.
- Issue a warning order.
- Make a tentative plan.
- Start necessary movement.
- Conduct a reconnaissance.
- Make final decisions and complete the plan.
- Issue an operations order.
- Rehearse.
- Supervise and refine.

RECEIVE AND ANALYZE THE MISSION

The troop commander will receive missions from squadron in the form of written, briefed, or radioed operation orders (OPORD) and fragmentary orders (FRAGO). Upon receipt of a mission, he will conduct a mission analysis to determine the *who*, *what*, *when*, *where*, and *why* elements of the mission and how much time is available until mission execution.

Mission. The mission analysis must identify the following:

- Specified tasks.
- Implied tasks.
- Essential tasks.
- Intent of the higher commander.
- Any constraints or limitations.

The items for analysis are derived from the squadron order as noted below.

- Paragraph 2, MISSION.
  - Mission Statement.

- Paragraph 3, EXECUTION.
  - Concept of the Operation.
  - Commander's Intent.
  - Maneuver.
  - Specific Instructions.
  - Coordinating Instructions.

- Execution matrixes.

- Thorough map reconnaissance.

List all specified and implied tasks not covered by unit SOP in a mission list. Identify those tasks that are vital to the squadron accomplishing its mission. These are essential tasks. Put the essential tasks in the form of a restated mission for the troop. The restated mission statement should be short and simple, and should cover *who, what, when, where,* and *why.*

One final note on this—keep the mission list at hand. It makes the job easier when preparing a course of action and assigning missions or tasks to platoons and sections.

Time. Because time is usually the most limited resource available, it must be used as efficiently as possible. Figure out what has to be done and what can be done in the time available. Use the backwards planning process to determine when critical events in the planning process must occur and stick to the timeline developed. Get the timeline out to subordinates in the warning order. Strive to use no more than one-third of the time available in the planning process at troop level. The remaining two-thirds is for subordinate leaders to plan and prepare.

Consider the following in terms of time when conducting mission analysis and preparing the troop warning order:

- Combat service support.

- Precombat inspections.

- Route reconnaissance (time/distance factors):
  - Assembly area to route start point.
  - Start point to command posts.
  - Command posts to release point.
  - Release point to line of departure.

- Subordinate's planning time.

- Platoon rehearsals:
  - Battle drill rehearsals.
  - Actions on contact.
  - Obstacle breaching/emplacement.
  - Movement formations.

Note. The preparatory actions above are triggered by receipt of a warning order at platoon level.

- Time to issue troop OPORD.

Note. One-third of the available time includes the time it takes to issue the troop OPORD.

- Troop rehearsals:
  - Backbriefs.
  - Rock drills.
  - Radio rehearsals.

- Time to move from—
  - Assembly area to line of departure.
  - Phase line to phase line.
  - Line of departure to objective.

- Effect of weather on movement.

Once the mission and time available have been analyzed, move to the next step in the trooping leading procedures.

ISSUE A WARNING ORDER

Issue the warning order promptly. Elements of the warning order include—

- Situation.
- Mission.
- Earliest time of movement.
- Time and place of OPORD.
- Attachments and detachments.
- Actions triggered:
  - Security.
  - Reconnaissance.
  - Combat service support actions.
  - Precombat inspections.
  - Sleep plans.
  - Rehearsals.

MAKE A TENTATIVE PLAN

Develop a course of action (COA). Consider the factors of METT-T when developing two, or better yet, three separate COAs.

- **Mission.** See *Receive and Analyze the Mission* paragraph on page 2-2.

- **Time.** In addition to the factors mentioned in the *Receive and Analyze the Mission* paragraph on page 2-2, the commander should consider the following details in relation to time when developing his tentative plan.

  - Time to move from—

    Line of departure to phase lines.
    Phase line to phase line.
    Line of departure to objective.

- Time for enemy to—
  - Attack.
  - Counterattack.
  - Reinforce.
- Effect of weather on movement and identification of friend or foe.
- Effect of NBC operations on movement.

• **Enemy.** Conduct intelligence preparation of the battlefield (IPB). IPB integrates enemy doctrine with weather and terrain to determine and evaluate enemy capabilities, vulnerabilities, and probable courses of action. IPB is a key part of preparing for battle. Much of the IPB workup can be found in Paragraph 1 of the squadron OPORD.

• **Terrain.** Analyze terrain in the area of operations. Consider the factors of OCOKA. (Arranged here in a more logical, systematic approach, based on IPB procedures and preparation of the modified combined obstacle overlay [MCOO].)

■ ***Obstacles.*** Identify all obstacles and restricted and severely restricted terrain throughout the sector or zone (a terrain sketch may be beneficial). Consider the following as obstacles:

- Water.
- Power lines.
- Bridges.
- Defiles.
- Slopes.
- Towns.
- Embankments.
- Railroads.
- Existing engineer obstacles.
- Wooded areas.

Terrain that flows around, or in some instances through or over, the obstacles identified is the primary maneuver corridor, or avenue of approach.

■ ***Avenues of Approach.*** Consider the following for each avenue of approach:

- What size force does it support? In what formation?
- Does it support long range observation and fields of fire? From where?
- Does it provide any cover and concealment? Where?
- Where does it lead?
  Through sector/zone?
  Enters/exits a flank?

The answers to these questions will help determine what terrain, if any, is key terrain.

■ ***Key Terrain.*** Key terrain is that terrain which control over would provide a clear advantage over another force. Seizing, securing, or even avoiding key terrain in sector/zone will figure prominently in the final plan. Identify what terrain is key and evaluate it in terms of observation and fields of fire along each avenue of approach.

■ ***Observation/Fields of Fire.*** Evaluate all terrain along the avenues of approach in terms of observation and fields of fire. Determine the following:

- What terrain provides the best observation of critical areas or along avenues of approach?
  At night?
  When visibility is obscured by smoke or fog?
- What terrain offers effective range of friendly and enemy weapon systems?
- What terrain provides the best fields of fire? For what size force?

■ ***Cover and Concealment.*** Evaluate all terrain along the avenues of approach in terms of cover and concealment. Determine the following:

- What terrain affords the best cover and concealment to the enemy and to the friendly forces?
- Where are likely hide positions that support potential enemy and friendly battle positions?
- What routes would a force use to move from hide positions to battle positions?

• **Troops Available.** Analyze the troops available to execute the mission. Determine the following:

■ ***Soldiers.***

- What is the strength of the platoons in the troop?
- Can all weapon systems be manned?
- How is the morale of the troop?
- What is the experience level of the troop?
- What training and skills are required? Are specialists such as engineers needed?
- What is the general health status and fatigue level of the soldiers?

■ ***Equipment.***

- How many weapon systems are operational and employable?
- Is special equipment or additional equipment needed to accomplish the mission?

START NECESSARY MOVEMENT

While developing his tentative plan, the commander may decide to reposition some of his forces before the operation starts so that he is prepared to meet mission requirements. If the entire troop needs to move, he will often execute the movement in accordance with a squadron plan. The amount of time he has to reposition his forces is determined by the readiness condition (REDCON) in effect.

Note. See Chapter 7 for a discussion of REDCON.

CONDUCT A RECONNAISSANCE

If sufficient time is available after determining his tentative course of action, the commander should make every effort to get out into his AO and look at the terrain. Ground reconnaissance is the norm; however, air transportation may be available. Staying on the ground gives a better feel for the terrain than does reconnoitering by helicopter, but it is slower. Although faster, aerial reconnaissance does not provide an appreciation of the terrain from a vehicle commander's point of view in terms of fields of fire and cover and concealment. If circumstances permit, subordinate leaders should accompany the commander to help perform the reconnaissance. It gives them an opportunity to become familiar with the area.

MAKE FINAL DECISIONS AND COMPLETE THE PLAN

Based on the new information gathered by his personal reconnaissance, the commander makes final adjustments to his plan and nails down the details. Once the plan is complete, he prepares the order. Troop orders need not be written and handed to subordinate leaders. Speak face to face with them if possible. Put the plan in the form of a standard five-paragraph field order. Jot it down in a notebook so every detail can be recalled.

ISSUE THE OPERATIONS ORDER

Assemble the orders group. Make sure everyone in the orders group is present. Issue necessary graphics to subordinate leaders and give them time to prepare their maps. Ensure the overlays are neat and accurate. Messy overlays with broad pen strokes cause confusion and waste valuable time. Once the subordinate's overlays are prepared, check them for accuracy. (A technique for ensuring accuracy is to include in the order a list of grid coordinates for key locations that might be ambiguous in the operational graphics. Even neat graphics, once copied from squadron down to platoon, may often deviate by as much as 500 meters.) Arrange the orders group in a semicircle from left to right in the order in which they will be addressed when giving specific instructions.

Issue the order. Use notes when issuing the order; do not rely on memory. If possible, issue the order while overlooking the AO. Keep in mind that the order will be translated a couple of times before the privates receive it, and it must make sense to them. Talk in their language. The order should be short and simple, and must be logical and easy to follow; be clear, organized, and concise so that no one becomes confused. Misunderstanding the commander's intent most often occurs in the transmission of orders to the platoon leader and his subordinates.

<u>Note</u>. See Appendix C for the operations order format and sample OPORD.

Conduct the confirmation brief. Immediately after the order is issued, call for questions from the subordinate leaders. After all questions concerning the order have been answered, begin the confirmation brief by each of the troop's subordinate leaders.

The confirmation brief is used by the commander to confirm that his intent and guidance for the conduct of the operation are clearly understood by everyone in the orders group before they are dismissed to begin their planning. The confirmation brief adjourns only when the commander is confident his subordinates understand their mission, his and the higher commander's intent, his concept of the operation, the scheme of maneuver, the time plan, and the type and location of the rehearsal.

REHEARSE

Rehearsals are of paramount importance before executing any plan. Rehearsals help in the following ways.

- Clarify the commander's intent.
- Expose combat, combat support, combat service support, or disconnected activities in the plan.
- Reinforce the scheme of maneuver and fire support plan.
- Focus on actions and decision points critical to mission accomplishment.
- Ensure subordinates explicitly understand their missions, how their missions relate to each other, and how each mission relates to the commander's plan.
- Outline conditions that, when present, would necessitate execution of branch plans.
- Provide feedback to the commander.

Rehearsals instill confidence in participants by—

- Giving participants faith in the success of their own plan as well as in their commander's plan.
- Providing subordinates with purpose, direction, and motivation.
- Enabling leaders to execute missions with speed, flexibility, and audacity.

There are generally seven types of rehearsals—full, key leader, terrain model, sketch map, map, radio, and backbrief. Preparation time and resources for each range from extensive to minimal. At troop level, the most common rehearsal types are backbrief, map, sketch map, terrain model, and radio.

Backbrief. During the backbrief, each subordinate leader briefs the commander of how he intends to accomplish his mission before he issues his OPORD to his respective unit. By having subordinates explain their intent and concept of operation, the commander can ensure their plans support his own. Flaws or potential problems with the plan may be revealed at this time.

Do not confuse the backbrief rehearsal with the confirmation brief the commander uses immediately after he issues an OPORD to determine how well his subordinate leaders understand the order.

Map/Sketch Map/Terrain Model. Using either a map and overlay of the same scale as used to plan and execute the operation, a large scale sketch map, or a terrain model, the commander and subordinate leaders move unit markers in a sequential (either by phase, event, or time), interactive, verbal execution of the operation. These type rehearsals are essentially conducted like a war game to show the planned sequence of action-reaction-counteraction to critical events or phases of the operation. This is a good opportunity to coordinate not only actions of the unit but also critical locations such as contact points, checkpoints, boundaries, battle positions, and hide positions among subordinate leaders.

Radio. Using existing communications networks (either FM or wire), the commander and subordinate leaders verbally and interactively execute critical portions of the operation. This technique can have obvious communications security disadvantages; if so, then only the essential, most critical portions of the operation are rehearsed. Radio rehearsals require few resources and little time and are best used in conjunction with other methods or to further refine the plan.

SUPERVISE AND REFINE

The last step in troop leading procedures must not be neglected. Failure to supervise the execution of the order, or to refine the plan as the situation changes, is the road to ruin. The importance of this step cannot be overemphasized. Actions during the final step may include, but are not limited to, higher commander and his staff being present in the subordinate unit staff planning process.

Section II. Situational Awareness

Situational awareness is always keeping a clear picture of the tactical situation, both mentally and graphically. This picture includes both the friendly and enemy situation and an understanding of the relevant terrain. Since the troop normally operates dispersed over wide frontages, it is essential that all leaders maintain situational awareness so they can make sound, quick tactical decisions. Situational awareness also permits the leaders to anticipate events and relate separate pieces of information to form logical conclusions. One of the critical outcomes of situational awareness observed by all leaders is a reduction of fratricide incidents.

BATTLEFIELD FRAMEWORK

How the commander has structured the battlefield impacts the troop commander's ability to maintain good situational awareness. A commander will structure the battlefield based on the conditions of METT-T. The framework of the battlefield can vary from a very rigid extreme with an obvious front and rear boundary and closely tied adjacent units, to a very dispersed and decentralized structure with no secure areas, unit boundaries, or definable front. Between these extremes is an unlimited number of possible variations. Maintaining situational awareness will become more difficult the more unstructured the battlefield is. Modern, highly mobile operations with small forces lend themselves to a less rigid framework which will challenge the ability to maintain a good picture of the battlefield.

PICTURING THE BATTLEFIELD

To have a good picture of the battlefield, all leaders must have virtually perfect knowledge of the friendly situation one level higher than their own. This means that all principal subordinate leaders must know the troop situation and the troop commander must know the squadron situation. All leaders must have accurate knowledge of the terrain, and they must know as much as possible about the enemy. The requirement to maintain a real-time picture of the battlefield one level higher does not relieve leaders of the requirement to understand the situation two levels higher. The difference is that a leader's understanding of the situation two levels higher than his own does not have to be as specific or in real time.

Most of the information passed between elements is in the form of reports over FM radio. The troop commander receives many reports as a result of his troop graphics. Good graphics require that the subordinate elements report periodically as they accomplish tasks. The troop commander must be aware of when subordinates report so he will know how current his visualization of the situation is. If an element does not report in a timely manner, the commander must quickly determine the situation of the overdue element.

Although many reports may not be addressed specifically to him, particularly on the squadron net, the commander must monitor them by eavesdropping on the nets as traffic is sent. How effectively he can accomplish this is, to some degree, experience dependent; however, there are techniques he can use to relate the information to his map and thereby track the tactical situation. The troop XO plays an important role in assisting the troop commander and platoon leaders in maintaining situational awareness. The troop XO provides leaders with periodic updates of friendly units to the front, flanks, and rear, based on traffic from the squadron operations and intelligence (OI) net.

The commander's map is the key to maintaining situational awareness (see Figure 2-1). He should plot all friendly position reports up to one level higher than his own. Information from spot reports should also be plotted. Using different colors for friendly and enemy elements allows quick distinction. To avoid cluttering the map, he should place a dot or symbol with a number on his map where the element is located. The same number is then written in the map margin (or beyond the area of operations) with the complete spot report or unit identification next to it. This notation should also include the time. As positions or reports are updated, the old symbol is crossed off and a new one with a corresponding notation is added. This system allows the commander to easily track and monitor the tactical situation. This system is augmented by a formal operations log kept in the troop TOC by the XO or NCOIC.

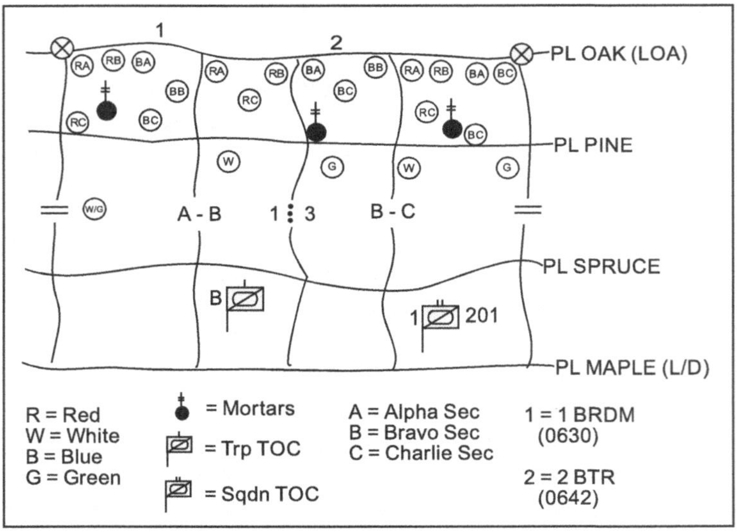

Figure 2-1. Sample situation map (troop commander).

BATTLE SPACE

Situational awareness, as previously discussed, is critical to successful reconnaissance and security missions; however, the troop commander's primary attention must be on his battle space. Battle space for the cavalry troop is determined by the location of individual platoons and sections, the range of direct-fire weapons, observation, sensors, and the terrain on which they are applied. This space is the immediate area over which the troop has influence. The troop commander must be aware of the general situation, but he must actively direct and manage all activity within his battle space. Most of the troop commander's command and control efforts focus on what is happening within his battle space and shaping his battle space to make it more efficient.

FRATRICIDE

Fratricide is a significant danger to all forces operating on a mobile battlefield where weapon system lethality is significantly greater than friend or foe identification ability. For this reason, situational awareness on the part of all leaders, particularly the commander, is critical not only to mission success but also to survival.

Under these types of conditions, it is critical that the commander know where other friendly elements are operating. With this knowledge of the situation, he must anticipate dangerous situations and take steps to either avoid or mitigate them. The commander must constantly be vigilant to changes and developments in the situation that may place his elements in danger. He must also ensure all subordinate unit positions are constantly sent to higher headquarters so all other friendly elements are aware of where they are and what they are doing. When the commander perceives a potentially dangerous situation, he must personally use the squadron command net to coordinate directly with the friendly element involved.

Section III. Communications

NETS

Several troop radio nets are used to circulate the volume of information in any combat operation. A troop command net links the troop commander with his subordinate units (see Figure 2-2). A troop fire support net links subordinate units with the fire support team (FIST) and mortar section (see Figure 2-3). Each net has an assigned radio frequency, but not necessarily a dedicated radio.

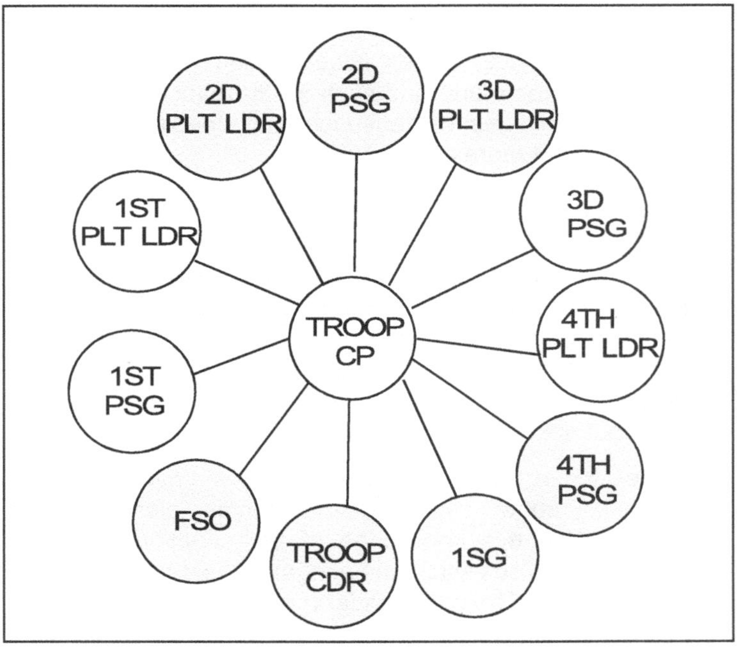

Figure 2-2. Troop command net.

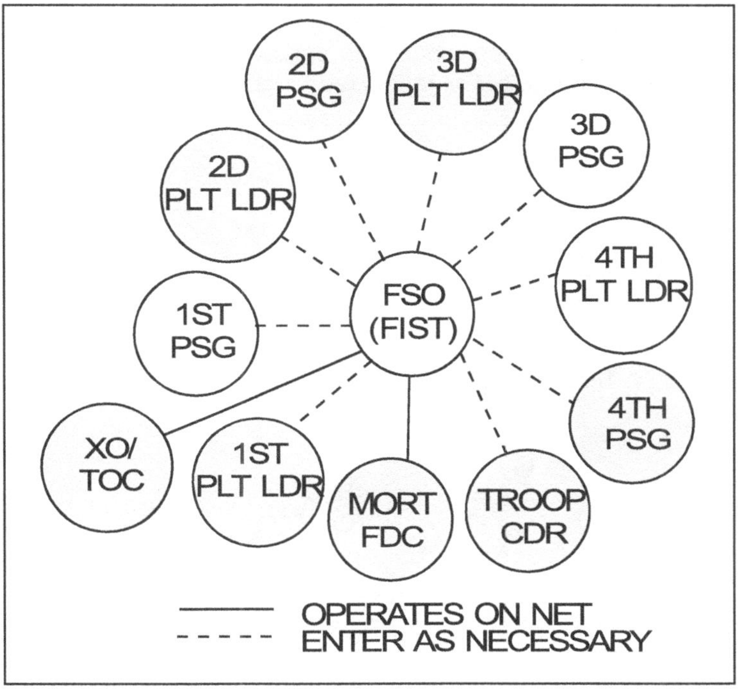

Figure 2-3. Troop fire support net.

In the troop CP, three additional radio nets are established to link the troop with the SCO and his headquarters. The squadron operations and intelligence (OI) net is used primarily to transfer enemy and friendly information by FM radio. All routine and recurring reports from the troop to squadron are transmitted on this net. Squadron headquarters uses this net to transmit information about current enemy and friendly situations within the squadron's AO. It is often used to issue FRAGOs. The troop commander and the XO are linked to the SCO by an FM squadron command net. This net is reserved strictly for use by the SCO, XO, S3, and CPs; and by subordinate troop, company, and battery commanders. In addition, an AM squadron command net allows the SCO to command and control the squadron when it is spread over distances that exceed the range of FM radios (see Figure 2-4).

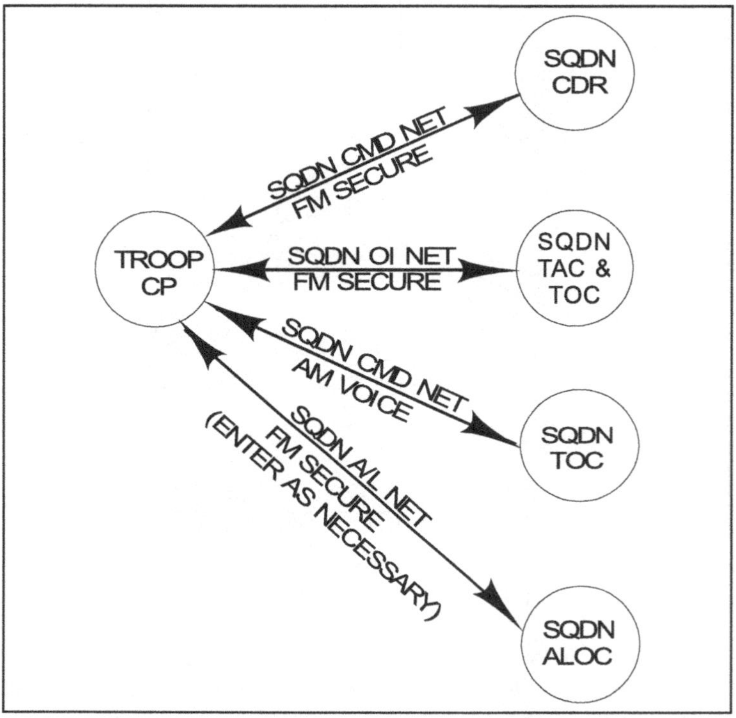

Figure 2-4. Troop command post external nets.

EAVESDROP

This tool is used by all cavalrymen. It is designed to give leaders time to think before they act. Most leaders have a hard time making decisions while they are talking on the radio. Decisions are usually better if the commander listens closely to what is said, and if he has uninterrupted time to consider information received. Eavesdropping is a method of communicating based on that principle. Everyone simply listens to the information sent from one leader to another, and unless necessary, stays off the radio net. In the troop, for example, reports from scout platoons are transmitted to the troop CP, where the XO reviews and records them. All others listen in and record the information themselves. Everyone stays off the net yet remains informed. If all or part of a transmission is missed, enter the net and quickly get the needed information from the troop CP.

FIXED CALL SIGNS

Fixed call signs are tools that establish immediate identity of subunits and leaders, thereby reducing the length of transmissions on the radio. These call signs eliminate the confusion often caused by the ever-changing signal operation instructions (SOI). Each platoon is referred to by a color, and leaders are referred to by numbers. These call signs do not change. An example of fixed call signs for a cavalry troop is outlined below.

Fixed Call Signs

Element	Call Sign
TROOP HEADQUARTERS	BLACK
FIST	Black 1
Medics	Black 2
TOC	Black 3
Supply	Black 4
XO	Black 5
Commander	Black 6
1SG	Black 7
Maintenance	Black 8
Mortars	Black 9
NBC	Black 10

1ST PLATOON (Scout)	RED
Plt Ldr 11	Red 1
Sec Ldr 12	Red 2
Sqd Ldr 13	Red 3
PSG 14	Red 4
Sec Ldr 15	Red 5
Sqd Ldr 16	Red 6
*Sec Ldr 17	Red 7
*Sqd Ldr 18	Red 8
*Sec Ldr 19	Red 9
*Sqd Ldr 10	Red 0
2D PLATOON (Tank or AT)	WHITE
Plt Ldr 21	White 1
TC 22	White 2
TC 23	White 3
PSG 24	White 4
3D PLATOON (Scout)	BLUE
Plt Ldr 31	Blue 1
Sec Ldr 32	Blue 2
Sqd Ldr 33	Blue 3
PSG 34	Blue 4
Sec Ldr 35	Blue 5
Sqd Ldr 36	Blue 6
*Sec Ldr 37	Blue 7
*Sqd Ldr 38	Blue 8
*Sec Ldr 39	Blue 9
*Sqd Ldr 30	Blue 0
4TH PLATOON (Tank or AT)	GREEN
Plt Ldr 41	Green 1
TC 42	Green 2
TC 43	Green 3
PSG 44	Green 4

*Light troop only

Section IV. Techniques of Tactical Control

CHAIN OF COMMAND

The troop commander cannot personally supervise the execution of all combat tasks he assigns. He must depend on a trusted team of subordinate leaders. The chain of command in the troop extends from the commander to the platoon leaders and section leaders. Within the platoons, it extends from platoon leaders and their platoon sergeants to section sergeants to vehicle/crew commanders. The troop XO and the first sergeant serve as the commander's principal assistants. They help coordinate the actions of subordinate units to ensure unity of effort. They also ensure the troop has what it needs to accomplish its assigned tasks.

ORDERS GROUP

The troop orders group consists of all subordinate leaders required to receive and execute the troop commander's orders. As a minimum, the orders group should include the XO, the first sergeant, the platoon leaders, the mortar section sergeant, the fire support officer (FSO), and the leaders of any units that are attached or under the troop's operational control (OPCON).

STANDING OPERATING PROCEDURES

A troop SOP is a written collection of standing orders. It establishes procedures for performing routine combat tasks. For example, the SOP tells subordinates what procedures to follow when the troop occupies an assembly area, breaches an obstacle, conducts a passage of lines, or reacts to air attack. It lists reports that are routinely sent to troop and squadron headquarters, and shows their correct formats. It also states procedures for communicating within the troop, obtaining indirect fire, or getting additional fuel or ammunition. The SOP is a valuable time saver for troop leaders. Most important, the SOP keeps everyone in step, reduces confusion, and provides the backbone of combat proficiency. Disciplined execution of the troop SOP is the hallmark of outstanding units.

TERRAIN INDEX REFERENCE SYSTEM

Terrain index reference system (TIRS) is used to maneuver the troop, rapidly report friendly locations without fear of compromise, and pass out control measures. TIRS is not used to report enemy locations.

Description

Each TIRS point is shown by a tick mark located on a grid line intersection. Each point is given a designator of one letter and two numbers, such as X56, placed in the upper right quadrant of the tick mark. TIRS point designation is SOP, and units determine which letters they will use. They may designate specific letters for specific unit sectors or AO. The TIRS point is then recorded as X56, located at PA2450 (four-digit grid).

TIRS point-to-point references designate kilometers. For example, 500 meters is given as "POINT FIVE," 1,000 meters as "ONE," and 3,500 meters as "THREE POINT FIVE."

For shifts from the TIRS point, use cardinal directions instead of left, right, up, or down. Using TIRS, "FROM X-RAY FIVE SIX—EAST ONE POINT SEVEN—NORTH POINT SEVEN" translates to "From the tick mark for TIRS point X56, shift east 1,700 meters and north 700 meters." When each TIRS point is placed on a four-digit grid intersection, the use of shifts makes the TIRS as accurate as the six-digit grid system. For an example of a TIRS point in use, refer to Figure 2-5.

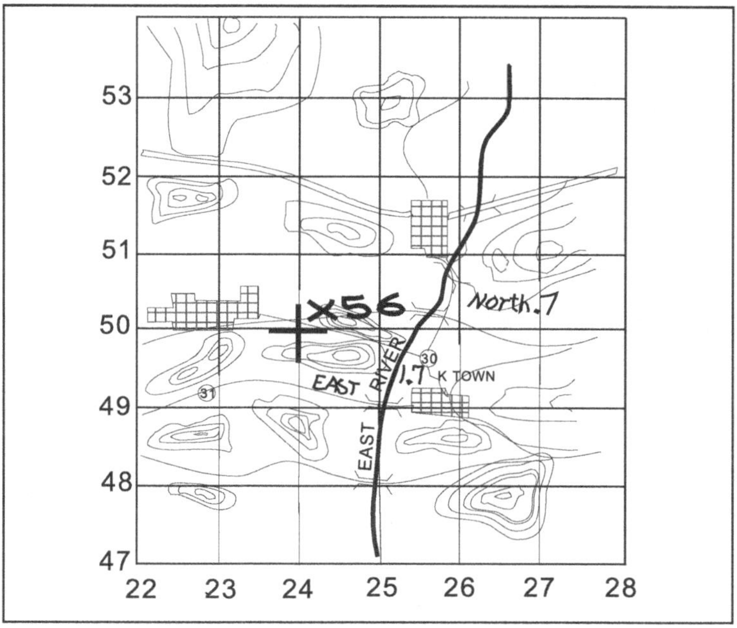

Figure 2-5. Placing TIRS on the map.

Application

The squadron normally issues the TIRS to be used for the operation as early as possible, perhaps with the warning order. The TIRS list should be issued to elements as an annex to a written OPORD.

The squadron should designate four to six TIRS points in each 10-kilometer square. The squadron TIRS should normally be sufficient for the troop to operate; however, if the troop designates additional TIRS, it should always ensure only squadron TIRS are referenced in its communications with higher headquarters.

TIRS should be used routinely to control combat operations. Use them—

- To identify BPs and to pass out control measures (such as LDs, PLs, and boundaries) quickly.
- To report friendly unit locations.

Passing control measures, such as BPs, sector boundaries, and PLs, are quick and accurate using TIRS.

Figure 2-6 shows BP B22. A troop commander could report "TROOP ESTABLISHED VICINITY ALFA ZERO SEVEN." To be more precise, he could report "TROOP BRAVO ESTABLISHED ALFA ZERO SEVEN WITH RED AT FROM ALFA ZERO SEVEN EAST POINT THREE—NORTH POINT FIVE—WHITE AT FROM ALFA ZERO SEVEN WEST POINT THREE—NORTH POINT EIGHT—BLUE AT FROM ALFA ZERO SEVEN EAST ONE POINT NINE—NORTH ONE POINT NINE—TO FROM ALFA ZERO SEVEN WEST POINT NINE—NORTH ONE POINT SEVEN—AND GREEN AT FROM ALFA ZERO SEVEN EAST ONE POINT ZERO—NORTH POINT FIVE." In the second transmission, the troop commander gives the precise location of the center of mass for all his platoons: red, white, blue, and green represent the 1st, 2d, 3d, and 4th platoons.

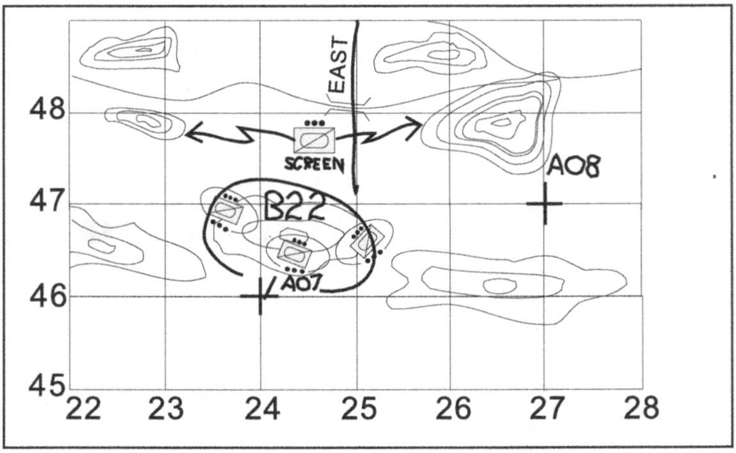

Figure 2-6. Reporting location using TIRS.

Using this transmission, the SCO orders Troop B to occupy a BP that is not on the troop commander's map: “BRAVO SIX SIX OCCUPY A BP EXTENDING FROM ALFA ZERO EIGHT WEST POINT SEVEN—SOUTH ONE POINT ZERO—TO FROM ALFA ZERO EIGHT EAST POINT SIX—SOUTH POINT FIVE—ORIENT NORTHEAST.”

Figure 2-7 shows how the SCO could quickly establish new phase lines and boundaries between troops.

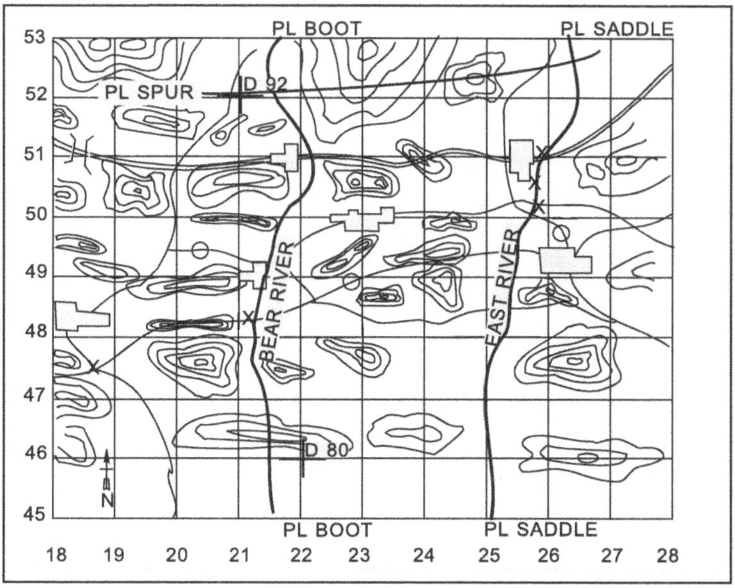

Figure 2-7. Establish graphic control measures using TIRS.

The SCO could transmit a new boundary between troops B and C by radio: “BRAVO SIX SIX AND CHARLIE SIX SIX BOUNDARY IS FROM DELTA EIGHT ZERO EAST POINT NINE—NORTH POINT EIGHT—TO FROM DELTA NINE TWO EAST TWO—SOUTH THREE—TO FROM DELTA NINE TWO EAST ONE POINT NINE—SOUTH POINT ONE.”

PL SPUR could be sent as "PL SPUR IS FROM DELTA NINE TWO WEST POINT NINE—TO FROM DELTA NINE TWO EAST FIVE POINT THREE—NORTH POINT SEVEN."

Section V. Command Guidance and Organizational Control

TROOP COMMANDER

The troop commander is mounted in an M1A1 tank in the heavy troop or a HMMWV in the light troop. His first priority is the command and control of the troop.

The commander must provide the troop with clear intent and guidance for the mission. Intent and guidance facilitate positive command and control over subordinate platoons yet provide flexibility in execution for subordinate leaders.

The commander positions himself where he can best command and control the troop. He may maneuver along with one of the tank/AT platoons or between the tank/AT platoons and scout platoons. It is imperative that the troop commander not become decisively engaged with fighting his vehicle. He must retain the ability to control the actions of his troop. The troop commander operates on both troop and squadron command radio nets. He must have the capability to communicate to subordinates as well as to the squadron commander.

TROOP TOC

The purpose of the troop tactical operations center (TOC) is to report information to the higher headquarters. The troop TOC is the critical link between squadron and troop. One of the primary purposes of the cavalry troop is to report, and the troop TOC is central to that function. The troop TOC operates under the direction of the troop XO who is responsible for assisting the commander in the command

and control of the troop. The troop TOC takes reports from subordinate units in the troop, processes the information, and reports the information to the higher headquarters. The TOC tracks the battle at the troop and squadron level and relays information to the commander and subordinate platoons pertaining to the friendly/enemy situation. The TOC is the net control station for the troop command net and operates on the squadron command and OI net. The troop TOC also monitors and operates as necessary on the squadron A/L and the troop fire direction net.

In the heavy troop the TOC vehicle is an M577. An M3 CFV moves with the heavy troop TOC to provide local security for the M577 and to serve as a command and control vehicle for the XO if the commander is out of action. The XO should not conduct TOC operations out of the CFV because his experience is needed in the TOC vehicle to handle the information flow between troop and squadron over the OI and command nets. If the commander is out of action, the XO should avoid, if possible, rushing forward to assume command of the troop because of the importance of the activities in the TOC. The XO should assume command of the troop from the TOC until the battle is over. However, if the XO does need to move forward to command and control the troop, a qualified replacement must be placed in the TOC to assume his duties. This individual may be the FIST chief, one of the platoon leaders, or the first sergeant. The correct replacement for the XO in the TOC should be based on the depth available at any one of the above mentioned positions.

The light troop TOC is a cargo HMMWV with a shelter.

The primary concern when positioning the troop TOC is based on the TOC's ability to communicate with the squadron and the subordinate elements of the troop. The TOC position should also be an area that provides good cover and concealment from enemy and limited access to civilian population. During reconnaissance or offensive operations, maneuver the TOC at least one terrain feature behind troop combat elements. During security or defensive operations, the TOC should be positioned in sufficient depth to avoid contact with the enemy yet maintain communications with the forward scout platoons. The TOC and the troop trains may collocate for increased local security.

FIST

The FIST is the critical link with the supporting artillery and responsible for coordinating indirect fires (both mortar and FA) for the troop. The team processes calls for fire from the platoons and allocates the appropriate indirect-fire system based on the commander's guidance for fire support. The FIST can also assist the squadron ALO with the employment of close air support.

The FIST operates on three radio nets: troop command, troop fire direction, and squadron FSE digital/voice. The FIST monitors at least one of the following nets: squadron command, squadron OI, and howitzer battery (supporting artillery headquarters in the heavy and light division).

The FIST vehicle also may serve as the alternate troop TOC. The FSO has ready access to the higher level situation and the radio systems to replicate the troop TOC if it becomes damaged or destroyed.

Command guidance to the FIST should include the following:

- **Purpose of indirect fires.** How does the commander intend to use field artillery and mortar fires to support his maneuver?
  - Screening.
  - Suppression.
  - Development of the situation.
  - Disengagement.

- **Engagement/attack criteria.** How many rounds and of what type and mix will be fired at a particular target? Which targets will be engaged with artillery and which with mortars?

- **Control of troop mortars.** If the FIST controls movement of troop mortars, how far forward of the scouts will the mortars be able to range? Where are the mortars going to move? When are the mortars going to move?

- **FIST movement.** The primary considerations when positioning the FIST are security of the team and the ability to communicate with the squadron FSE, howitzer battery, or the DS artillery. The FIST is not the forward observer team for the troop; the troop has 19Ds who act as forward observers. The five techniques to maneuvering the FIST are—

  - Maneuvers with the commander.
  - Maneuvers with or near the mortar section.
  - Maneuvers with the TOC.
  - Maneuvers alone to maintain communications.
  - Maneuvers with the scouts or tanks to directly control fires or to use the ground laser designator (GLD).

Note. See Chapter 8 for a more in-depth discussion of troop fire support command and control techniques.

TROOP MORTARS

The troop mortars provide responsive indirect-fire support to the troop. In the light troop they are equipped with three cargo HMMWVs and two 120-mm towed mortars. The heavy troop is equipped with two M106 mortar carriers with 107-mm mortars. The mortars operate on the troop command and fire direction nets.

Command guidance for the mortars should include the following:

- **Purpose of mortar fires.** How does the commander intend to use mortar fires to support his maneuver?
  - Screening.
  - Suppression.
  - Development of the situation.
  - Disengagement.

- **Engagement/attack criteria.** How many rounds and of what type and mix will be fired at a particular target?

- **Positioning.** How far forward of the scouts will the mortars be able to range?

- **Movement.** Where will the mortars move and what will the mortars orient their movement on, for example, checkpoints, a route, mortar firing positions? Who will control the movement of mortars—the mortar section sergeant, the FIST, or the troop XO?

SCOUT PLATOONS

The scout platoons, HMMWV or Bradley equipped, perform reconnaissance and security missions for the troop. They may also perform missions within an offensive or defensive role in certain METT-T situations. To enhance command and control and to provide scout platoons the freedom of action required in cavalry operations, the troop commander should give the following guidance to scouts, regardless of what missions they are performing.

- Engagement, disengagement, destruction, and bypass criteria.

- Tempo of the operation. Tempo is dictated by the higher commander's intent for the operation, and is a function of the enemy situation as it relates to the survivability and firepower of the systems conducting the operations (METT-T). Tempo for the scout platoon determines the rate at which they conduct their mission and develop the situation. High tempo operations, for example, could be characterized by very aggressive reconnaissance focusing on mounted avenues of approach and likely enemy defensive positions. Engagement criteria in high tempo operations would be unrestricted, allowing the scouts to rapidly develop the situation by fire and maneuver. Low tempo operations could be characterized by scouts conducting a very detailed reconnaissance and maximizing the use of stealth to develop the situation.

The scout platoon leader and platoon sergeant operate on both the troop command and platoon command nets. As required, they will operate on the troop fire direction net. For platoon organization for combat, see FM 17-98, Chapter 2, Section VI.

TANK/AT PLATOONS

The tank/AT platoons are the principal killers in the troop. They assist the troop commander in the rapid development of the situation during operations by fire and maneuver. The troop commander should give the platoons the following guidance to facilitate their command and control during battle.

- Location in the troop formation and distance behind the scout platoon.

- Engagement, disengagement, and destruction criteria. When and where will they be employed and against what size force? Will they be used to suppress in support of another maneuver force, or destroy by fire and maneuver?

- Tempo of the operation. (Same as for the scout platoons above.)

The platoon leader and platoon sergeant operate on troop command and platoon command nets. As required, they will operate on the troop fire direction net.

TROOP TRAINS

The troop trains provide logistical support and casualty evacuation to the troop. The trains are under the direction of the first sergeant and consist of the troop combat and field trains. The first sergeant monitors the troop command and squadron A/L nets. He directs resupply, treatment and casualty evacuation, and maintenance/recovery operations for the troop.

The troop trains may be consolidated or echeloned. Either way, the trains should move one terrain feature behind the trail combat elements of the troop. If echeloned, the field trains move with the squadron field trains under the direction of the squadron HHT commander. The troop combat trains consist of the first sergeant in his vehicle, a maintenance contact team with recovery vehicle, and a medical evacuation vehicle. The field trains are made up of the supply sergeant and the remaining members of the maintenance team.

Command guidance to the troop trains should include the following:

- Trains organization. Should they move consolidated or echeloned?

- Movement guidance. Where do the trains move in the troop formation and what is their orientation for movement? Use checkpoints, routes, or trains locations to orient the movement of the trains.

Note. See Chapter 9 for a more in-depth discussion of the TTP for combat service support.

Section VI. Tactical Movement Formations

During tactical operations, the troop normally moves in one of four movement formations: troop column, troop line, troop vee, or troop split-vee.

TROOP COLUMN

Use the troop column when moving on a designated route, when speed is essential, and when contact with the enemy is not expected. This formation moves the troop quickly and efficiently from one place to another. Control of the troop while moving in column is eased, but the troop is vulnerable to enemy air or ground attack.

The troop usually moves at a designated speed and with a set distance between vehicles when moving in column formation. The march speed and distance between vehicles are designated by the troop SOP, based on the situation.

Note. See Chapter 7 for a discussion of column formation.

TROOP LINE

The troop line formation can be used when conducting defensive or retrograde operations in the heavy and light troops (see Figure 2-8). The formation may also be used by the light troop during reconnaissance operations when the troop is scrambled.

Figure 2-8. Troop line formation.

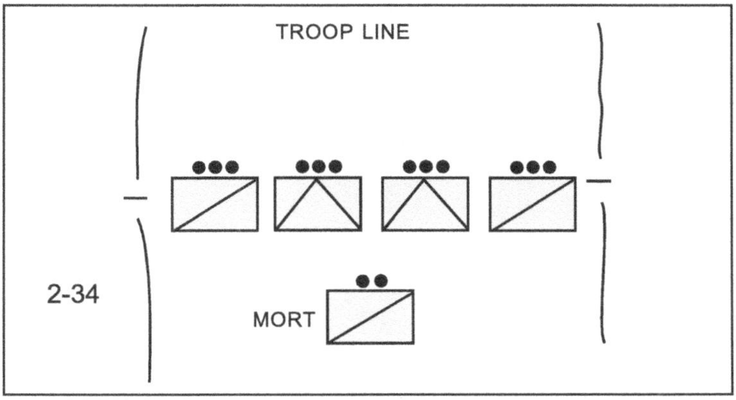

TROOP VEE

The troop vee formation is used to conduct reconnaissance or security missions (see Figure 2-9). The purpose of the vee formation is to keep maximum reconnaissance forward; the two tank or AT platoons are positioned in depth, ready to attack or defend any where in the troop zone/sector. This formation allows the troop to mass combat power at the decisive point (see Figure 2-10).

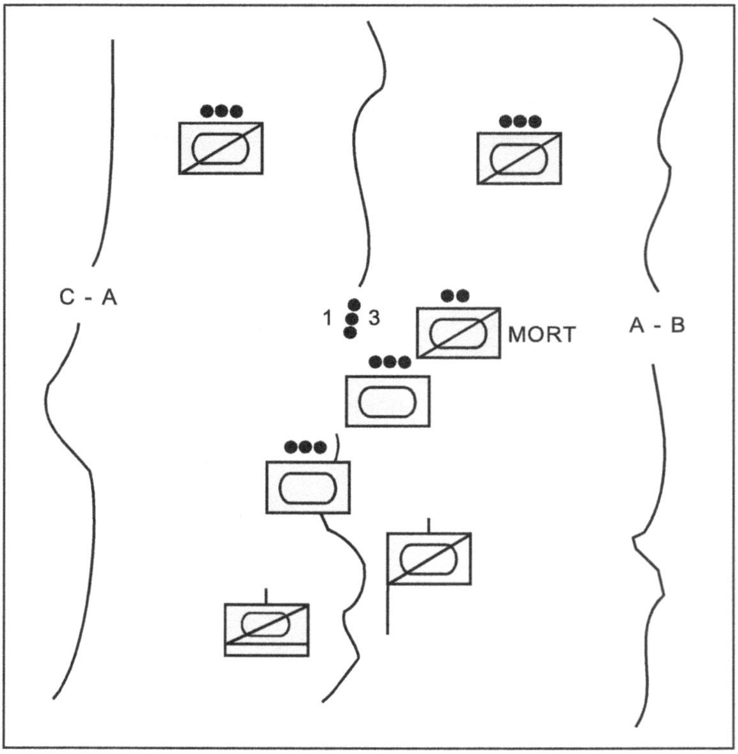

Figure 2-9. Troop vee formation.

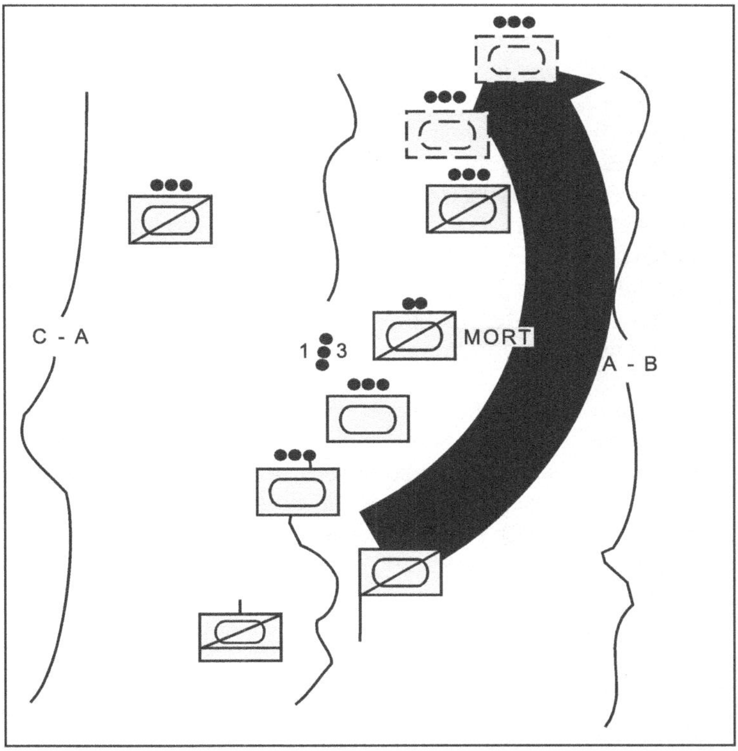

Figure 2-10. Troop commander maneuvers his tank platoons (troop vee).

In the troop vee formation, the two scout platoons work abreast in their respective zone/sector. The two tank/AT platoons are collocated in depth of the scouts, and they maneuver in the troop zone/sector.

Heavy Troop. Normally the distance between the scouts and tanks varies from 1 kilometer to 4 kilometers. The tanks should be close enough to respond quickly, but far enough back to respond to enemy contact based on the commander's guidance for engagement. Generally, do not position the tank platoons to overwatch the scout platoons. When the scouts make contact with the enemy, the tanks should be far enough in depth to maintain their ability to maneuver. The scouts are usually responsible for providing their own security as they move.

Light Troop. Normally the distance between the scouts and TOWs varies between 500 meters to 1,500 meters. TOW vehicles should respond quickly to enemy contact in order to assist in the scout platoon's development of the situation. Because of the reduced fire power in the HMMWV scout platoon, the TOW systems may provide the scouts with additional security in the form of overwatch during reconnaissance operations.

Position the mortars in or near the center of the troop zone/sector to provide indirect-fire support across the troop front. Keep them positioned to fire about one third to two-thirds of their range (about 3 kilometers to 3.5 kilometers, terrain dependent) beyond the scouts, so the scouts can engage enemy forces at long range with indirect fires. (Line of sight is the determining factor in range forward of the scouts. In densely wooded terrain, the scout's visibility may be only 100 to 200 meters forward of their front line trace, so mortar range need not always be 3,000 meters forward of the scouts.)

TROOP SPLIT-VEE

The troop split-vee is a variation of the troop vee, and can be used in reconnaissance and security missions (see Figure 2-11). In the split-vee formation, the scout platoons work abreast and forward of the troop. The two tank/AT platoons work in depth of the scout platoons, but each tank/AT platoon initially follows a single scout platoon in its zone/sector. The scout platoons conduct their reconnaissance or screening mission within their zone/sector, and the tank/AT platoons key their movement on the progress of the scout platoons. Give the tank/AT platoon leaders command guidance on where they should move in relation to the scout platoons. Tank/AT platoon leaders eavesdrop on the troop command net to monitor the progress of the scouts and to move their platoons to maintain their orientation with the scouts based on the troop commander's guidance.

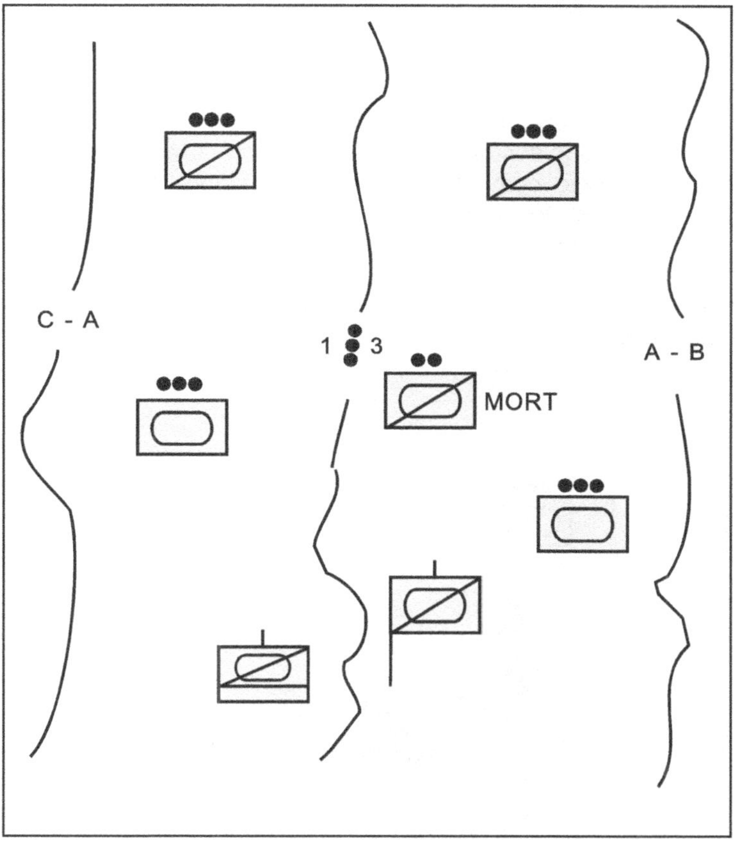

Figure 2-11. Troop split-vee formation.

Develop a habitual relationship between a scout platoon and a tank/AT platoon. First and second platoons work together, and third and fourth platoons work together. This builds teamwork between the scout and tank/AT units, and gives them confidence in each other.

All four platoons still work for the troop commander in this formation. The purposes of having a tank/AT platoon work with a scout platoon are to have combat power spread across the entire troop front, to keep the tanks/TOWs in a position to respond rapidly in support of a scout platoon, and to have scouts available to reconnoiter routes and positions for use by the tanks/TOWs. The troop split-vee is often used in close or restrictive terrain that reduces the tank/AT platoon's mobility so that it cannot quickly maneuver across the troop zone/sector. Generally the troop commander is the only one who commits the tank/AT platoons to conduct a hasty attack, to counterattack, or to move into overwatch positions. If he must mass the combat power of both tank/AT platoons, the troop commander can still shift the tank/AT platoons across the troop zone/sector during the mission (see Figure 2-12).

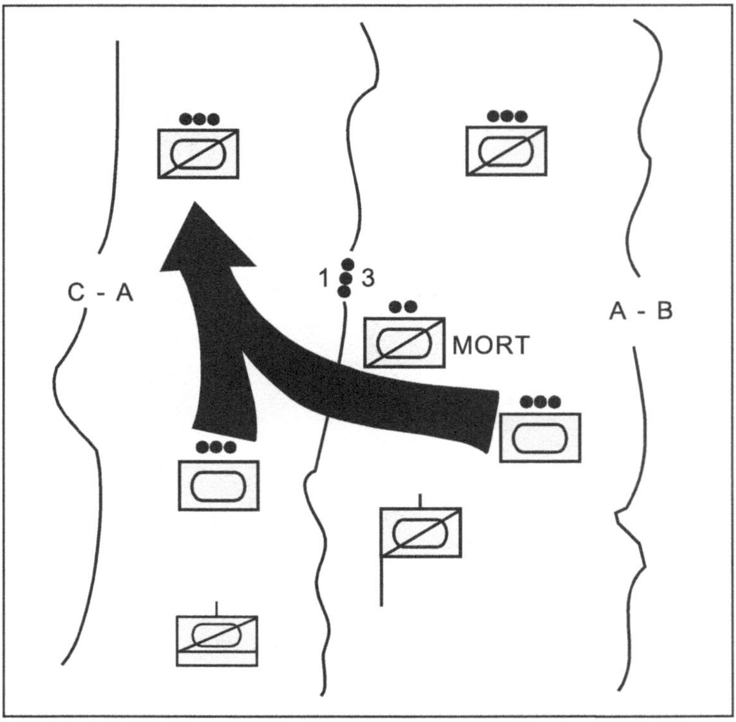

Figure 2-12. Troop commander maneuvers his tank platoons (troop split-vee).

LIGHT TROOP SCRAMBLE

The light troop scramble is a task organization that allows the troop commander to use a mix of weapon systems to enhance acquisition and destruction of the enemy under reduced visibility conditions and in restrictive terrain, where massed TOW fires are not possible. This task organization can also be used to respond to multiple but separate taskings of a troop usually associated with operations other than war, for example, manning checkpoints and area security missions. The troop commander may maneuver his troop in a vee, split-vee, column, or line formation when task organized into a scramble.

The troop is task organized into four like platoons of seven vehicles each. Five scouts and two AT vehicles are in each platoon (see Figure 2-13). The platoon may task organize and operate two ways:

- TOWs integrated with the scout sections (see Figure 2-14).
- TOWs maneuvering as a section to overwatch the movement of the scouts (see Figure 2-15).

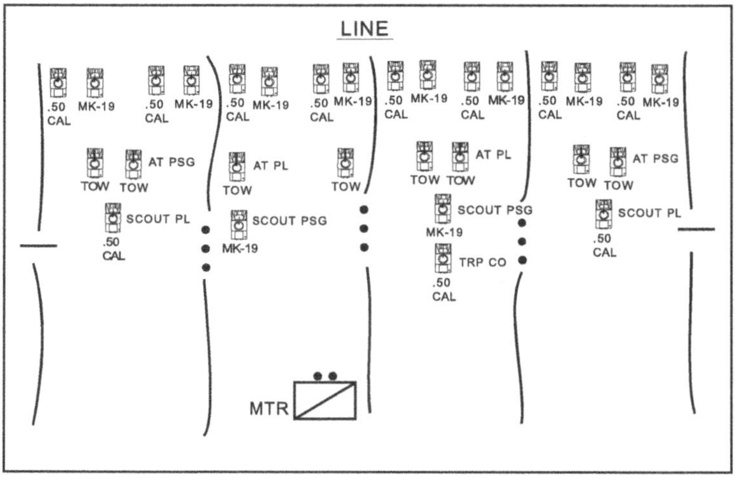

Figure 2-13. Light troop scramble.

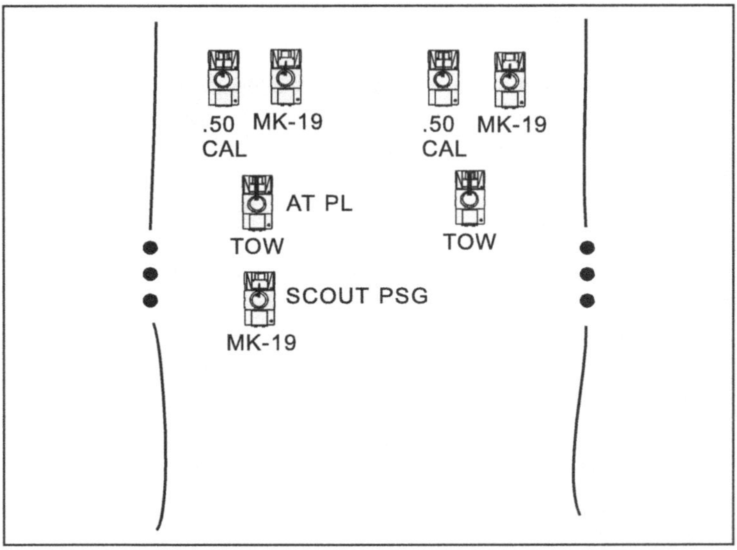

Figure 2-14. TOWs integrated with scout section.

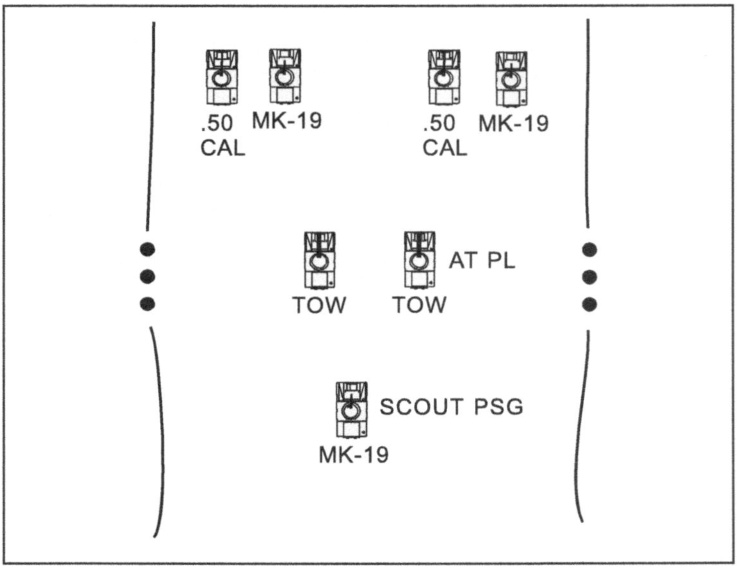

Figure 2-15. TOWs maneuvering as a section to provide overwatch.

Chapter 3

Reconnaissance

Reconnaissance is a mission undertaken to obtain information about the activities and resources of an enemy or about the meteorologic, hydrographic, or geographic characteristics of a particular area. Reconnaissance produces combat information. Combat information is a by-product of all operations acquired as they are in progress. Reconnaissance, however, is a focused collection effort. It is performed before and during other combat operations to provide information used by the commander to confirm or modify his concept.

Cavalry is the corps or division commander's principal reconnaissance organization. Cavalry troops perform reconnaissance using a combination of mounted and dismounted techniques. The cavalry troop is uniquely organized, trained, and equipped to perform the crucial task of reconnaissance for other combined arms forces. Reconnaissance is the cavalryman's specialty.

CONTENTS

Page

Section I. Purpose, Fundamentals, and Capabilities

PURPOSE

Cavalry troops conduct reconnaissance forward of another friendly force to provide information about the terrain and enemy within the area. The reconnaissance allows the follow-on force an opportunity to maneuver freely and rapidly to its objective. Reconnaissance keeps the follow-on force from being surprised or interrupted, and protects it against losing soldiers and equipment on the way to the objective. The cavalry troop performs three types of reconnaissance—route, zone, and area.

FUNDAMENTALS

Successful reconnaissance operations are planned and performed with six fundamentals in mind.

- Maximum reconnaissance force forward.
- Orient on the location or movement of the reconnaissance objective.
- Report all information rapidly and accurately.
- Retain freedom to maneuver.
- Gain and maintain enemy contact.
- Develop the situation rapidly.

Maximum Reconnaissance Force Forward. In reconnaissance, every pair of eyes makes a difference. Do not keep scouts in reserve. This does not mean that every scout should be forward in a strictly linear sense, but actively employed in the conduct of the reconnaissance.

Orient on the Location or Movement of the Reconnaissance Objective. From the IPB (intelligence preparation of the battlefield) the S2 will identify gaps in the squadron's knowledge of the enemy and terrain it will be operating in. These gaps together with the commander's guidance form the PIR (priority intelligence requirements) and direct the reconnaissance efforts of the troop. Focus the efforts of the troop in selecting a course of action and a scheme of maneuver that continuously orients the troop on its reconnaissance objective based on the commander's guidance. The troop's focus may be a terrain feature, a specific area, or an enemy force.

Report All Information Rapidly and Accurately. Higher commanders base their decisions and plans on the battlefield information cavalry troops gather during reconnaissance. Combat information loses value as it ages, so the fresher the better. Scouts report exactly what they see. Troop commanders and executive officers may summarize or clarify reports to the squadron, but must never delete raw information reported by the scouts. Information that seems unimportant to the troop may be extremely valuable at higher echelons.

Retain Freedom to Maneuver. Cavalry troops must be able to maneuver on the battlefield to continue the reconnaissance mission. If the troop becomes decisively engaged, the reconnaissance stops. Use of proper movement techniques and overwatch helps prevent decisive engagement. IPB provides the troop commander information that allows him to anticipate battlefield events and to retain freedom of maneuver.

Gain and Maintain Enemy Contact. Contact is any condition ranging from a surveillance sighting to engaging in close combat. Surveillance is often sufficient and is the preferred method of maintaining contact. When necessary or required, troops will use fire and maneuver to maintain contact with an enemy force. Once contact is gained, however, it is not lost unless ordered by higher headquarters.

Develop the Situation Rapidly. During reconnaissance operations cavalry troops frequently and repeatedly encounter situations that require action to determine what the troop must face. These situations may be terrain oriented, obstacles, or enemy. Terrain or obstacle situations require close reconnaissance, bypass, hasty breach, if necessary, and marking. If an enemy force is encountered, the troop will determine the size, composition (What is the enemy force made up of—tanks, personnel carriers?), disposition (Is the enemy force in an offensive or defensive posture? Is he dug in? What is his orientation?), and activity. IPB provides the threat situational information that guides the effort. Reconnaissance techniques, often in the form of drills, are used while developing the situation.

Troops must develop the situation rapidly to get inside the enemy's decision cycle and force him to react to the actions of the troop rather than the troop reacting to the enemy. This requires the troop to rapidly execute development of the situation at the platoon level through the execution of rehearsed battle drills.

CAPABILITIES

The ability of a cavalry troop to conduct reconnaissance is a function of the enemy situation, the terrain the troop is operating in, and the type of troop conducting the reconnaissance.

- Heavy Troop
  - Can reconnoiter up to a 10-kilometer wide zone.
  - Can reconnoiter up to two routes simultaneously.
  - Can conduct reconnaissance at the rate of about 1 kilometer per hour, depending on the terrain.
  - When faced with a heavy-equipped threat, will conduct either aggressive or stealthy reconnaissance, depending on the higher commander's guidance.
- Light Troop
  - Can reconnoiter up to a 10-kilometer wide zone.
  - Can reconnoiter up to two routes simultaneously.
  - Can conduct reconnaissance at the rate of about 1 kilometer per hour, depending on the terrain.

- When faced with a heavy-equipped threat, will conduct stealthy reconnaissance and aggressive reconnaissance when reinforced with heavy forces.
- When faced with a light-equipped threat, will conduct either aggressive or stealthy reconnaissance, dependent on the commander's guidance.

Section II. Planning Considerations, Methods, and Techniques

The purpose of this section is to outline the planning considerations, methods, and procedures a cavalry troop uses to execute reconnaissance missions.

PLANNING CONSIDERATIONS

When planning a reconnaissance mission, the troop commander must take into account his unit's capabilities and limitations and consider the following:

- Time available from mission receipt to completion.
- Threat size, composition, disposition, and will to fight.
- Terrain and weather effects on the troop's ability to maneuver.
- Tempo of the operation.
- The squadron commander's (SCO) intent and guidance, as follows:
  - What are the SCO's focus and desired endstate of the reconnaissance?
  - What triggers the squadron's employment of the tank/AT company?
  - What does the SCO want the troop to destroy, fix, and bypass?
- Task organization or reinforcements.
- Critical tasks to be accomplished by the troop. Specifically, identify which critical tasks may be deleted during the reconnaissance.

Based on the considerations above, the troop commander determines the following:

- What is the focus of the reconnaissance? What critical tasks must be accomplished within the constraints of time and terrain?
- What specified or implied missions are associated with the squadron endstate?

Note. If given a limit of advance (LOA) for the reconnaissance mission, the troop commander should plan to screen along the LOA (see Chapter 4).

- How will the troop deal with enemy contact? What are the troop's criteria for engagement, destruction, and bypass?
- Under what situations does the commander see employment of his tank (heavy troop) or antitank (AT) (light troop) platoons?
- How will the commander use indirect fires from mortars and artillery to support his maneuver?
- Who controls the troop's attachments and how are they integrated into the reconnaissance?

The troop may receive attachments from higher headquarters. These assets may be maintained under troop control or tasked down to platoons for their use in the execution of the platoon's specified tasks. Examples include attached engineers, ground surveillance radar (GSR), or a chemical reconnaissance element.

- **Engineers.** If an engineer platoon is attached to the troop, the commander may elect to keep them under his control and treat them as a maneuver platoon. He may assign them to conduct the reconnaissance of the route, while the scout platoons move just ahead and reconnoiter terrain on either side. If the troop receives a squad or section, the troop commander may elect to task it to the scout platoon conducting the reconnaissance of the route.

- **GSR**. If a GSR squad or section is attached, the troop commander may elect to task organize them to a platoon or keep them under troop control. During reconnaissance operations, GSRs may be focused on flank avenues of approach into the troop zone with their movement controlled by the scout platoon leader or commander. In either case the commander should give clear guidance for positioning and orientation of the reconnaissance.
- **Chemical reconnaissance element.** If a chemical reconnaissance element (squad, section, or platoon) is task organized to the troop, the commander may maintain the element in reserve to reinforce a scout platoon that comes in contact with a contaminated area, or he may task organize the element down to one of the scout platoons.

METHODS OF RECONNAISSANCE

There are three methods of conducting reconnaissance at the cavalry troop level: *dismounted, mounted, and reconnaissance by fire.* The troop commander may use any method or combination of methods to accomplish the reconnaissance mission under the restrictions placed on him by METT-T (mission, enemy, terrain, troops, and time available) and the higher commander's intent and guidance. A fourth method of reconnaissance, *aerial reconnaissance*, may be executed by an air cavalry troop conducting a coordinated reconnaissance forward of the ground troop.

Dismounted Reconnaissance

The troop commander may direct scouts to conduct dismounted reconnaissance when—

- Time is not a limiting factor.
- Detailed information is required.
- Stealth is required.
- Enemy contact is expected or has been achieved through visual means.

- Scout vehicles cannot move through an area because of terrain.
- Security is the primary concern.

Dismounted reconnaissance permits the cavalry troop to collect the most detailed information about the terrain and enemy within a given zone, area, or route. However, dismounted reconnaissance is also the most time-consuming of all the reconnaissance methods. The cavalry troop, heavy or light, is limited in the number of dismounted scouts it can employ at any time.

Mounted Reconnaissance

The troop commander directs scouts to conduct mounted reconnaissance when—

- Time is limited.
- Detailed reconnaissance is not required, or mounted method affords the same opportunity to collect information as the dismounted method.
- Enemy locations are known.
- Enemy contact is not likely.
- An air cavalry troop is conducting a coordinated air reconnaissance.

Reconnaissance by Fire

When conducting reconnaissance by fire, the troop places direct and/or indirect fire on positions the enemy is suspected of occupying. This action causes the enemy to disclose his presence by moving or by returning fire. The troop commander may use reconnaissance by fire when—

- Time is critical.
- Natural or man-made obstacles that could be overwatched by an enemy force are encountered.
- A suspected enemy position fits the situational template.
- Bunker complexes that may or may not be occupied are encountered.
- Enemy locations are known.

The disadvantage of the reconnaissance-by-fire method is that the troop will lose any element of surprise it may have had. However, reconnaissance by fire may reduce the chance of some portion of the troop being caught in an enemy kill zone. Reconnaissance by fire may not always provide the desired effect. A well-disciplined force will resist the inclination to move when probed by weapon fires.

When indirect-fire situations exist, the troop commander ensures scouts are in a position to observe the target area. Once the decision is made to use reconnaissance by fire, weapons should be used in the following priority.

- Indirect-fire systems.
- Machine gun.
- 25-mm chain gun or MK-19.
- TOW or tank cannon fire.

Reconnaissance by fire does not mean the indiscriminate use of direct and indirect fires at all woodlines and hilltops in the hopes of causing the enemy to react. Not only will the enemy recognize this ploy for what it is and not react to it, but also it wastes valuable ammunition.

Aerial Reconnaissance

Air cavalry troops perform aerial reconnaissance when time is critical. Aerial reconnaissance is often coordinated closely with a ground reconnaissance troop. The air and ground forces complement each other. The air element can move forward of the ground unit and reconnoiter key pieces of terrain or restrictive terrain, allowing the ground troop to concentrate its efforts in other areas or to increase the tempo of its reconnaissance. The air troop provides the ground troop with added security by clearing the ground forward of the ground unit, thereby facilitating movement of the ground force and quickening the pace of the operation. The ground troop can move rapidly mounted to the areas of interest within its area or zone of operations and have the time to dismount and collect detailed information (see Figure 3-1).

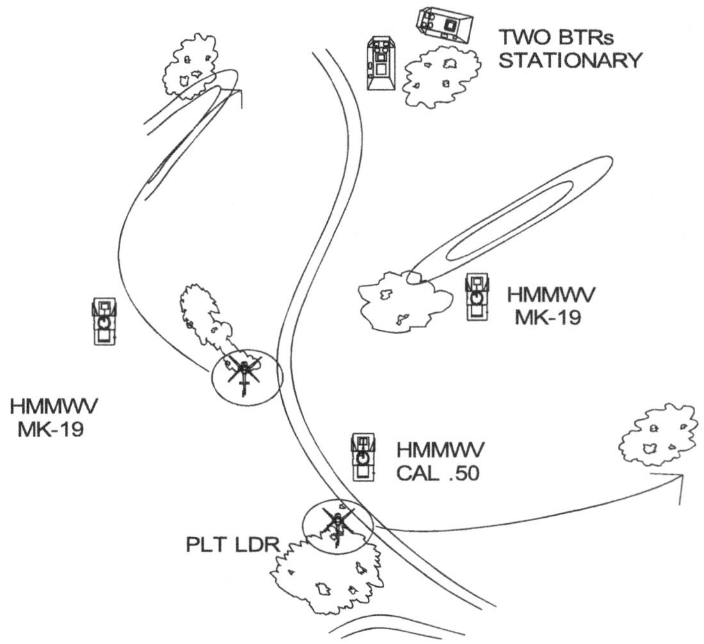

- Air cavalry platoon identifies two stationary BTRs along the woodline.
- Air cavalry platoon leader sends spot report to air cavalry troop commander. Troop commander tells platoon leader to develop situation and get in contact with ground platoon leader in zone.
- Air cavalry platoon leader sends information to ground platoon leader and assists ground scout section in developing situation.
- Air cavalry troop commander and ground troop commander choose a course of action.
- This coordination of air-ground supports squadron commander's intent, which increases troop survivability and squadron operating tempo of reconnaissance.

Figure 3-1. Air scouts conduct coordinated reconnaissance of zone with ground troop.

Aggressive Versus Stealthy Reconnaissance

The cavalry troop uses various combinations of reconnaissance methods. The method or combination of methods chosen by the troop and its scout platoons is based on the higher commander's guidance and METT-T. The method or methods chosen will characterize the reconnaissance as either aggressive or stealthy.

The stealthy approach to reconnaissance is more time-consuming. It emphasizes avoiding detection and engagement by the enemy. To be effective, a stealthy approach must rely primarily on dismounted reconnaissance and maximum use of covered and concealed terrain.

Aggressive reconnaissance is characterized by the speed and manner in which the reconnaissance force develops the situation once contact is made with an enemy force. A troop conducting aggressive reconnaissance uses firepower from direct- or indirect-fire systems and maneuvers to rapidly develop the situation. The troop will primarily use mounted reconnaissance and reconnaissance by fire when conducting aggressive reconnaissance.

Cavalry troops must be trained to conduct both aggressive and stealthy reconnaissance. The troop commander, scout platoon leader, and squad leaders will most likely use many different methods over the course of a reconnaissance operation. The constraints of commander's guidance, the opposing threat, the capabilities of the troop, the terrain the troop is operating in, and the time available to complete the mission (METT-T) all have an impact on the method or methods chosen by the troop leadership.

TECHNIQUES

Actions on Contact

Actions on enemy contact are a series of steps the troop takes when it encounters an enemy force or situation that warrants/demands action. Actions on contact are important because they allow the troop to maintain its tempo of operation by rapidly developing the situation and taking action before the enemy can gain the initiative and force the troop to react. At platoon level, actions on enemy contact consist of four steps.

- Deploy and report.
- Develop the situation.
- Choose a course of action.
- Recommend or execute a course of action.

Note. See FM 17-98 for an in-depth discussion of platoon actions on contact during reconnaissance.

While the platoon that makes contact executes actions on contact, the commander must continue to maneuver the remainder of the troop to ensure a clear picture of the enemy situation across the entire troop front. The following steps demonstrate the actions taken by the platoon in contact and the corresponding actions at the troop level.

Deploy and Report

Platoon Action. The elements of the scout platoon that make initial contact with the enemy immediately deploy to terrain that affords them both cover and good observation. If necessary, scouts return fire to suppress the enemy, and then deploy to their positions. The scout making contact sends a contact report to his platoon leader. The platoon leader forwards the report to the troop commander. Once the scout in contact is in a good covered and concealed position, he sends his initial spot report, which is forwarded through the platoon leader to the troop commander (see Figure 3-2).

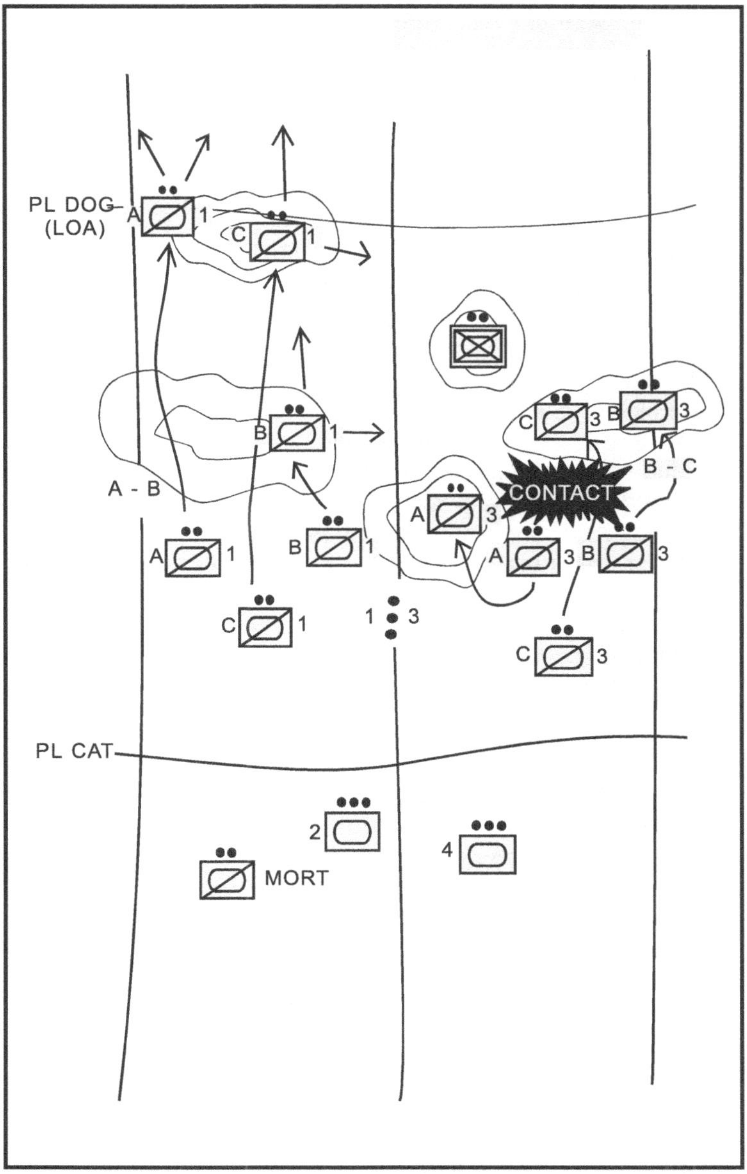

Figure 3-2. Scouts deploy and report then begin to develop the situation.

Troop Action. All other platoon leaders and platoon sergeants monitor the contact report. The troop commander assesses the information and moves, if necessary, to a position where he can observe the action. However, the troop must not lose focus of the reconnaissance mission.

Develop the Situation

Platoon Action. Next, the platoon in contact defines what it is up against. Scouts use dismounted and mounted reconnaissance to determine the enemy's size, composition, and orientation, and the exact location of weapon systems. The platoon may also use reconnaissance by fire to determine the enemy's tactical intentions. The reconnaissance-by-fire technique should, however, be conducted with indirect-fire assets when possible to avoid revealing the scouts' position. The scout platoon will search for minefields, wire, antitank ditches, and other obstacles that could force a friendly unit into a fire sack or cause it to flank itself to the enemy. To determine if the enemy can be supported by any other forces, the scouts search for enemy flanks and scour all adjacent terrain. They identify good counterattack routes into the flanks or rear of the enemy. Once the platoon leader determines the extent of the situation, he forwards a follow-up spot report.

Troop Action. The troop commander will most likely tell the scout platoon not in contact to continue its reconnaissance to a designated LOA to develop the situation across the entire troop front. By doing this, the troop can determine if there are any other enemy forces with which the troop must be concerned. The scout platoon not in contact will establish hasty observation posts along the LOA oriented on likely enemy locations or avenues of approach. The tank platoons will continue to monitor the troop command net and prepare to act based on the commander's intent/guidance and the enemy situation.

Choose a Course of Action

Platoon Action. Now that the platoon leader knows what he is up against, he considers two or three possible courses of action and selects the one that best meets the commander's intent/concept of the operation, is within his capabilities, and allows the troop to resume its reconnaissance mission as soon as possible. The possible courses of action open to him, based on the commander's intent/concept, might be hasty attack, bypass, hasty defense/screen, or support of a hasty attack by another platoon(s). Some courses of action will quickly be ruled out because they do not meet the commander's intent and guidance for the operation.

- **Hasty Attack.** The platoon leader can conduct a hasty attack if he has enough combat power to defeat the enemy quickly. In most cases, the scout platoon does not have the capability to mass enough combat power to defeat an enemy in prepared positions. In addition, the scout platoon leader may not want to risk battle losses that would reduce his effectiveness and his ability to complete the mission.

- **Bypass.** If the platoon does not have enough combat power to conduct a hasty attack, or if it wants to remain undetected and continue the reconnaissance mission, the scout platoon can bypass the enemy. The platoon leader must receive the troop commander's permission to bypass. If he has permission to bypass, the platoon leader must leave scouts in contact with the enemy force unless ordered to break contact by the troop commander. The platoon leader will know from the troop commander's guidance in the operations order if this is a viable course of action.

- **Hasty Defense/Screen.** If the platoon cannot conduct a hasty attack and cannot bypass, it establishes a hasty defense or screen. The platoon will conduct a hasty defense if it can defend against an enemy force. If the enemy contact exceeds the platoon's capability to conduct a hasty defense, it may elect to establish a screen and maintain contact through observation. The platoon concentrates on maintaining contact with the enemy and fixing it in place with indirect or possibly direct fire until additional support comes from the troop.

- **Support a Hasty Attack by Another Platoon(s).** The platoon in contact may become the support element for a hasty attack by a tank/AT platoon(s).

Troop Action. The troop commander continues to maneuver the troop and assists the platoon leader as required. He maintains situational awareness.

Recommend/Execute a Course of Action

Platoon Action. If the course of action the platoon leader chooses meets the commander's guidance for actions on contact, the platoon leader executes the course of action. If the situation dictates that the platoon execute a hasty defense/screen, or support a hasty attack by another platoon, the platoon leader reports this information to the troop. The report is brief. He updates his spot report (Blue 1) with any additional information, tells the troop commander what he is doing about the situation, and recommends the course of action that he thinks best suits the situation. The platoon in contact has then completed its actions on contact.

Troop Action. The commander must approve or disapprove the recommended course of action, based on its effect on the troop and squadron mission.

If the scout platoon is required to establish a hasty defense/screen, the troop commander assumes responsibility for continuing to develop the situation. The commander can do this best in a position from which he can influence troop actions. Based on what the scouts have found and reported and what the commander can see, he must weigh the advantages and disadvantages of each course of action, and choose the one that best meets the troop's mission requirements.

- **Hasty Attack**. The troop can conduct a hasty attack, which throws the weight of one or both tank/AT platoons against the enemy and destroys him quickly, then continue its reconnaissance. The commander may coordinate the efforts of the scout platoon and tank/AT platoon(s) in the hasty attack or delegate the responsibility to the scout platoon in contact. The hasty attack at troop level should have two elements: the *assault/attack-by-fire* element and the *support* element.

  Steps to the execution of a hasty attack are as follows:

  - The scouts determine that the enemy force encountered cannot be bypassed based on the enemy disposition and composition. The enemy encountered meets the commander's intent for destruction.

  - The scout platoon leader recommends hasty attack to the troop commander and identifies a good attack-by-fire or assault position for the tank/AT platoons.

  - The troop commander approves the scout platoon leader's recommended course of action and issues orders to execute a hasty attack.

  - The troop commander decides to use either attack by fire or assault as his method of destruction. Heavy troops equipped with tanks can perform attack by fire or assault. Because of the defensive nature of the TOW, light troops tend to use attack by fire when conducting hasty attack.

  - The scout squad or section moves to link up with the tank/AT platoons and guides them to an assault/attack position. The assault element reports when set in assault positions, and if time is available, reconnoiters the attack-by-fire position or the attack axis.

- The troop commander moves to the scouts who are positioned in overwatch of the enemy contact. He also positions the FIST in overwatch to assist in controlling indirect fires.

- The troop commander designates the scouts in overwatch as the support element. He establishes direction of fire and methods of control for direct-fire weapon systems; for example, left and right limits for fires, and signals for the initiation of direct fires and the shifting of fires as the assault element begins its attack.

- The troop commander directs the FIST to fire for effect on enemy positions when given the signal (AT MY COMMAND). The FIST reports when guns are ready.

- The troop commander issues the fire command for the support force and indirect-fire systems and directs the attack-by-fire or the assault element to execute its movement. The troop commander lifts and shifts direct and indirect fires as necessary to cover the movement of the assault element and to seal off the enemy withdrawal (see Figure 3-3).

- **Bypass**. The troop may bypass the enemy force and continue the reconnaissance to further develop the enemy situation throughout the depth of the zone while maintaining an element of surprise. The decision to bypass is based on the higher commander's intent for the operation. The troop should leave an element in contact with the enemy force; however, the more elements in contact with the bypassed enemy mean the less reconnaissance will be conducted.

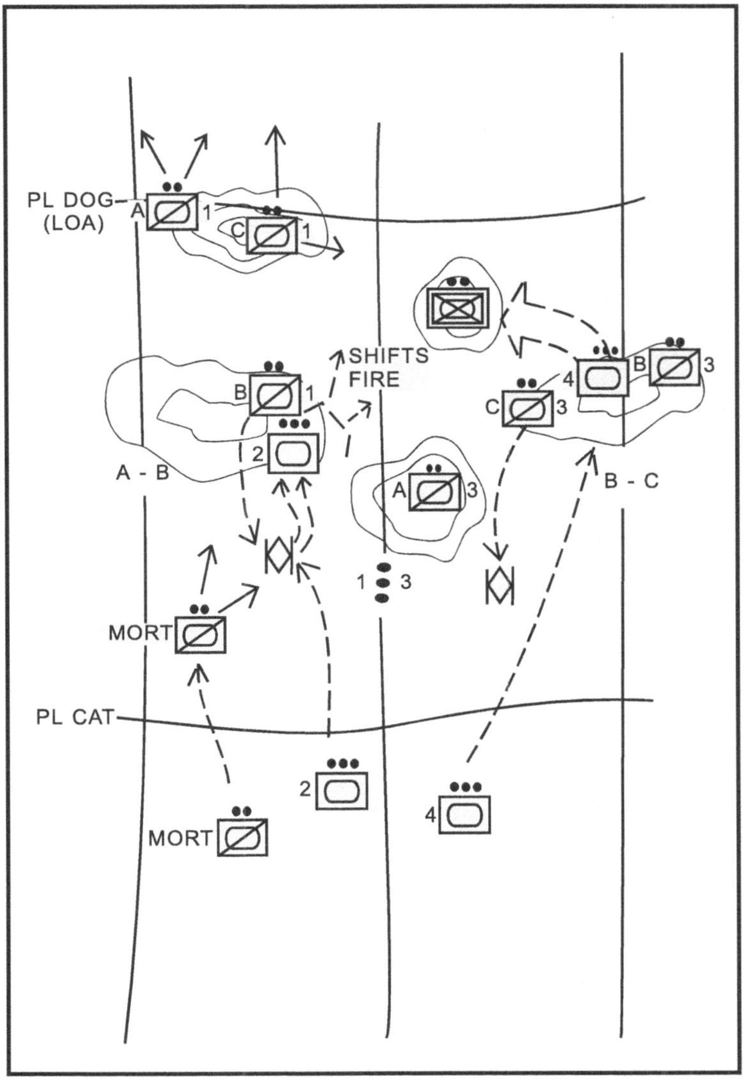

Figure 3-3. Troop positions for hasty attack.

- **Hasty Defense**. If the troop cannot conduct a hasty attack or bypass the enemy, it may establish a hasty defense to fix the enemy while waiting for further orders. If this is the troop's course of action, the squadron commander has the responsibility to develop the situation further.

The overriding considerations in selecting a course of action are the intent of the squadron commander and the troop's ability to complete the mission with minimum losses.

DANGER AREAS

Obstacles

Obstacles encountered during reconnaissance are treated as enemy contact, because all obstacles are assumed to be covered by enemy fire. Upon encountering an obstacle, scouts deploy to covered positions, report, and then begin to develop the situation. The troop commander may move his tank platoon(s)/AT platoon(s) forward to overwatch the obstacle while the scouts conduct dismounted and mounted reconnaissance to determine the following:

- Is the obstacle defended by the enemy? If so, how many enemy soldiers are there and where are they?
- What are the extent and composition of this obstacle? How deep, wide, steep, or long is it?
- Can the obstacle be bypassed, or will breaching be necessary?

Once the scouts have completed the reconnaissance of the obstacle and recommended a course of action (either bypass, hasty attack/breach, or hasty defense), the troop commander must choose the course of action that best meets the commander's intent and guidance for the mission. If the commander's decision is to bypass or breach the obstacle, the troop must mark the obstacle and any lane created through the obstacle (see Figure 3-4).

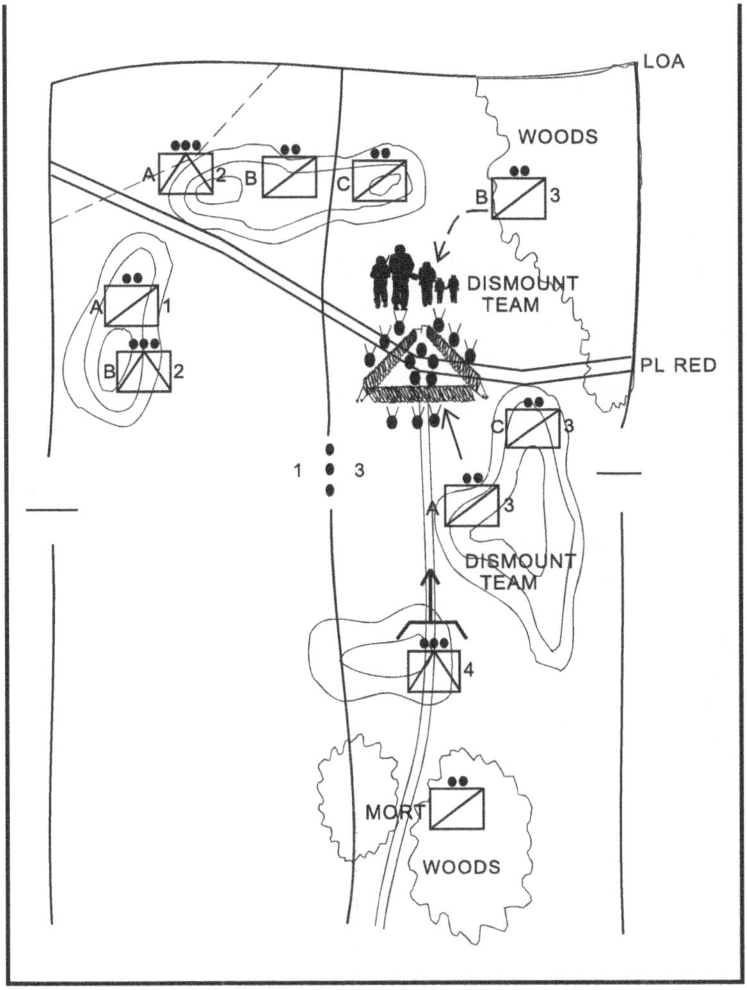

Figure 3-4. The troop develops the situation at the obstacle.

Open Areas

Open areas are dangerous for cavalry troops because they permit the enemy to observe or engage units as they move through them. When reconnoitering a route, zone, or area that includes an open area, do not try to force the reconnaissance through it. Treat it like a danger area. Reconnoiter the flanks using good covered and concealed routes. The troop should try to clear the far side of the open area before attempting to reconnoiter the open area itself. Open areas within the troop's zone or area of operations may force the troop commander to abandon the use of platoon boundaries to facilitate the reconnaissance to the flanks and rear of the open area (see Figure 3-5).

Built-up Areas

In most areas of the world the troop can expect to conduct reconnaissance of built-up areas (BUA). BUAs range from small hamlets to large towns, even portions of cities. BUAs are very dangerous for mounted units and should be avoided when possible. If the situation permits, the troop should conduct a reconnaissance of BUAs from a distance, and then bypass them. However, it may be necessary to move through a town as part of the troop's reconnaissance mission.

The troop's ability to conduct reconnaissance of a BUA is constrained by the lack of dismounted scouts. The troop is capable of finding the following in BUAs:

- Enemy vehicles.
- Enemy command and control facilities.
- Obstacles such as rubble, blown bridges, and craters.
- Logistics elements.
- Bypasses within the BUA.

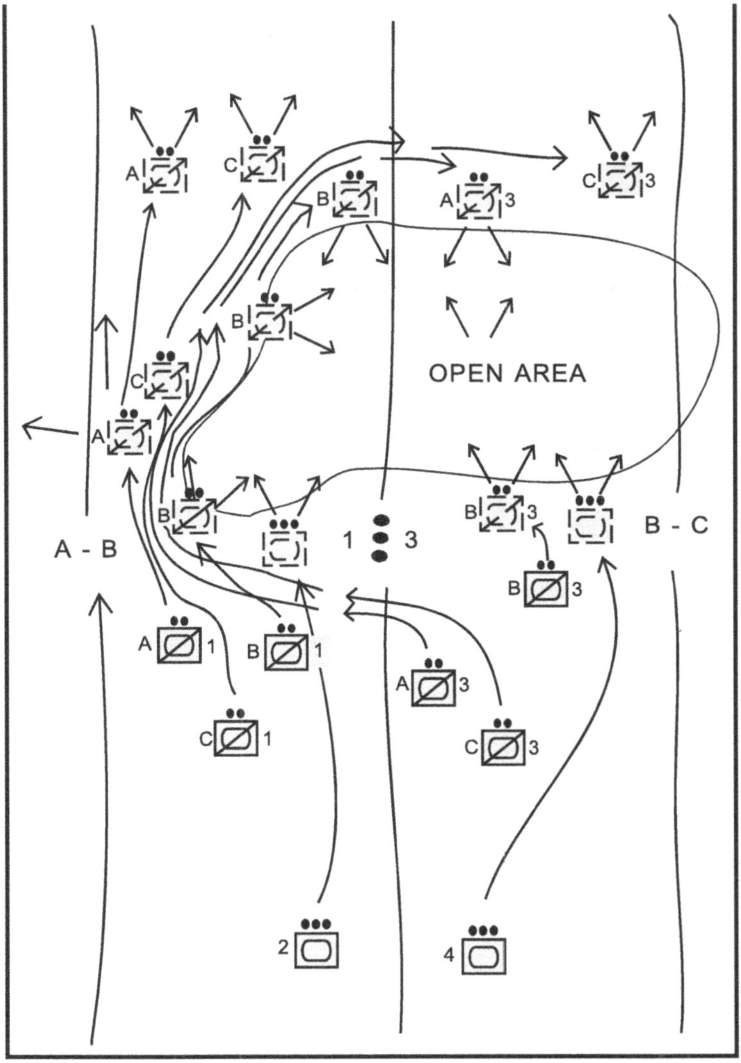

Figure 3-5. The troop reconnoiters to the flanks and rear of an open area.

If the troop is not augmented, it is not capable of performing the following in BUAs:

- Reconnoitering each building and street.
- Reconnoitering underground structures (basements and sewers).
- Determining the strength of dug-in enemy units.
- Determining the detailed disposition of dismounted enemy units.
- Conducting hasty attacks against dismounted enemy units of larger than platoon size.

The following are techniques and considerations for performing reconnaissance in BUAs:

- Dirt roads, alleys, and curves in paved roads are excellent places to employ mines. Structures in BUAs are excellent places to employ booby traps.
- Ask the S2 for detailed street maps of all major BUAs the troop might encounter during an operation. These are more useful than standard 1:50,000 military maps.
- Clearly define platoon zones. Do not divide responsibility for a street between platoons.
- Observe BUAs from outside for signs of enemy activity prior to entering them.
- Main roads through towns are normally accompanied by open areas such as parks, traffic circles, and medians. The enemy may use these same open areas for logistics elements, concentrations of armored vehicles, artillery positions, and combat support vehicle locations.
- Collateral damage constraints may limit the ability to employ some types of weapon systems.

LIGHT TROOP EXAMPLE

The squadron is conducting a zone reconnaissance. The troop zone contains a large town, 2 to 3 kilometers square, and the squadron main supply route, ROUTE GOLD, passes through the town. The troop will have to reconnoiter enough of the town to ensure there are no enemy who can disrupt movement along ROUTE GOLD.

On approaching the town, the troop has options on how to enter it. The techniques used will largely depend on the amount of time available for the mission. If time is limited, the scouts will halt short of the BUA, just off ROUTE GOLD, and observe the town from multiple vantage points before entering. If more time is available, the scouts will move well off the route and observe the BUA before entering. In any case, remember if enemy is present, he will probably cover the main road with fires. Once the platoon is inside the BUA, the main road may provide the best axis for the reconnaissance.

As the scout platoon observes the town, they look for any signs of enemy presence. The antitank platoons move into overwatch positions. The mortars establish a firing position in support of the scouts. Scouts reconnoiter to the flanks and rear of the BUA to find a bypass and establish flank and far side security (see Figure 3-6).

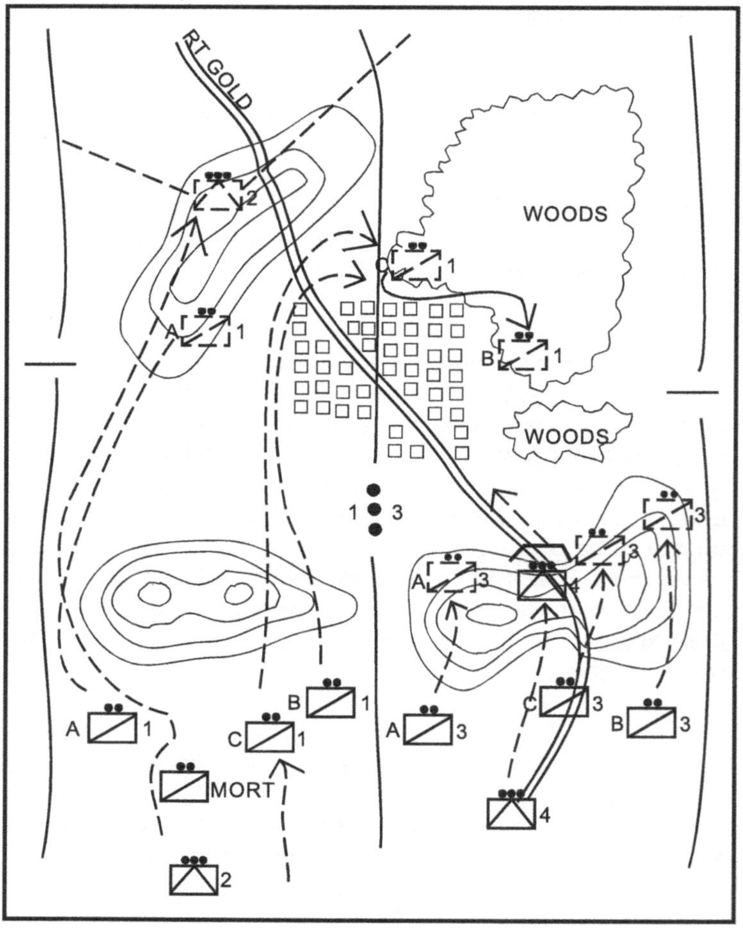

Figure 3-6. The troop securing the BUA with one scout platoon, ATs in overwatch, and one scout platoon preparing to reconnoiter the BUA itself.

Once the area surrounding the BUA is secured, the scouts move mounted to the last concealed position and dismount, searching for suitable entry points into the BUA. *If there is enemy contact*, the overwatching vehicles suppress with direct and indirect fires while the scouts move to a covered and concealed position. At this point the troop commander must decide to maintain contact and bypass, or continue to probe for entry points into the BUA. The determining factors for the commander's decision are the intent of the higher commander and the level of threat in the BUA. In most instances, cavalry troops maintain contact and bypass BUAs occupied by enemy forces. The enemy forces are handed off to follow-on forces for destruction.

If the troop does not encounter enemy contact on its initial reconnaissance of the outskirts of the BUA, the scout platoons continue their search for suitable entry points into the town. Once the scouts determine those entry points into the town are clear, they move into the BUA and reconnoiter nearby buildings (moving through side and back yards, alleys, and other off-street paths) and surrounding areas. The dismounted teams from the scout platoon do not enter or clear each building, but look for obvious signs of enemy presence. HMMWVs may be brought forward to overwatch the dismounted team's movement with caliber .50 machine guns or MK-19s (see Figure 3-7). The AT platoons should continue to overwatch from outside the BUA. After the scouts complete the reconnaissance of the BUA, the troop continues its reconnaissance in zone, ensuring that bypasses around the BUA are marked.

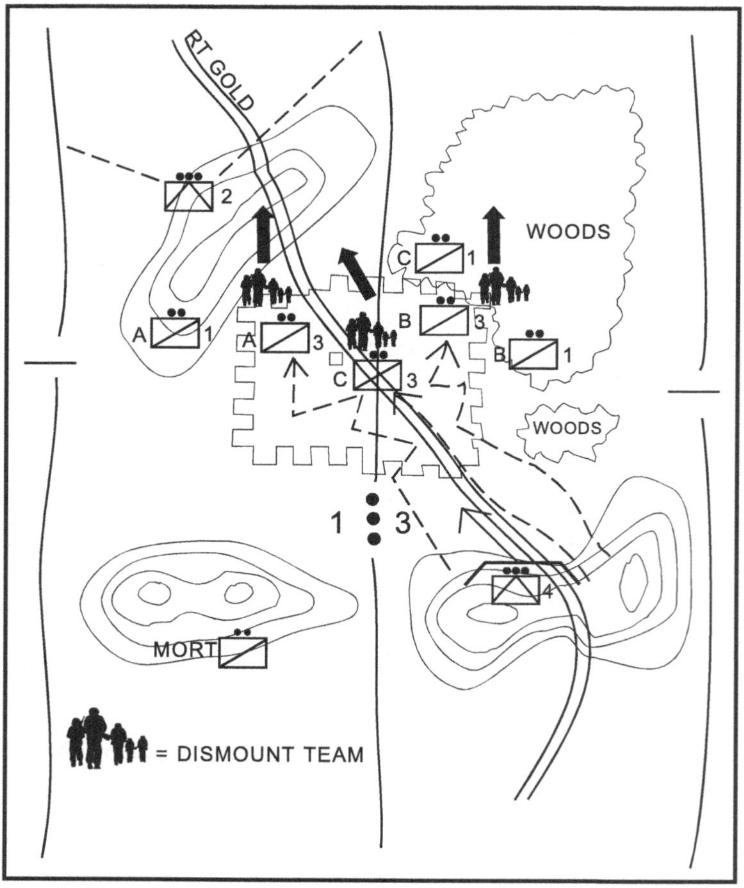

Figure 3-7. Scout section reconnoitering in a built-up area.

Section III. Route Reconnaissance

Route reconnaissance is a directed effort to gain detailed information about a specific route and the terrain on either side of the route that the enemy could use to influence movement along the route. Depending on the terrain surrounding the route, a route reconnaissance mission may be accomplished by a single scout platoon, or it may require the entire troop to cover the terrain and accomplish the required tasks.

A cavalry troop may be assigned the mission to conduct a zone reconnaissance (see Section IV of this chapter), but also have the task of reconnoitering a route that is within its zone. In this case, the troop mission would be zone reconnaissance. However, one of the subordinate units in the troop, scout platoon, or attached engineers may be given the route reconnaissance mission by the troop commander.

CRITICAL TASKS

During a route reconnaissance, the following critical tasks must be accomplished unless the SCO directs the troop to do otherwise:

- Reconnoiter and determine trafficability of the route.
- Reconnoiter all terrain the enemy can use to dominate movement along the route.
- Reconnoiter all BUAs along the route.
- Reconnoiter all lateral routes in the area of responsibility.
- Inspect and classify all bridges along the route.
- Locate fords or crossing sites near all bridges along the route.
- Inspect and classify all overpasses, underpasses, and culverts.

- Reconnoiter all defiles along the route; clear all defiles of enemy and obstacles within its capability, or locate a bypass.

- Locate mines, obstacles, and barriers, and within its capability, clear the route.

- Locate a bypass around BUAs, obstacles, and contaminated areas.

- Report route information.

- Find and report all enemy that can influence movement along the route.

TECHNIQUES

The SCO directs the troop to conduct a route reconnaissance as a mission or as a specific task in another mission. This section discusses route reconnaissance in the context of an assigned troop mission.

The troop commander considers several factors in formulating his concept.

- Start point, release point, and designation of the route.

- Mission to be performed at the start point and after reaching the release point.

- Time the mission is to start, and if required, be completed.

- Critical points along the route identified as checkpoints.

- Any constraints or restrictions.

- IPB information on the route, to include current enemy situation.

IPB provides critical information on the enemy and terrain. Enemy threats may be encountered in two basic forms.

- Ambushes along the route in close or restricted terrain or even tied to obstacles along the route.

- Attack by long-range direct or indirect fires from dominating terrain along the route.

Analysis of the terrain provides an indication of danger areas and the nature of the potential threat. From this information, the commander determines how much terrain on each flank of the route must be reconnoitered and the organization for combat. Any constraints or restrictions may also influence how much terrain is reconnoitered.

The troop commander adds control measures to provide an adequate framework for the mission. He places a boundary on both sides of the route far enough out to provide reconnaissance of the dominating terrain. A line of departure is placed perpendicular to the route short of the start point, allowing sufficient space to deploy into a tactical formation. An LOA is placed far enough beyond the release point to enclose dominating terrain that overwatches the release point. The commander uses additional phase lines to maintain a coordinated effort throughout the troop. Boundaries and phase lines are drawn along recognizable terrain features. The terrain features should be visible from both ground and air to facilitate air-ground integration. The commander uses other control measures as necessary for flexibility in the maneuver and execution of the mission (see Figure 3-8).

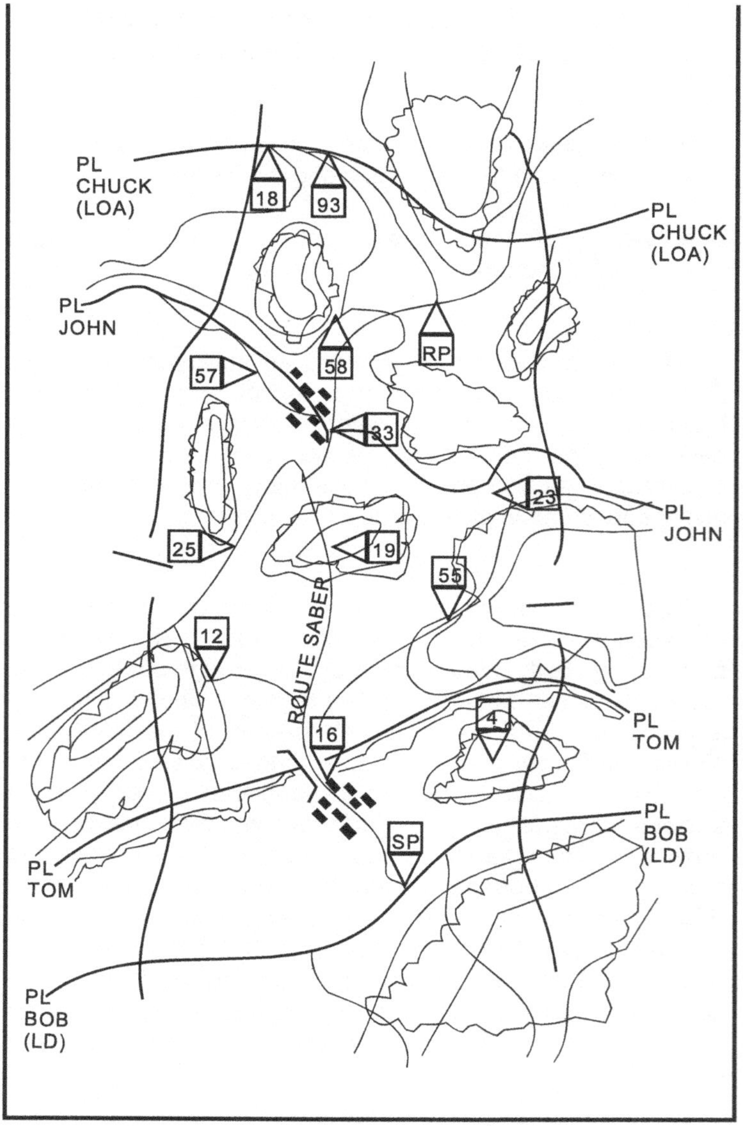

Figure 3-8. Troop graphic control measures for a route reconnaissance.

The cavalry troop normally performs a tactical road march to the line of departure and deploys to execute the reconnaissance of the route. Based on the amount of intelligence known about the enemy, the commander will determine how much security is required for the move forward to the line of departure. Also the commander should consider the effect his final disposition of forces will have on the troop's follow-on mission.

One technique for accomplishing all tasks as rapidly and safely as possible is as follows: Draw a 2.5-kilometer to 3-kilometer boundary on either side of the route. Near the start point, draw a line of departure from one boundary to the other. Next, draw an LOA about 3 kilometers forward of the release point from one boundary to the other. Add phase lines between the line of departure and LOA, as needed, to help control the progress of the troop. To control movement on the route, establish checkpoints. Use checkpoints to identify specific areas or locations to be reconnoitered. These checkpoints will serve as references to orient reconnaissance efforts.

One scout platoon moves out early across the line of departure to reconnoiter the terrain or BUAs on either side of the route. The other scout platoon reconnoiters the route itself, trailing 1 kilometer to 2 kilometers behind the lead scout platoon. This provides a good measure of security for the troop and the platoon working to reconnoiter and classify the route. The mortar section follows the lead scout platoon and remains positioned to range from 3,000 to 3,500 meters forward of the lead scout elements. The mortar section avoids moving on the route and stays on the flanks near the route. The tank/AT platoons follow and support behind the second scout platoon. They key their movement on the advance of the scout platoons. The tank/AT platoons should be kept back in depth to retain flexibility, because they must be ready to react anywhere in the troop zone. The troop trains follow about 2 kilometers behind the tank/AT platoons, and bound from one covered and concealed position to another. The troop CP displaces generally along the route using terrain that affords effective and continuous communications with troop elements and the squadron. The troop commander positions himself to observe the actions of the scout platoons.

EXAMPLE OF A TROOP ROUTE RECONNAISSANCE

The 1st platoon (scout) crosses the LD (PL BOB) using the two-section organization, and reconnoiters the terrain on either side of ROUTE SABER up to PL TOM. The platoon looks for the threat around the outskirts of the village, and explores routes into the village that intersect ROUTE SABER. The scouts also search the woods near checkpoint 4 for the enemy. The platoon scans the terrain north of the stream, and then moves up. One section locates a fording site west of the bridge at checkpoint 16, and the other section swims the stream on the east. The platoon makes a quick visual inspection of the bridge, then searches for the threat in the forested terrain near checkpoints 12 and 55. The mortar section crosses the LD next, about 2 kilometers behind the 1st platoon (scout), and establishes a firing position on the outskirts of the village. The 3d platoon (scout) crosses the SP when the 1st platoon (scout) crosses PL TOM.

The 3d platoon (scout) reconnoiters and classifies the route through the village, then moves up and classifies the bridge just north of town. The troop commander bounds closely behind the 3d platoon (scout) and observes their actions. The 2d and 4th platoons (tank) follow along the flank of ROUTE SABER about 1 kilometer behind the 3d platoon (scout), then move into the outskirts of the village. They observe the dominating terrain on the far side of the stream while the 3d platoon (scout) reconnoiters the bridge. The troop CP and trains remain in concealed positions south of the LD (see Figure 3-9).

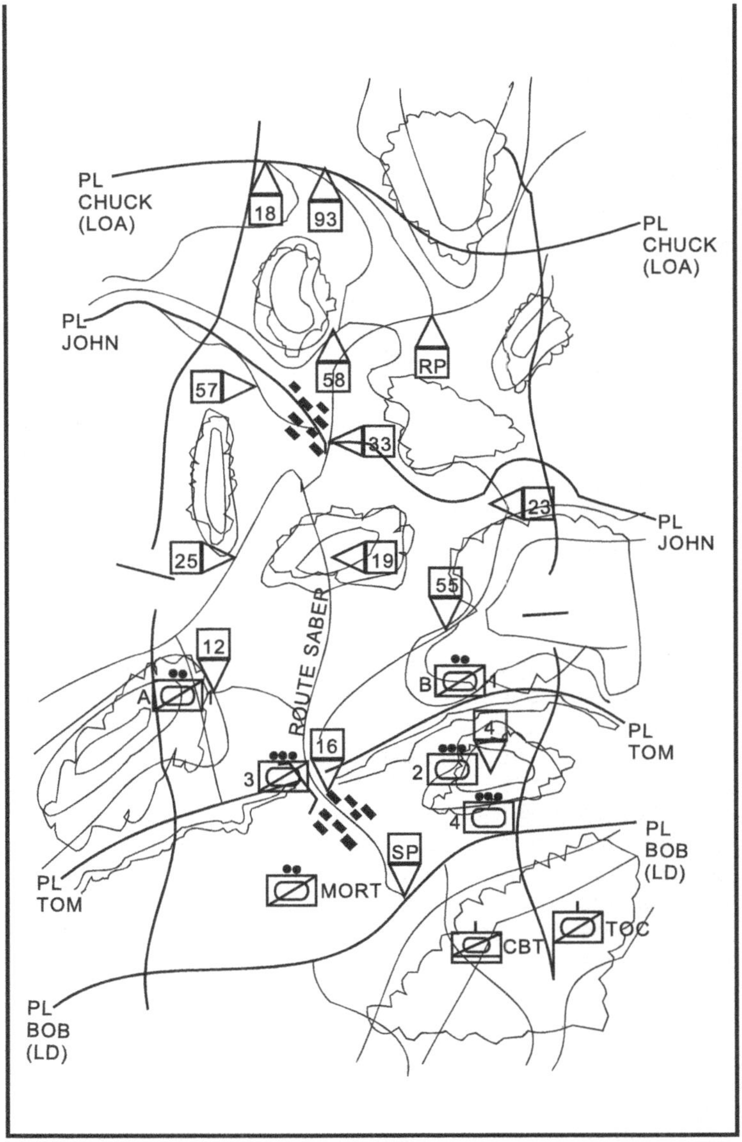

Figure 3-9. Route reconnaissance (part one).

The 1st platoon continues to reconnoiter in zone up to PL JOHN. The platoon searches the wooded areas for the threat, and explores the high-speed lateral routes to the east and west of ROUTE SABER. The platoon clears the shoulders and makes a visual inspection of the defile on ROUTE SABER at checkpoint 19, then moves forward to search the outskirts of the village at checkpoint 33. As with the first village, the platoon looks for threat forces along the roads into the village. The mortar section follows the 1st platoon and establishes a firing position on the edge of the woods southwest of checkpoint 19. The 3d platoon continues classifying the route along ROUTE SABER. At the defile near checkpoint 19, scouts using mine detectors check the route to ensure it is free of mines. The 2d and 4th platoons cross the bridge at checkpoint 16, and then move into concealed positions in the woods on either side of ROUTE SABER near checkpoint 19. The troop CP moves to a concealed position on high ground near checkpoint 12. The troop trains move into covered and concealed positions in the village south of checkpoint 16 (see Figure 3-10).

As the 1st platoon crosses PL JOHN and approaches the RP, the lead scout squad of Bravo section is hit by cannon and heavy machine-gun fire from concealed positions near checkpoint 93. The vehicle has its left track blown off and sustains external damage. The gunner and one scout are wounded by shell fragments. The PSG immediately returns fire and deploys to cover then reports the contact to the platoon leader who forwards a contact report to the troop commander. Once the PSG determines the location of the enemy force, he sends a spot report to the platoon leader and a request for immediate suppression to the troop FSO. After the call for fire is sent forward, the PSG directs his remaining scout squad to search for additional threat forces in the woods north of the RP and along the high-speed route heading northeast (see Figure 3-11). As mortar fire falls on the enemy position, the 1st platoon leader quickly moves up near checkpoint 58 to observe the threat. He directs his section to search the woods north of checkpoint 57, and to move around to the west flank of the threat and find out

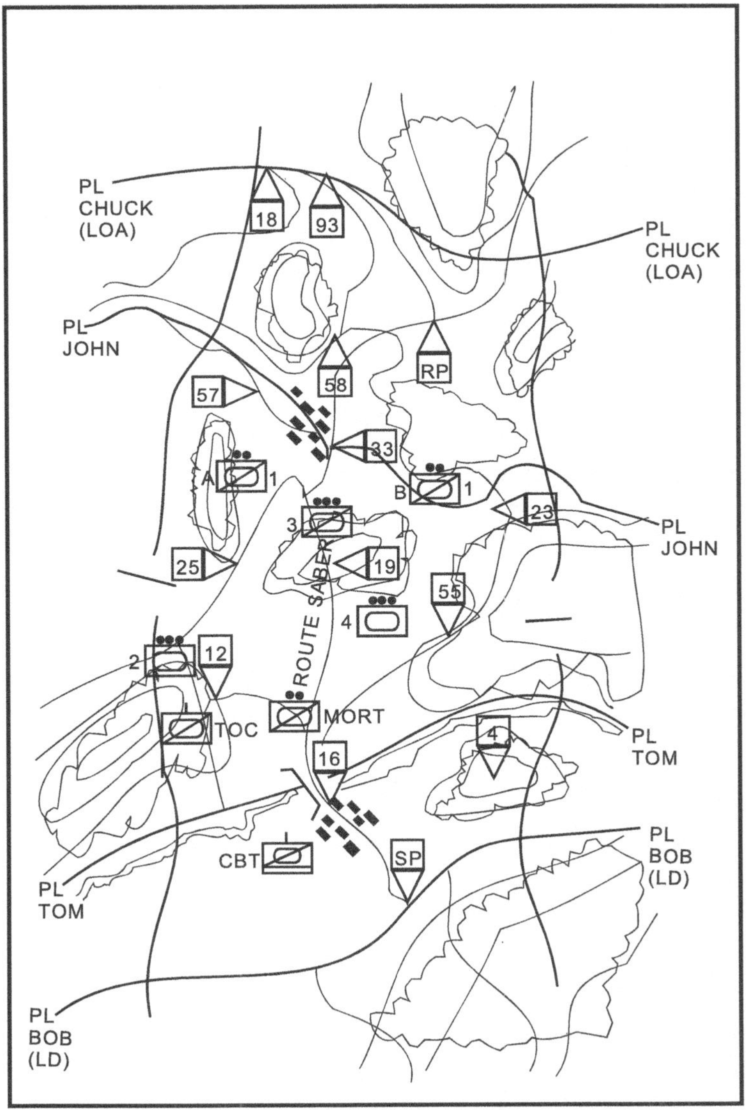

Figure 3-10. Route reconnaissance (part two).

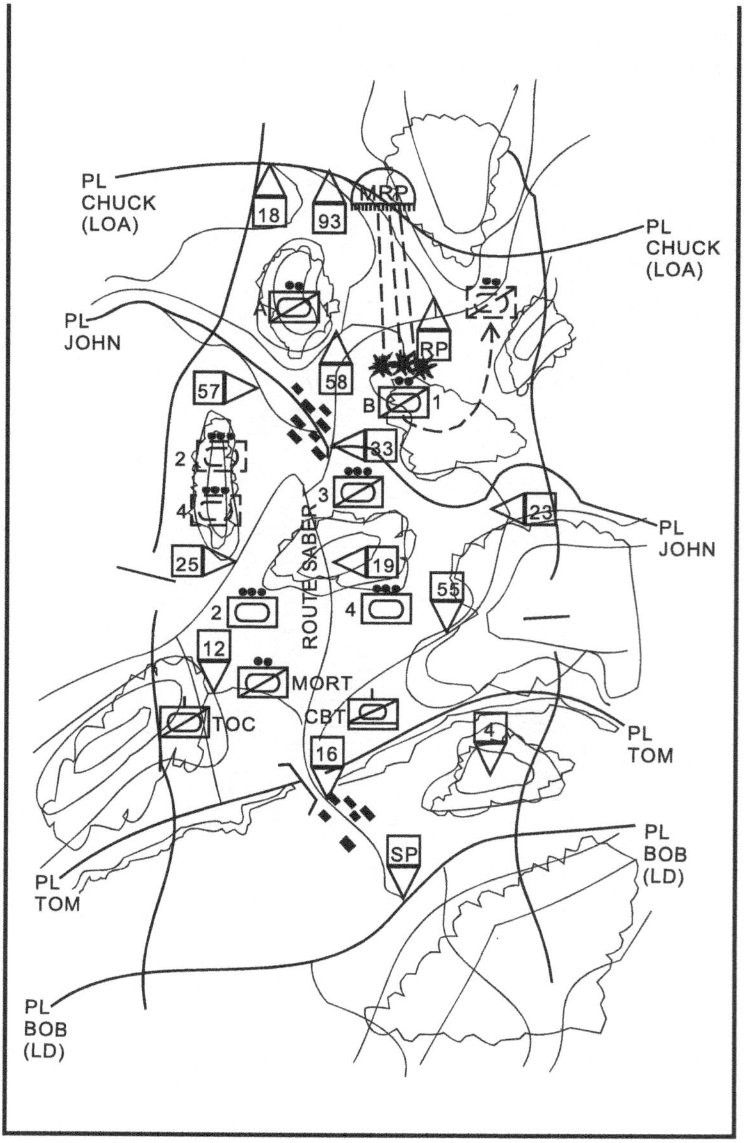

Figure 3-11. Route reconnaissance (part three).

what the platoon is up against and whether or not the threat force is supported by other elements in the village at checkpoint 18. He then reports to the troop commander. The troop commander immediately moves forward to join the 1st platoon leader and to see the situation for himself. En route, the commander tells the 2d and 4th platoons to move into concealed positions in the village at checkpoint 33, and to prepare to counterattack. The 3d platoon continues to reconnoiter the route through the village at checkpoint 33, and then moves into covered positions along ROUTE SABER just south of checkpoint 33. The troop CP holds in place and reports the situation to squadron. The 1SG quickly leads the M88 and the medic vehicle forward to treat and evacuate the injured scouts and their damaged vehicle.

Dismounted scouts from the 1st platoon determine that the threat force consists of three BMPs and one T-64 in prepared positions. Threat crews remain mounted and continue to slew turrets, scanning for more friendly elements. The threat reinforced platoon is oriented southeast toward the route. Their western flank is free of obstacles and unobserved. The troop commander decides to conduct a hasty attack (see Figure 3-12).

When the tank platoons complete their hasty attack, the troop commander gets permission to advance beyond the LOA and reconnoiter the enemy position. The 3d platoon returns to complete the reconnaissance of ROUTE SABER. As the platoon checks for mines along the route to the RP, they discover the route is mined just short of the RP. They find a bypass around the obstacle and forward a bypass report (Blue 10) to the troop XO. The platoon then organizes into two sections and moves out to occupy OPs overlooking the lateral routes at the east and west flanks of the troop near PL CHUCK. The troop CP moves into position on high ground near checkpoint 58. The 1SG directs activities of the troop trains, to include evacuation of injured soldiers and inoperable vehicles, security and evacuation of injured prisoners, requests for casualty replacements and resupply.

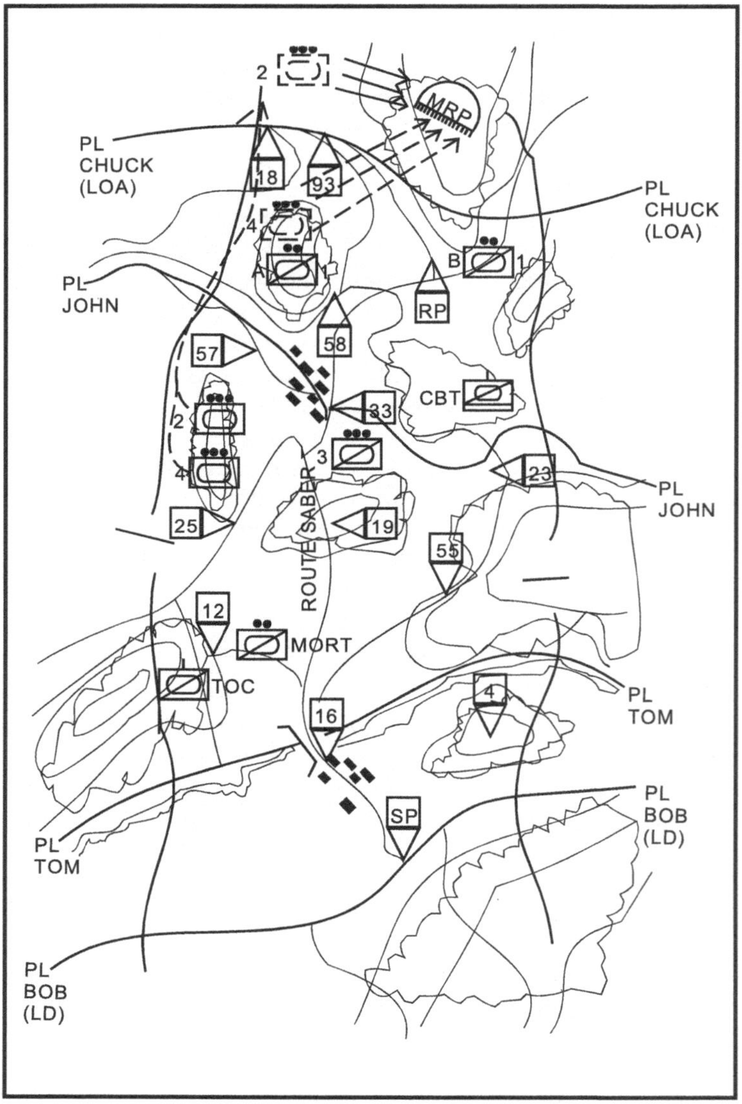

Figure 3-12. Route reconnaissance (part four).

Section IV. Zone Reconnaissance

Zone reconnaissance is the directed effort to obtain detailed information concerning all routes, obstacles, terrain, and enemy forces within a zone defined by boundaries. Obstacles include chemical and radiological contamination. A zone reconnaissance is assigned when the enemy situation is vague or when information concerning cross-country trafficability is desired. It is appropriate when previous knowledge of the terrain is limited or when combat operations have altered the terrain.

CRITICAL TASKS

Zone reconnaissance is a deliberate, time-consuming process. During a zone reconnaissance, the troop accomplishes the following critical tasks unless specifically directed otherwise by the higher commander:

- Reconnoiter all terrain within the zone.
- Inspect and classify all bridges within the zone.
- Locate fords or crossing sites near all bridges in the zone.
- Inspect and classify all overpasses, underpasses, and culverts.
- Locate and clear all mines, obstacles, and barriers in the zone within its capability.
- Locate a bypass around BUAs, obstacles, and contaminated areas.
- Find and report all enemy forces within the zone.
- Report reconnaissance information.

TECHNIQUES

The cavalry troop is usually assigned a zone reconnaissance as part of a squadron zone reconnaissance mission.

A zone reconnaissance mission is very time-consuming. Any commander who orders this mission must remember the number and complexity of the tasks to be accomplished. If he wants a faster tempo, he must prioritize reconnaissance tasks for the troop.

The cavalry troop can effectively reconnoiter a zone from 6 to 10 kilometers wide. If stretched any farther, the troop quickly exceeds its ability to accomplish the critical tasks.

Because the enemy situation is vague and knowledge of the terrain is limited to what is shown on maps or aerial photos, platoons will face unexpected situations around every corner; therefore, the course of action selected must also provide a good measure of protection for the troop as it executes the mission. When considering techniques of conducting a zone reconnaissance, the scheme of maneuver has to be flexible. The troop commander must convey his intent to subordinates so they can act quickly and without orders. He should employ forces, tank or AT platoons, in depth to give the troop flexibility and responsiveness.

When the troop receives a zone reconnaissance mission, the zone is usually identified by lateral boundaries. The line of departure and a reconnaissance objective or LOA are specified. The commander should divide the troop zone into two platoon zones for the scout platoons. Use caution when drawing the boundary. Make sure it is on easily identifiable terrain that is not a high speed avenue of approach. The number of critical tasks within the troop zone, and terrain restrictions should guide the commander in determining the location of the scout platoon boundaries. Doctrinal distances are not always the best solution if one platoon will be overtasked.

Add phase lines every 5 to 8 kilometers on easily identifiable terrain to control progress of the troop through the zone. Place contact points near the intersection of the boundary and all phase lines. Ensure the contact points provide at a minimum the potential for concealment from enemy observation to allow the exchange of information between flank units. Use checkpoints to focus the efforts of scout platoons and to assist in maneuvering the troop (see Figure 3-13). Use TIRS on maps as described in Chapter 2.

Use a troop vee or split-vee to conduct zone reconnaissance. As scout platoons reconnoiter the zone, the tank/AT platoons follow and support behind the scouts, keying their movement off the scouts' forward progress. The tank/AT platoons are kept far enough in depth to retain flexibility, and yet remain responsive to situations developed by the scout platoons. The distance from the scouts is determined by the terrain and enemy situation. Therefore, the distance the tanks/ATs move from the scouts may vary during the course of the operation. Control tank/AT platoon movement by one of the following methods:

- Give movement guidance to the platoon leaders. Have them key movement off the scouts.
- Move the platoons by CPs, battle positions, and hide positions.

Move the mortar section center of the troop zone to range forward of the scouts. The range forward of the scouts is determined by the scouts' ability to acquire the enemy. The mortars may move under the control of the FIST or eavesdrop on the troop net and key their movement off the progress of the scout platoons. Regardless of who controls their movement, the TOC and the FIST track the mortar position constantly to ensure they can support the scouts.

Note. If the mortar section is busy firing missions and computing fire data, they may not be able to stay abreast of the friendly situation. Therefore, the FIST or XO should be ready to control the mortars displacement.

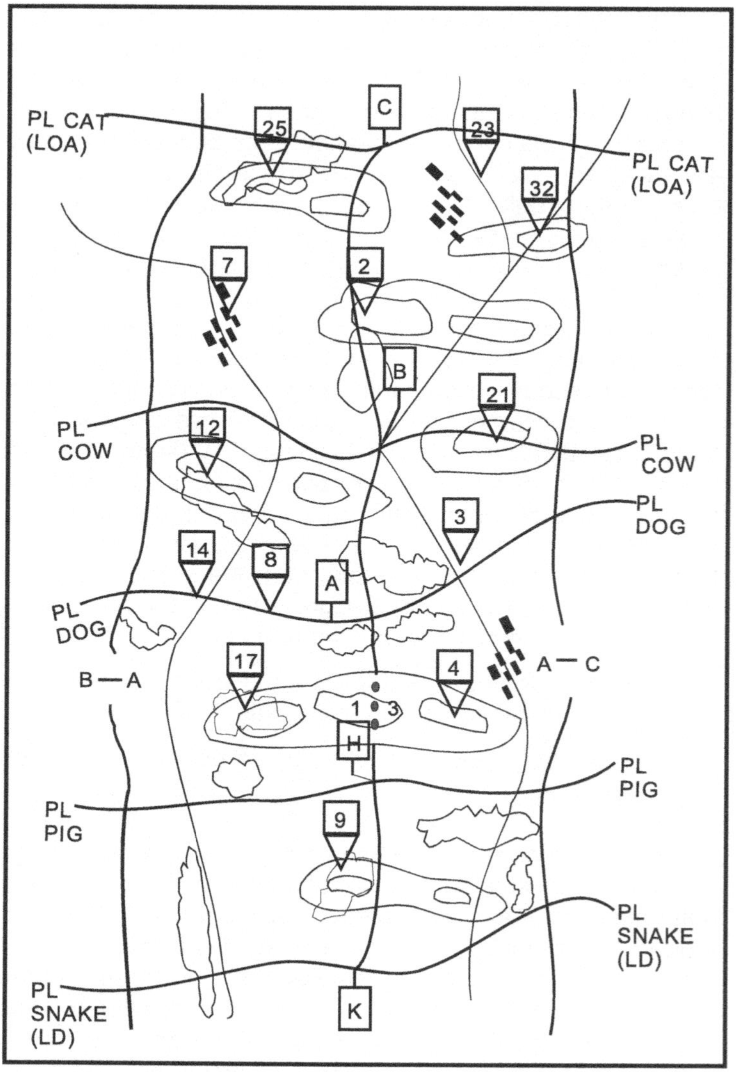

Figure 3-13. Zone reconnaissance graphic control measures.

The XO displaces the troop CP through the zone using terrain that affords effective and continuous communications with troop elements and the squadron. The 1SG moves the troop trains about 2 kilometers behind the tank platoons and bounds the trains from one covered and concealed position to another. The troop commander positions himself well forward to observe the actions of one scout platoon or the other. His position usually depends on where he expects initial enemy contact or problems to occur.

LIGHT TROOP EXAMPLE

The troop commander deploys the troop in a split-vee formation. The 1st and 3d platoons (scout) cross the LD, PL UTAH, abreast and reconnoiter the terrain in troop zone up to PL OHIO. The platoons determine both the open terrain and the trafficability of the route within their zone. Some hill slopes are greater than thirty percent and dense woods inhibit armor movement. Except for a marshy area north of checkpoint 9, the open terrain is dry and firm. Both routes will support heavy armor movement. The troop commander follows the 1st platoon. The mortar section eavesdrops on the troop command net to follow the progress of the scout platoons, and coordinates its moves with the troop FSO accordingly. As the scout platoons cross PL OHIO, the 2d and 4th platoons (antitank) cross the LD. They eavesdrop on the troop command net and key their movement on the progress of the scout platoons. The troop CP and the trains hold in place.

The scout platoons continue reconnaissance up to the river (PL BAMA). They take a close look at the dominant terrain near checkpoint 17 and at the village near checkpoint 4. As the platoons approach the river, scouts move into covered and concealed positions, dismount, and visually search the dominant terrain on the north side of the river. The 1st platoon inspects the bridge at checkpoint 14 and determines that it will support up to 40 tons. Scouts also verify that a good fording site exists near checkpoint 8. The 3d platoon confirms there is a fording site with a rock bottom just east of the blown bridge at checkpoint 3. It will support heavy armored traffic (see Figure 3-14).

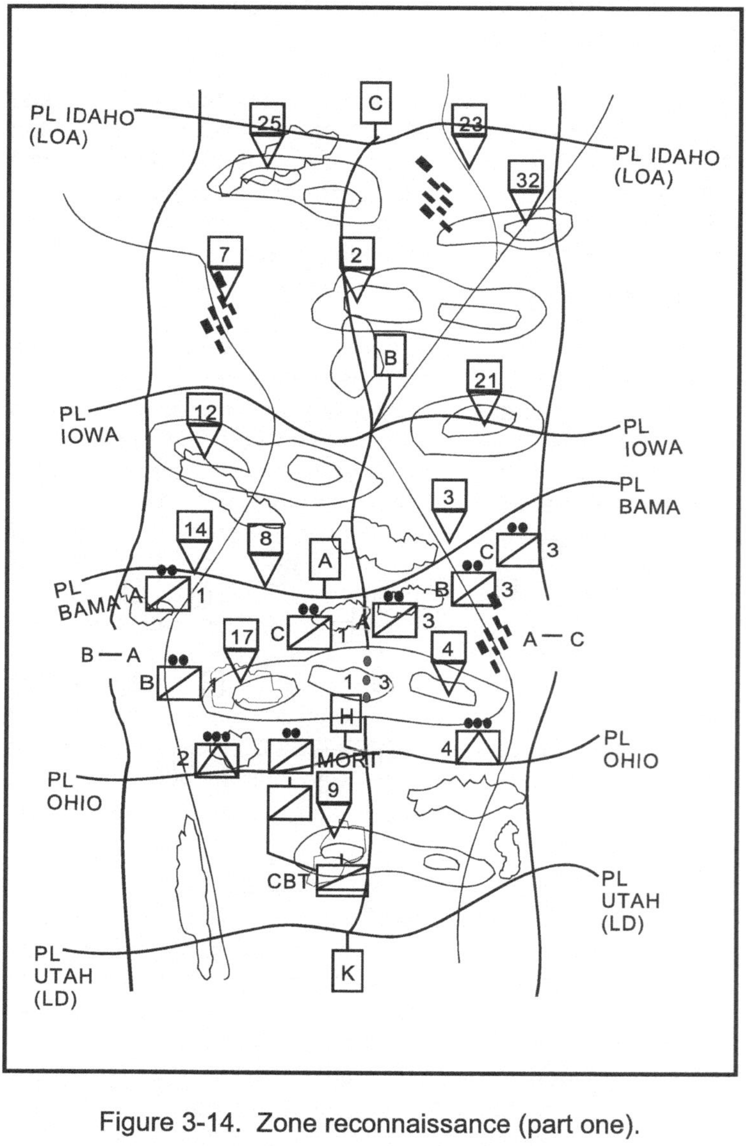

Figure 3-14. Zone reconnaissance (part one).

As the scout platoons continue reconnaissance towards PL IOWA, the 1st platoon observes two stationary BRDM-2s on the high ground near checkpoint 12. They look like a security outpost. The platoon engages with immediate suppression from the troop's 120-mm mortars. As the rounds impact, the outpost withdraws behind the ridgeline to the north. The 3d platoon determines the dominant terrain near checkpoint 21 is clear of threat forces. The mortar section crosses the river at checkpoint 8. The 2d platoon follows, crossing at checkpoint 8, while the 4th platoon uses the ford near checkpoint 3. The troop CP bounds forward to high ground near checkpoint 17. The troop trains cross the river and move into the woods on the northern bank.

The scout platoons continue reconnaissance while moving toward the troop LOA. As the 3d platoon crosses the ridge (PL IOWA), Bravo section receives heavy machine-gun fire from concealed positions south of checkpoint 2. The platoon immediately deploys, reports contact to the troop commander, and calls for indirect fire to suppress the threat (see Figure 3-15).

Under this protection, the platoon, using primarily a dismounted technique, reconnoiters to the flanks and rear to develop the situation. The troop commander acknowledges the report, and moves immediately to link up with the 3d platoon leader. The troop commander orders the 1st platoon to continue its reconnaissance to locate the enemy's flank and coordinate with the FIST to switch indirect fires to the squadron howitzer battery. The 3d platoon is organized in a three-section, three-vehicle configuration; the platoon leader is with the Charlie section. The platoon leader sends his Bravo section forward on the enemy's left and his Alpha section on the right to determine if the enemy is mutually supported by other threat forces from the flanks or rear. The Charlie section sergeant and a two-man dismount team move in closer and determine that the enemy consists of an MRP with three BTR 70s and one BMP-2. The Alpha scout section and scouts from 1st platoon identify the right flank of the threat and discover the flank is unprotected by obstacles and is exposed to direct fires.

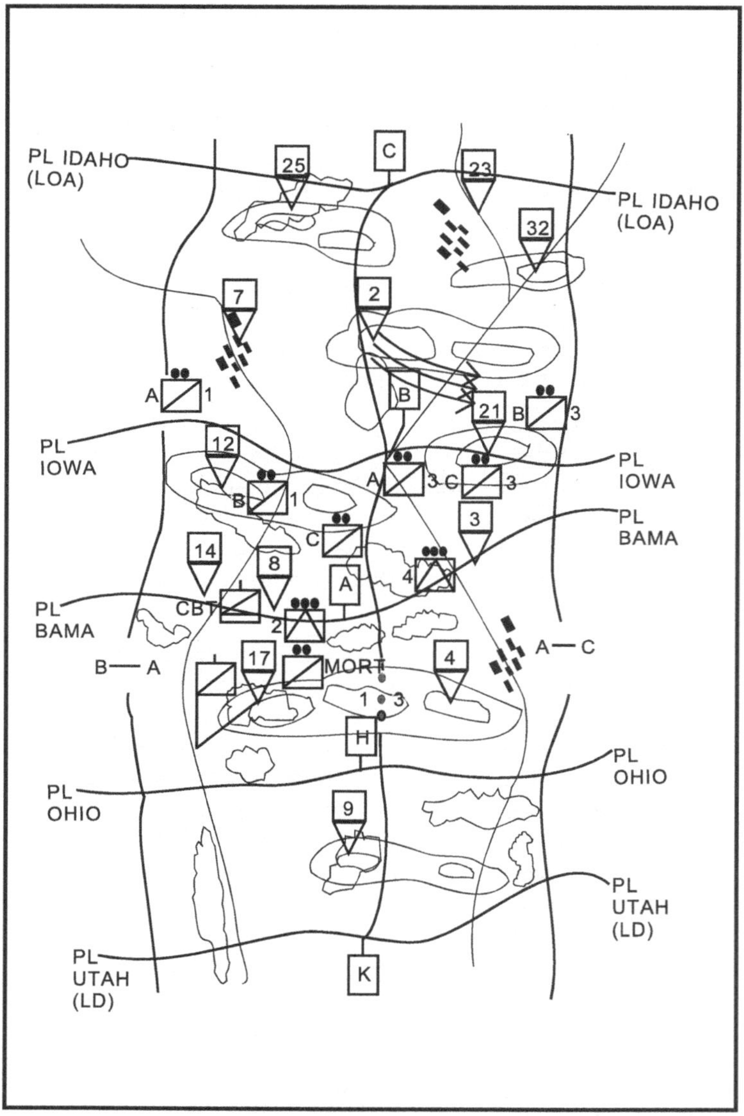

Figure 3-15. Zone reconnaissance (part two).

While en route to the 3d platoon's zone, the troop commander tells the 1st platoon to continue reconnaissance forward to PL IDAHO and determine if other enemy forces are providing mutual support to the enemy in the 3d platoon's zone. At PL IDAHO, he tells the platoon to establish a screen. The troop commander tells the 4th platoon to move over and join the 3d platoon Charlie section in overwatch of the enemy MRP, and prepare to overwatch with fires. The troop commander tells 2d platoon to prepare to conduct a hasty attack of the enemy MRP. The 1st platoon scout squad moves to checkpoint 12, links up with the 2d platoon, and guides them into position to attack the threat platoon.

The 2d platoon reports they are set in the assault position. The 4th platoon reports they are set in the overwatch position with the 3d platoon Charlie section. The 3d platoon leader reports that his dismount teams from his Alpha and Bravo sections are out of the target area and set. The 1st platoon leader reports he is set in a screen along PL IDAHO. The troop commander orders the FSO to fire for effect on the enemy position at his command. The commander issues orders and guidance for limits of fire and signals for the initiation, shifting, and lifting of fires. The troop commander issues the fire command for indirect fire over the command net and follows with a fire command for direct fire for the 4th platoon and the Charlie section of 3d platoon. The indirect fire from the squadron's howitzer battery and the troop mortars impact on the enemy position and is immediately followed by the fire from four TOWs, MK-19s, and M2 machine guns. The commander then issues the orders for 2d platoon to initiate its move to the attack-by-fire position followed by the order for the support element (4th platoon, Charlie section of 3d platoon) to shift fires. The troop FSO shifts fires to the rear of the enemy position. The 2d platoon moves forward and engages the enemy target from the flank to complete the destruction of the enemy force (see Figure 3-16).

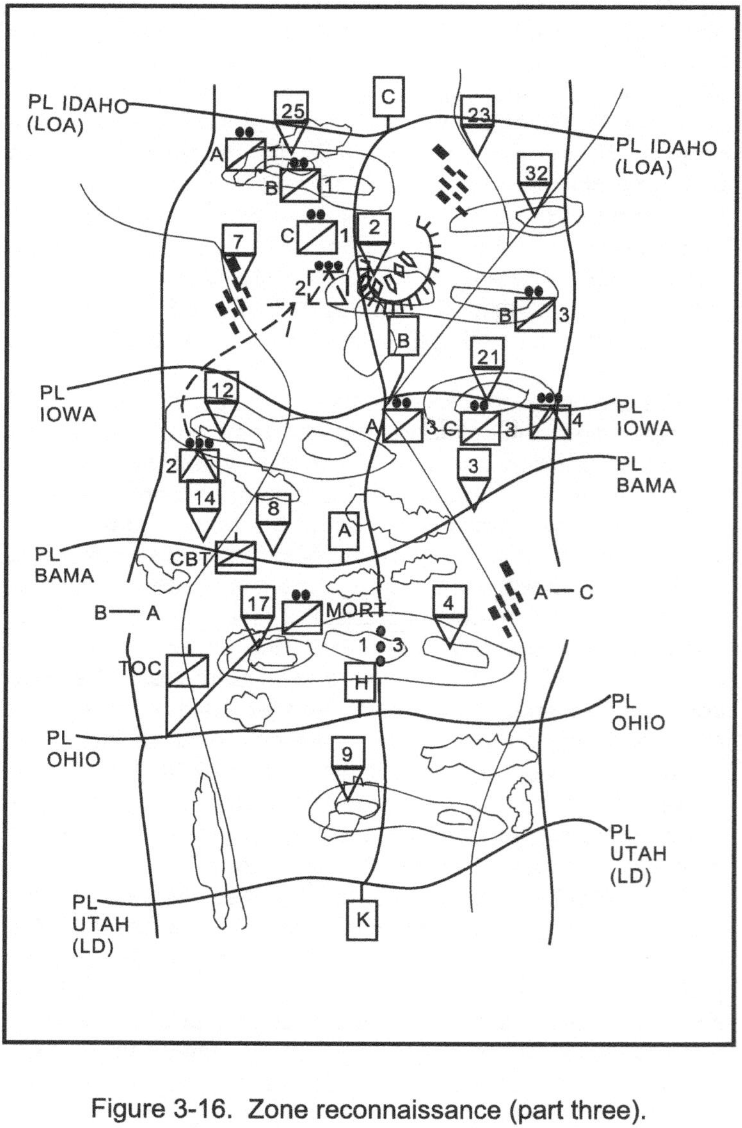

Figure 3-16. Zone reconnaissance (part three).

The antitank platoons consolidate in covered and concealed positions west of checkpoint 2. Scout dismount teams from the 3d platoon move in quickly, capture two wounded prisoners, and search the vehicles and personnel. The 3d platoon leader orders one scout squad to stay and secure the prisoners, and then completes his reconnaissance up to the LOA. The 1SG leads the medics to the location of the 3d platoon scout squad and the EPWs. The XO collects and transmits final reconnaissance reports to the squadron.

AIR-GROUND COORDINATION

Cavalry troops will often conduct a coordinated reconnaissance with an air cavalry troop. Because of the air cavalry troop's ability to maneuver faster and look deeper than ground cavalry, they are often placed to the flanks or forward of ground cavalry. The air cavalry troop should focus its reconnaissance efforts on areas that impede ground cavalry movement.

The air cavalry troop quickens the pace or tempo of the ground reconnaissance because it provides the ground unit with added security during movement and increased situational awareness once contact is made. The air cavalry troop can gain initial contact with the enemy or a critical piece of terrain. Once the air troop develops the situation with his assets, he can pass off the information/contact to the ground troop (see Figure 3-17).

- Ground troop and air troop coordinate zone reconnaissance operations.
- Air troop divides the zone into platoon zones.
- Air troop will use checkpoints to identify critical areas. Checkpoints should enhance ground troop movement throughout the zone.
- Platoon on the right reconnoiters forward of ground troop, identifies platoon ford site, and conducts route reconnaissance.
- Platoon on the left reconnoiters forward of ground troop, supports the other platoon in conducting route sweep, checks lateral routes in the zone, and looks for sites along the obstacle (river/road).
- ACT commander and ground troop commander identify critical areas where air-ground coordination is required or needed.
- ACT commander positions himself close to the ground troop commander (if needed).

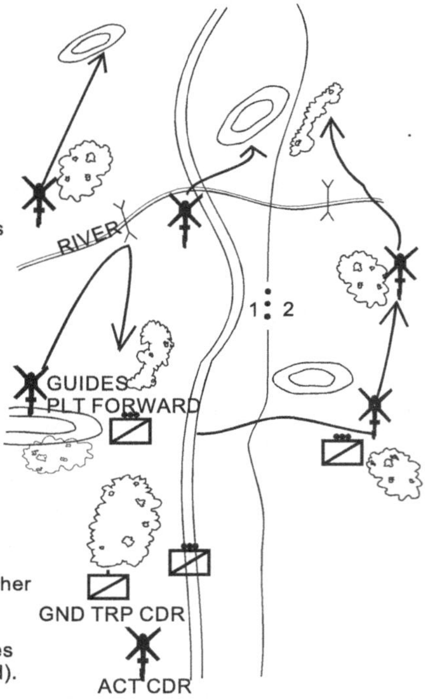

- If the ford site or bridge site is blocked, ACT elements reconnoiter the site forward of the location. They also find a bypass to the obstacle and lead ground troops to bypass the location. The ACT will start the screen forward of ground elements.

Figure 3-17. Air-Ground troop coordinated reconnaissance.

Section V. Area Reconnaissance

An area reconnaissance is a specialized form of zone reconnaissance. It is a mission conducted to gain detailed information about terrain features and enemy forces within a specified area or point that other forces intend to occupy, pass through, or avoid. A commander usually calls for area reconnaissance before he sends his forces into or near an area to avoid being surprised by actual terrain conditions or unexpected enemy forces.

CRITICAL TASKS

During an area reconnaissance, the following critical tasks must be accomplished unless the SCO directs the troop to do otherwise:

- Reconnoiter all terrain within the area.
- Inspect and classify all bridges within the area.
- Locate fords or crossing sites near all bridges within the area.
- Inspect and classify all overpasses, underpasses, and culverts.
- Locate and clear all mines, obstacles, and barriers in the area within its capability.
- Locate a bypass around BUAs, obstacles, and contaminated areas.
- Find and report all enemy within the area.
- Report reconnaissance information.

TECHNIQUES

An area reconnaissance is conducted like a zone reconnaissance. When the troop receives an area reconnaissance mission, the assigned area is identified as the terrain inside a solid, continuous boundary. Planning the movement to the area is the first step. Select the route(s), establish a march order on each route, and specify a start point, checkpoints, and a release point. Use a movement technique that keeps the troop moving quickly and securely. If possible, avoid contact with the enemy while en route. Report and bypass. Be careful when approaching the area to begin the reconnaissance. Use common sense. Avoid known enemy forces outside of the area where reconnaissance will be conducted; start somewhere else.

As in a zone reconnaissance, enclose the area within a troop zone. Draw a line of departure, an LOA, and lateral boundaries. Divide the troop zone into two platoon zones. Add phase lines along identifiable terrain to control movement through the area. Place contact points at the intersection of the platoon boundary and all phase lines. Place TIRS on the map. To identify specific areas or features, use checkpoints for reference (see Figure 3-18).

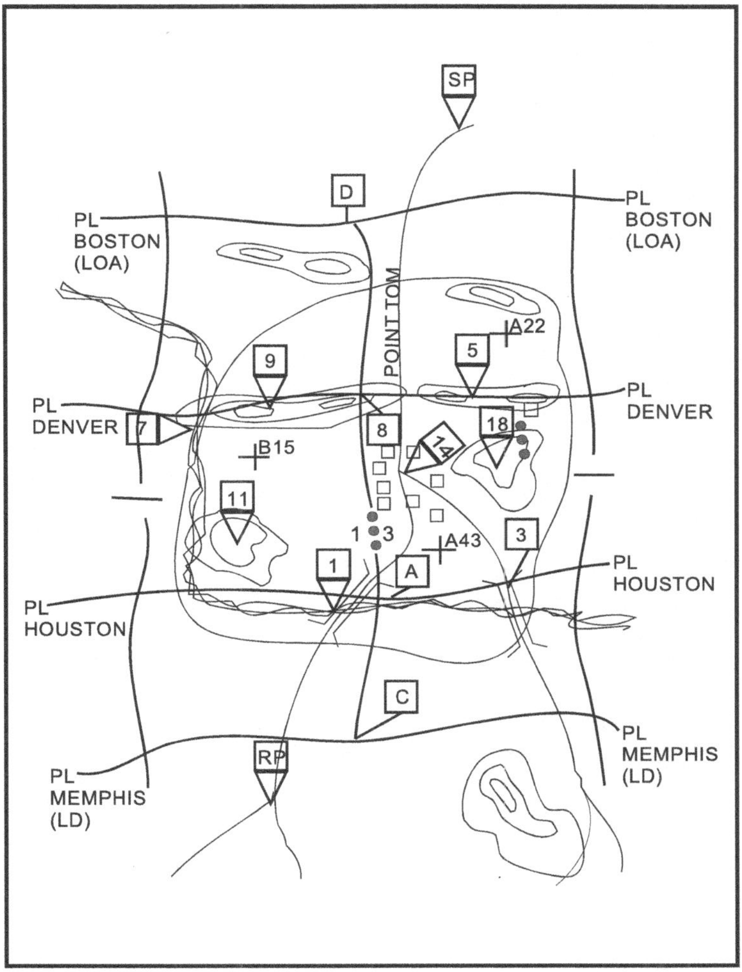

Figure 3-18. Area reconnaissance.

Use a troop vee or split-vee to conduct an area reconnaissance. Scout platoons deploy abreast from the release points to accomplish all the reconnaissance tasks. They move out across the line of departure first. The mortar section moves through the center of the troop zone, remaining in position to range from 3 to 3.5 kilometers forward of the lead elements in the scout platoons. The mortar section could also follow one of the scout platoons, depending on where initial enemy contact is expected. The tank platoons are kept back far enough in depth to retain flexibility, but remain responsive to situations developed by the scout platoons. They key their advance on the progress of the scout platoons. The tank platoons are told how far behind the scout platoons they should stay as the troop deploys across the line of departure. The distance between the tank platoons and the scout platoons will usually be adjusted as the terrain or enemy situation changes. The troop CP displaces through the zone using terrain that affords effective and continuous communication with troop elements and squadron. The troop trains follow about 2 kilometers behind the tank platoons and bound from one covered position to another. The troop commander positions himself well forward to observe the action of one scout platoon or the other. The location usually depends on where initial enemy contact or problem situations are expected.

Chapter 4

Security

Security is an essential part of all offensive and defensive operations. Cavalry provides security for the commander along an exposed front, flank, or rear of the main body where a threat may exist. Surveillance is continuous during security operations. Even during security missions that involve fighting the enemy, the scouts' primary task remains gathering information. Scouts do this by establishing OPs, conducting patrols, and performing reconnaissance.

Counterreconnaissance is an inherent task in all security operations. Counterreconnaissance is not a mission. It is the sum of actions taken at all echelons to counter enemy reconnaissance and surveillance efforts through the depth of the area of operations. Counterreconnaissance denies the enemy information about friendly units. It is both active and passive and includes combat action to destroy or repel enemy reconnaissance elements.

CONTENTS

Section I. Purpose, Fundamentals, and Capabilities

Security includes screening operations, guard operations, covering force operations, area security operations, convoy security operations, and route security operations. The squadron performs screen, guard, and route security missions. Covering force operations are normally the mission of a cavalry regiment. Separate brigades or task organized divisional brigades may perform cover operations as well.

Both the heavy and light cavalry troops perform two security missions (screen and convoy security) independently or as part of their parent squadron. Both also participate in guard, cover, and route security missions as part of their parent squadron or regiment. Either troop will normally perform reconnaissance, screen, defend, delay, attack, or a combination of these missions in support of their parent squadron or regiment (see the applicable section of Chapters 3, 5, and 6 for specifics).

PURPOSE

Security operations are designed to obtain information about the enemy and to provide reaction time, maneuver space, and protection to the main body. Security operations are characterized by conducting continuous reconnaissance to reduce terrain and enemy unknowns, gaining and maintaining contact with the enemy to ensure continuous information, and providing early and accurate reporting of information to the protected force.

FUNDAMENTALS

Five fundamentals are common to all cavalry security operations.

Orient on the Main Body. As a security force, the troop will be operating at a specified distance from a main body, between it and a known or suspected enemy force. If the main body moves, the troop also moves. The troop commander must know how the main body commander intends to maneuver his forces and where he wants the troop in relation to his movement. The troop commander maneuvers his troop to positions where he can provide the needed security.

Perform Continuous Reconnaissance. The troop's security and the security of the main body come in large measure from knowing everything about the terrain and the enemy within the troop's area of operations (AO). Reconnaissance and continuous patrolling go hand in hand with security operations. Cover all the ground in the AO. Determine what the terrain will allow the troop *and* the enemy to do.

Provide Early and Accurate Warning. Early and accurate warning of enemy approach is the cornerstone of security operations. The main body commander needs as much time as possible to shift and concentrate his forces to meet and defeat an unexpected enemy attack. Put observers in positions that afford long-range observation of expected enemy avenues of approach. Use aeroscouts and ground surveillance radar (GSR), if available, to enhance their ability to see. Place remote sensors in the ground to monitor avenues of approach that cannot be easily observed. If possible, send dismounted or mounted patrols forward of OPs to extend their ability to see, providing additional reaction time for the main body commander.

Provide Reaction Time and Maneuver Space. All security operations are designed to provide reaction time and maneuver space for the main body so it can deal effectively with an unexpected enemy attack. The troop provides early warning to the main body commander and sometimes has to fight hard to buy time and space so the main body's combat power can be concentrated on defeating the enemy.

Maintain Enemy Contact. Once visual or physical contact with the enemy is gained, do not allow him to break the contact. Maintain enemy contact and continue to report his activities until told to stop.

CAPABILITIES

Capabilities of the heavy troop include—

- Screen up to a ten-kilometer-wide sector.
- Maintain continuous surveillance of up to six battalion-size avenues of approach.*
- Can establish up to 12 short-duration OPs.

Capabilities of the light troop include—

- Screen up to a ten-kilometer-wide sector.
- Maintain continuous surveillance of up to six battalion-size avenues of approach.*
- Can establish up to 16 short-duration OPs.

*The maximum six long-duration OPs either cavalry troop can occupy is a function of personnel required to perform the following tasks at each OP:

- Man the actual OP.
- Maintain radio communications with the OP and with the platoon leader.
- Provide local security for the vehicles.
- Conduct dismounted patrols as required.
- Conduct resupply.
- Perform maintenance.
- Sleep/rest.

The ability to perform the above tasks simultaneously for periods in excess of 12 hours requires at least 9 to 10 personnel (collocation of two scout squads for M3-mounted scouts, or three scout squads for HMMWV-mounted scouts).

Section II. Screen

Screen is the most common security mission heavy and light cavalry troops conduct. Both troops conduct screen missions for their parent squadrons or other forces to—

- Provide early warning of enemy approach.
- Provide real-time information, reaction time, and maneuver space to the protected force.
- Destroy enemy reconnaissance elements within their capability (perform counterreconnaissance).
- Impede and harass the enemy.

The screen mission provides the least amount of protection of any security mission, and is appropriate when operations have created extended flanks, when gaps between forces exist and cannot be secured in force, or when required to provide early warning over gaps that are not considered critical enough to require security in greater strength. A commander normally assigns cavalry this mission when he needs time to respond to an unexpected enemy attack, and cannot afford to commit other forces to the task.

The screen mission is defensive in nature. As such, both heavy and light cavalry troops screen the front, flanks, and rear of a stationary force but only to the flanks or rear of a moving force. Screening operations are not performed forward of a moving force. Zone reconnaissance or movement to contact is the appropriate cavalry troop mission suited to the requirements of the offensive force.

CRITICAL TASKS

A screen mission has certain critical tasks that guide planning. To achieve the intent of a screen mission, the troop must accomplish the following critical tasks:

- Maintain continuous surveillance of all battalion-size avenues of approach into the troop sector under all visibility conditions.
- Destroy or repel all enemy reconnaissance patrols (if within the unit's capability and directed by the higher commander).
- Locate lead elements of enemy order of battle and determine the direction of movement of each.
- Maintain contact with, report the activity of, and impede and harass the enemy while displacing.

EXTENDED SCREENS

Both the heavy cavalry troop and the light cavalry troop normally screen a sector up to ten kilometers in width. However, METT-T may dictate an extended screen across frontages in excess of the norm. Either troop's ability to accomplish its critical tasks, or its ability to screen in depth, can diminish rapidly as frontages increase.

Examples of extended screens are described below.

- Heavy cavalry troop screens 20 kilometers of southern bank of unfordable river crossed by four bridges in sector.
- Light cavalry troop screens 25 kilometers of desert terrain, from dominant ridge.

DEPTH

Depth is also important in a screen. The term *"screen line"* is descriptive only of the forward trace along which security is provided. Depth allows an enemy contact to be passed from one element to another without requiring displacement. Depth is advantageous to—

- Destroy an enemy reconnaissance patrol without compromising critical OPs.
- Prevent an enemy from penetrating the screen line too easily.
- Prevent gaps from occurring when OPs displace or are lost.
- Maintain contact with moving enemy without compromising OPs.
- Prevent enemy templating of the screen line.

Depth is achieved primarily by positioning OPs, particularly where there are limited avenues of approach. Tank platoons, AT platoons, the mortar section, and attached elements positioned behind the screen line establish local security and provide surveillance. The degree to which depth can be attained is a function of many factors, which include—

- Higher commander's intent and concept as expressed in—
  - Graphical trace of the screen line (LOA).
  - Engagement criteria.
  - Destruction criteria.
  - Displacement/disengagement criteria.
- Width of the sector.
- Depth of the troop sector.

- Terrain and avenues of approach it will support.
- Attachments and detachments.

Screening is largely accomplished by establishing a series of OPs and conducting patrols to ensure adequate surveillance of the assigned sector. Screens are active operations. Stationary OPs are only one part of the mission. Employing patrols (mounted and dismounted), aerial reconnaissance, ground-based sensors, intelligence from space-based sensor systems, and OPs relocated on an extended screen ensure that continuous overlapping surveillance occurs. Inactivity in an immobile screen promotes complacency.

PLANNING CONSIDERATIONS/GUIDANCE TO SUBORDINATES

The enemy situation is often vague when planning a screen. The troop should develop plans that are flexible enough to react to any enemy course of action, particularly the worst case. Planning considerations for a screen should include a detailed description of how contact with the enemy reconnaissance will be gained then how and where it will be destroyed. Planning should also cover the method of displacement once the main body of the enemy force has been identified and how that force will be handed off to the main body in the main defensive belt. Because of the need for flexibility, screen operations will often begin to inherit the characteristics of defense or delay missions. To cover the displacement of scouts, some elements of the troop may be required to execute missions such as delay or defend.

Screen operations at troop level usually occur in four phases.

- Movement to and occupation of the screen line.
- Surveillance and counterreconnaissance.

- Gaining and maintaining contact with the enemy main body displacement of screen.
- Rearward passage of lines.

Higher command guidance should address each phase of the operation and cover at least the following:

- Location/orientation/width of the screen.
- Duration of the screen.
- Method of movement to and occupation of the screen line.
- Location and disposition of the friendly force being screened.
- Engagement/destruction criteria.
- Displacement/disengagement criteria.
- Follow-on missions.
- Positioning and orientation guidance for GSRs (if attached).

The following must be considered when developing and completing the plan and executing the screen mission:

Time Screen Must Be Established. The time the screen must be set and active will influence the troop's method of deploying to and occupying the screen line.

Movement to the Screen Line. If the screen mission is the result of a previous tactical maneuver such as zone reconnaissance, the troop will essentially be postured to begin screening from present positions. This situation occurs frequently, and may be the result of a FRAGO to halt at a specified phase line.

If the troop is not currently sited on the screen line, obviously, deployment to the screen line must occur before actually beginning the screen mission. Time determines the method of occupying the screen line. Thorough analysis of METT-T will determine which deployment technique or combination of techniques best meets mission requirements.

Trace and Orientation of Screen Line. The initial screen line is depicted as a phase line and often represents the forward line of own troops (FLOT). As such, the screen line may be a restrictive control measure for movement (limit of advance); coordination/permission would be necessary to move beyond the line to establish OPs or to perform reconnaissance. When occupied, OPs are sited on or behind the phase line. OPs should be given specific orientation and observation guidance.

Initial OP Locations. The squadron or troop commander may determine tentative initial OP locations to ensure effective surveillance of the sector and designated named area(s) of interest (NAI). At a minimum, the troop commander designates a primary orientation for the scouts during the conduct of the screen. Scouts, once set on the screen line, will report their location to the troop TOC and verify they are in compliance with the commander's orientation guidance. The scouts who occupy each OP always retain the responsibility to modify the location to achieve the commander's intent and guidance for orientation. The OPs are positioned along or behind the screen line.

OPs are generally categorized based on their expected duration of employment as either—

- Short-duration (less than 12 hours)

  or

- Long-duration (more than 12 hours).

OPs may be either mounted or dismounted. Mounted OPs maximize use of vehicular optics, weapon systems, and speed of displacement, but are more readily detected by the enemy. Dismounted OPs provide maximum stealth at the expense of speed of displacement, and vehicle-mounted optics and weapons.

A heavy cavalry troop can occupy up to 12 short-duration OPs (one per scout squad, six per scout platoon). For extended periods of time, the heavy troop can occupy six OPs (one per scout section, three per scout platoon).

A light cavalry troop can occupy up to 16 short-duration OPs (one per scout squad, eight per scout platoon). For extended periods of time, the light troop can occupy six OPs (one per scout section, three per scout platoon).

Width and Depth of the Screened Sector. The troop sector is defined by lateral boundaries extending out to a limit of advance (the initial screen line), forward of a rear boundary. The troop sector is established by the squadron or unit being screened. The troop rear boundary may be a squadron phase line and may serve as a battle handover line (BHL) to control passing of responsibility for the enemy to the protected force. The troop's ability to gain depth decreases as screened frontage increases.

Locations of Subsequent Screen Lines. The squadron or troop commander uses additional phase lines to control the operation. These phase lines may serve as subsequent screen lines. Displacement to the subsequent screen lines is event driven.

Scout Platoon Sectors. Assign clear responsibility of identified avenues of approach and designated NAIs. The nature of a screen normally requires both scout platoons to deploy abreast.

Tank/AT Platoon Sectors. Position the tank (heavy troop) or AT (light troop) platoons in the scout platoons' sectors. They may occupy hide or battle positions along avenues of approach. The tank/AT platoons remain responsive to the troop commander. They are the primary direct-fire killing asset.

Force to be Screened. The troop must orient on the force it is securing. If the main body is moving, the troop must move to maintain the screen's position relative to the main body.

Reinforcements. Any unique requirement posed by the mission may require assets not organic to the troop. GSR and engineers are common attachments at troop level.

- **GSR.** During screen operations, GSR is used to augment scout OPs and to add depth to the screen. GSRs should be attached to scout platoons, and the commander should provide the scout platoon leader with positioning and orientation guidance.
- **Engineers.** If engineers are attached to the troop, the troop commander should assign them with priority of mission and priority of effort in support of commander's guidance. During screen operations, engineers will normally dig survivability positions for scouts and tanks. Also engineers may emplace obstacles in support of the counterreconnaissance battle or assist the troop with displacement of the screen once contact has been established with the enemy body.

Special Requirements or Constraints. Specify all requirements for observing any NAI identified during the intelligence preparation of the battlefield (IPB). Task subordinate platoons as required. Specify the following in the troop OPORD:

- Engagement criteria.
  - What size force will scouts engage and destroy (if any)?
  - Where will this action occur?
  - What size force will the tank or AT platoons engage and destroy?
  - Where will this action occur?

- Disengagement criteria.
  - What event will cause scouts to displace from the initial screen line?
  - How will scouts maintain contact with the enemy while displacing?
  - What event will cause the tank or AT platoons to displace to subsequent or alternate positions?

Indirect Fire Planning. Fire planning integrates artillery and mortar fires. Position the troop mortars to fire up to two-thirds of their maximum range forward of the initial screen line. A wide sector may require the troop commander to position them to provide effective coverage of the most likely avenue of approach determined by IPB. The troop FSO plans artillery fires to adequately cover any gaps in mortar coverage.

Direct Fire Planning. Based on his analysis of the terrain, the troop commander determines where to engage the enemy (engagement areas). He also determines the location of battle positions that provide observation, fields of fire, and cover and concealment that support each engagement area.

Positioning of C2, CS, and CSS Assets. The troop commander positions himself to observe the most dangerous enemy avenue of approach. The troop TOC positions itself in depth to provide continuous control and reporting during initial movements. After the screen line has been reestablished in depth following displacement from the initial screen line, the TOC can reposition. Combat trains position behind masking terrain close enough for rapid response. They are best sited along routes providing good mobility laterally and in depth.

Patrol Requirements. Patrols may be required to cover gaps between OPs. The troop commander tasks the scout platoon leaders to perform specific patrols.

Coordination. The troop commander coordinates his concept with air cavalry troop (ACT) commanders who may be operating the same ground, flank troop commanders, tank company (heavy cavalry) and AT company (light cavalry) commanders, and other unit commanders as appropriate.

TECHNIQUES

Graphics

The following is a list of common graphic control measures used to control screen missions:

- Boundaries.
- Phase lines.
- Checkpoints.
- Contact points.
- NAIs.
- OPs.
- Mortar firing positions.
- Battle positions.
- Hide positions.
- TIRS.
- Unit symbols.

Movement to the Screen Line

In deploying to the screen line, the troop commander must deal with the competing requirements to establish the screen quickly to meet mission requirements and to provide the necessary level of security for the troop in doing so. The troop moves to the screen line using one of three basic methods—a tactical road march, zone reconnaissance, or movement to contact.

Tactical Road March. The troop conducts a tactical road march to a release point behind the screen line. From the release point, platoons deploy to occupy initial positions. This method of deploying to the screen line is the fastest, but least secure. It is appropriate when enemy contact is not expected and time is critical, or when an air cavalry troop is conducting zone reconnaissance forward of the ground troop.

Movement to Contact. The troop conducts a movement to contact from a line of departure to the initial screen line. This method is slower than a tactical road march, but more secure. It is appropriate when enemy contact is likely, time is limited, or when an air cavalry troop is conducting zone reconnaissance forward of the ground troop.

Zone Reconnaissance. The troop conducts a zone reconnaissance from a line of departure to the initial screen line. Given adequate time, this method is preferred as the troop can clear the zone of any enemy and enables platoons to become thoroughly familiar with the terrain. The troop can reconnoiter potential subsequent OP locations, battle and hide positions, and mortar firing positions, for example, as they move to the screen line. A zone reconnaissance is appropriate when time is available and information about the enemy or terrain is unknown.

Security Drill

A security drill is a series of rehearsed actions (battle drills) a scout platoon or cavalry troop takes to maintain contact with the main body of an advancing enemy force. It is used when collapsing the screen line to subsequent OP positions or when transitioning from a screen mission to a delay or defend mission.

At platoon level, OPs gain contact with the enemy main body, then report and prepare to displace to a subsequent position. When the enemy force reaches the OP's break point (point where the OP must displace or his position/movement will compromise him to the enemy), the OP passes off the responsibility to track the enemy to another OP in depth. The platoon displaces its OPs to subsequent positions in depth while maintaining contact with the enemy.

At troop level, the security drill combines the collapse of the initial screen line with the actions of organic tank or AT platoons. Scout platoons may perform platoon security drills initially, consolidating some or all of their combat power at a battle position to aid execution of a troop-level engagement.

At platoon and troop levels, conduct of security drills is tempered by the commander's overall concept, intent, and scheme of maneuver. Enemy actions (events) drive security drill execution (response) (see Figure 4-1).

STATIONARY SCREEN

Take a close look at the high-speed avenues of approach into the sector. Divide the sector into two platoon sectors. Make sure the platoon boundary is on easily identifiable terrain. Do not split avenues of approach with a platoon boundary or place the boundary on a road. Place NAIs, TIRS, or checkpoints beyond the screen line to focus surveillance from OPs. If needed, add additional phase lines to control withdrawal of the troop at 5- to 8-kilometer intervals. Place contact points at the intersection of the platoon boundary and all phase lines. Place TIRS on the map or overlay as described in Chapter 2.

Deploy the scout platoons abreast and establish a series of OPs along or behind the initial screen line, as terrain allows, but never forward of it without permission. Make it clear to the scout platoon leaders which avenues of approach (depicted as NAIs or checkpoints) they are to observe. If unobservable areas between OPs need to be routinely checked, have the scout platoon leaders prepare patrol plans for approval and subsequent execution.

Position the mortar section to fire 3 to 4 kilometers forward of the initial screen line, oriented on the expected enemy avenue of approach. Establish subsequent firing positions for the mortar section back through the sector. Position each tank (heavy) or AT (light) platoon in one of the scout platoon sectors, or have the tank/AT platoons consolidate. Position the tank/AT platoons in hide positions or battle positions in depth behind the scout platoons, oriented on the expected enemy avenue of approach. Establish subsequent positions for the tank/AT platoons back through the sector to support the scheme of maneuver.

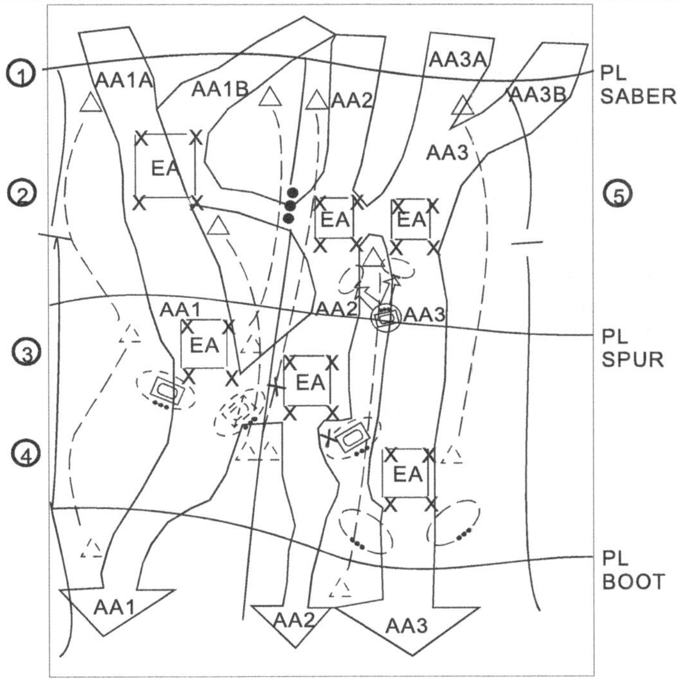

INDEX:

- (1) Initial screen line.
- (2) Scout platoon engagement area dependent on commander's intent.
- (3) Subsequent OPs.
- (4) Troop consolidates fire power of tanks and scouts to execute engagement while maintaining contact.
- (5) Scouts maintain contact and execute engagement with tank platoon.

Figure 4-1. Scout platoon/cavalry troop security drill.

Have the XO position the command post on terrain that affords good FM radio communications with troop elements and squadron headquarters. If possible, the command post should be positioned behind the subsequent screen positions. This allows the TOC to remain in position during the initial collapse of the screen line. Establish tentative subsequent command post sites back through the troop sector. The XO retains the authority to adjust the actual TOC location to maintain effective communications.

The 1SG first positions the troop trains within 5 to 8 kilometers of the initial screen line, then he establishes subsequent locations for the trains to bound back through the troop sector. The troop commander positions himself well forward where he can best observe and control the actions of the troop (see Figure 4-2).

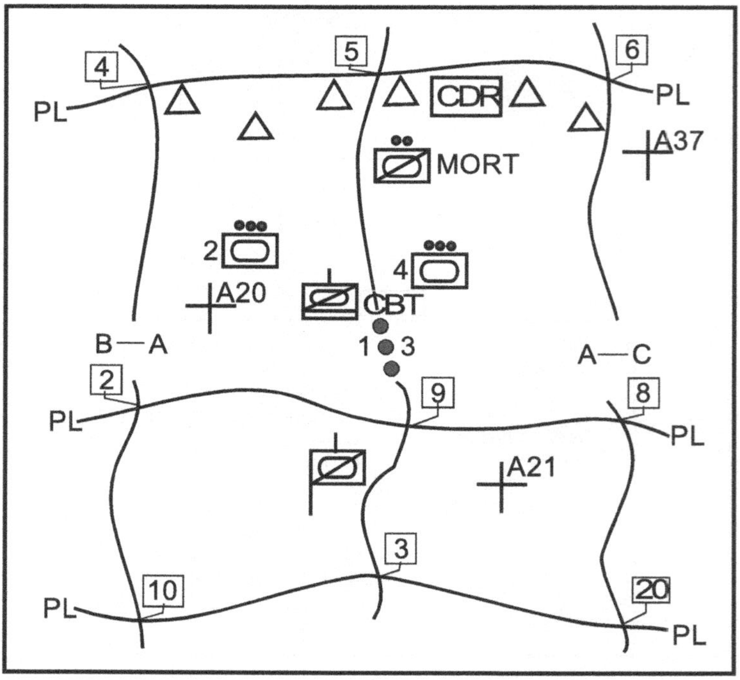

Figure 4-2. Troop screen positioning.

MOVING FLANK SCREEN

The troop may conduct a moving flank screen by itself or as part of the squadron. It may be tasked to screen the exposed flank of the squadron while the squadron conducts a movement to contact, a hasty attack, or a zone reconnaissance (see Figure 4-3). The troop may also participate in a squadron mission to screen or guard another combined arms force (see Figure 4-4).

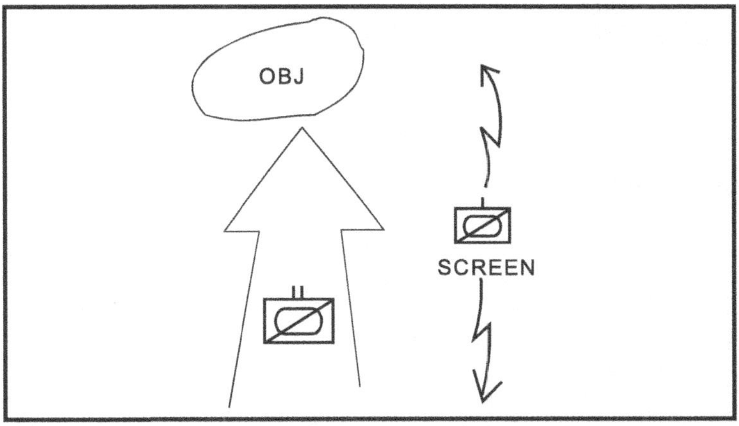

Figure 4-3. Squadron movement to contact; troop flank screen.

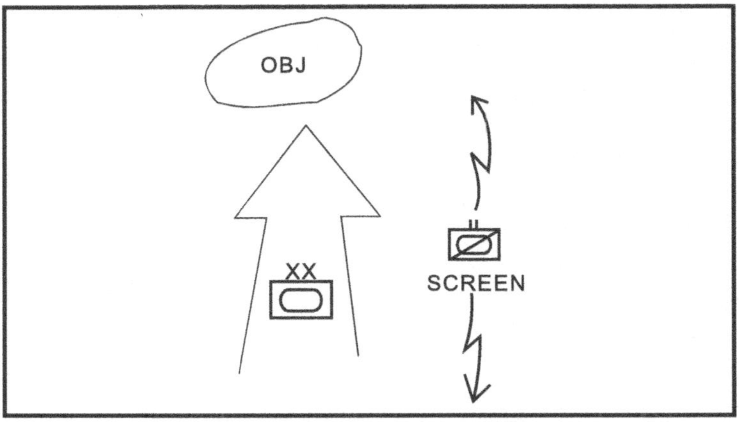

Figure 4-4. Squadron flank screen.

A moving flank screen uses the same techniques as when screening a stationary unit. Position the scout platoons to maintain continuous surveillance on the avenues of approach, the mortar section to cover likely avenues of approach with indirect fire, and the tank/AT platoons in depth to destroy or repel enemy reconnaissance units.

Applying Graphic Control Measures

Because of the inherent dual orientation of a moving flank screen (direction of movement versus orientation of the screen line), control of the operation poses numerous challenges. Control measures must facilitate both orientations (see Figure 4-5).

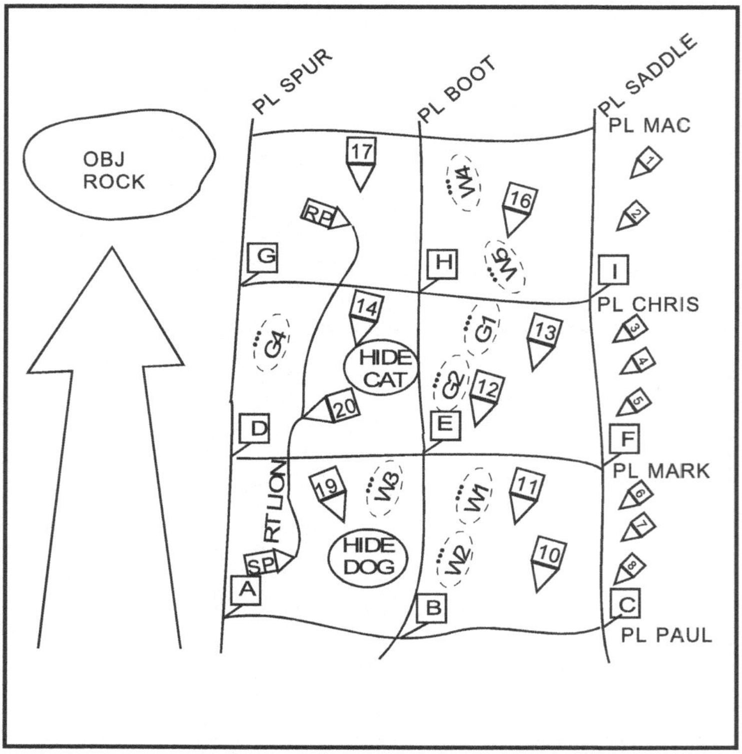

Figure 4-5. Troop moving flank screen graphics.

Follow the procedures below when applying graphic control measures to a moving flank screen.

- Use phase lines to control the scout platoon's movement (placed perpendicular to the screen line). Plan to use these phase lines as on-order boundaries for subordinate platoons if enemy contact is gained. Place phase lines no more than 5 kilometers apart (corresponding to the width of a scout platoon's screen frontage). Do not divide avenues of approach with them.

- Use additional phase lines rearward of (parallel to) the initial screen line to control retrograde movement (toward the protected force). Plan to use these phase lines as subsequent (on-order) screen lines.

- Use objectives, checkpoints, or axes of advance to control the movement of the tank/AT platoons for movement. Position and plan to use these objectives or checkpoints as battle positions for the tank/AT platoons if the scouts make enemy contact at the screen line. Plan subsequent (on-order) battle positions between the screen line and the protected force.

- Use mortar firing positions or checkpoints to control movement of the mortar section. Position these firing positions rearward of the screen line where they allow the mortars to fire two-thirds maximum range forward or cover likely avenues of approach. Plan subsequent (on-order) mortar firing positions between the screen line and the protected force.

While this number of graphic control measures and required planning may seem excessive, they provide maximum flexibility in terms of mission execution. The troop commander can issue simple FRAGOs to adjust the plan to the enemy situation.

Repositioning the Screen

The troop must reposition to stay oriented on the force it is securing. Movement along the screen line is determined by the speed of the protected force. Movement is conducted by one of three techniques—continuous marching, bounding by platoons (alternately or successively), and bounding by OPs (alternately or successively).

Continuous Marching. This technique is appropriate when the protected force is moving quickly and contact is not likely. It is the least secure movement technique.

Deploy both scout platoons abreast with the two tank or AT platoons, the mortar section, and the remainder of the troop in depth (between the screen line and the protected force). The trace of the screen line is essentially the route of advance for the two scout platoons in column. Have the tank/AT platoons and the remainder of the troop move along a designated route or axis of advance (see Figure 4-6).

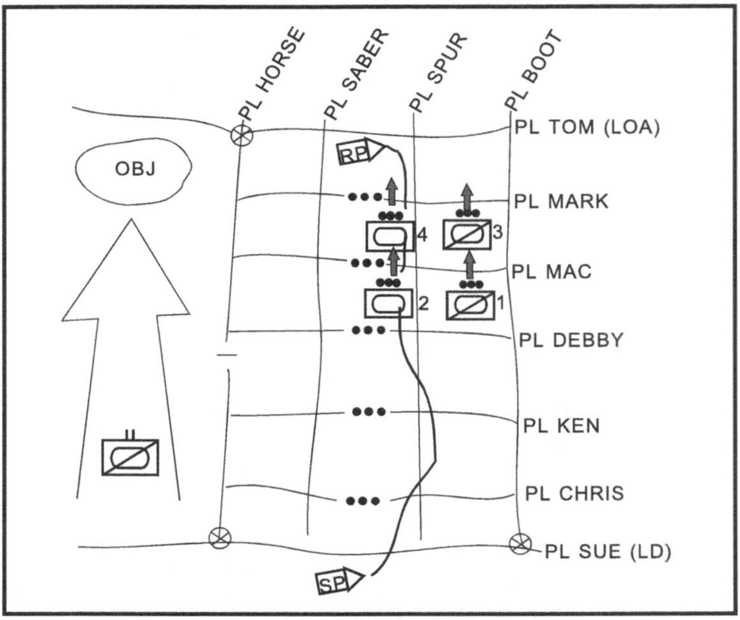

Figure 4-6. Troop moves by continuous marching.

Bounding by Platoons. This technique is appropriate when the protected force requires greater protection than afforded by continuous marching, is not moving quickly, or knows enemy contact is possible. Bounding platoons alternately may leave temporary gaps in the screen line as they move. Bounding platoons successively is more secure but slower than bounding platoons alternately.

Deploy both scout platoons abreast with the two tank/AT platoons, the mortar section, and the remainder of the troop in depth (between the screen line and the protected force). Alternately bound one scout platoon around (to the rear of) the other to assume new positions along the screen line (see Figure 4-7), or successively bound the scout platoons along the screen line (see Figure 4-8). Have the tank/AT platoons occupy designated positions sequentially or alternately.

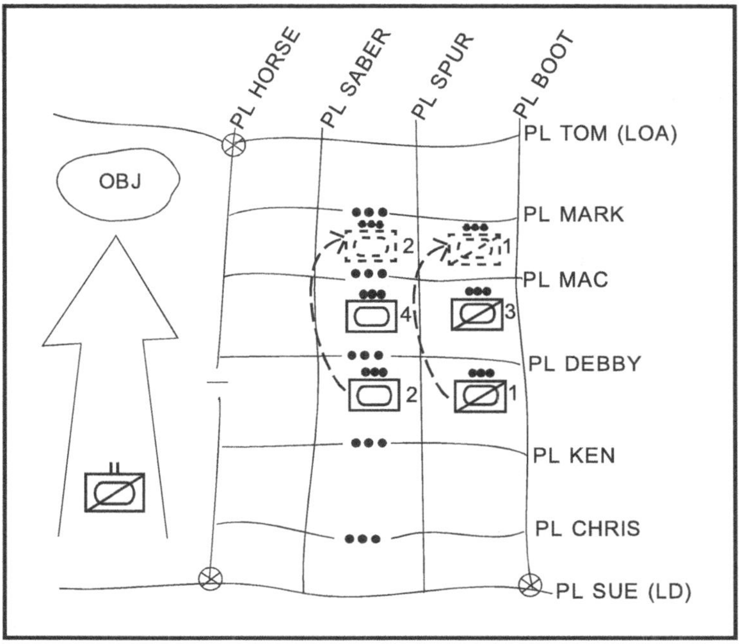

Figure 4-7. Troop moves by alternately bounding platoons.

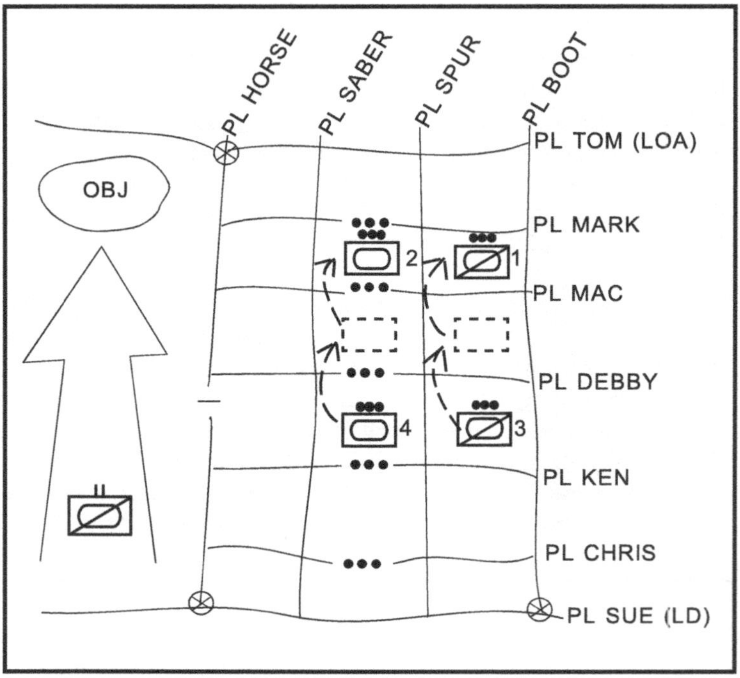

Figure 4-8. Troop moves by successively bounding platoons.

Bounding by OPs. This technique is appropriate when the main body is moving slowly, contact is possible, and maximum security is required. Bounding OPs alternately will disrupt the integrity of the scout platoons as OPs bound to their next position. Bounding OPs successively is easier for the scout platoons to control.

Deploy both scout platoons abreast with the two tank/AT platoons, the mortar section, and the remainder of the troop in depth (between the screen line and the protected force). Alternately bound the rearmost OP around (to the rear of) the other OPs (both platoons) to assume a new position along the screen line (see Figure 4-9), or successively bound the OPs along the screen line (see Figure 4-10). The number of OPs on the screen line may be reduced, as two or more may be bounding at any given time. The rate of advance of the protected force will determine this. Have the tank/AT platoons occupy designated positions sequentially or alternately.

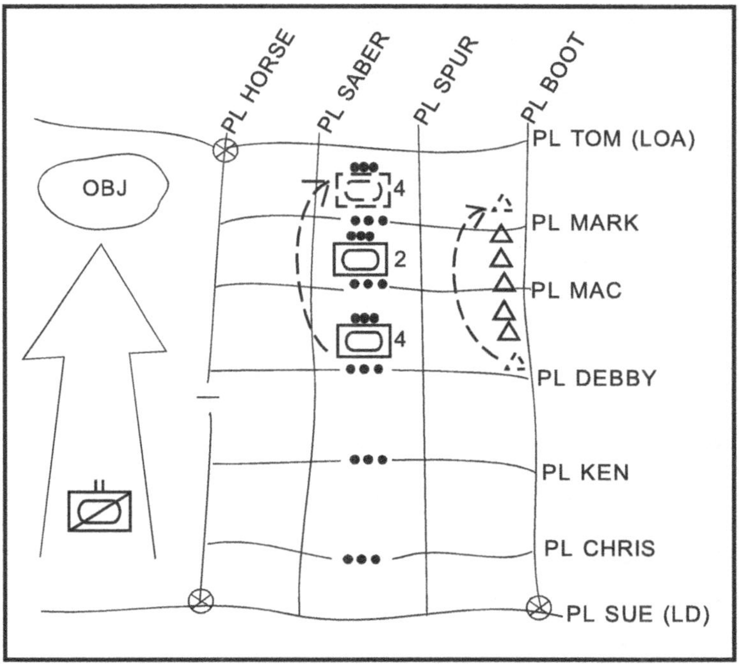

Figure 4-9. Troop moves by alternately bounding OPs.

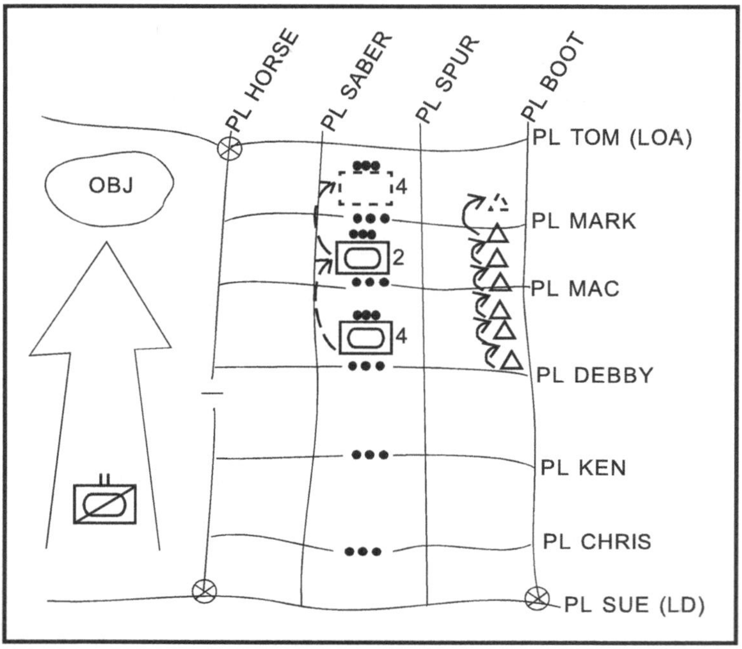

Figure 4-10. Troop moves by successively bounding OPs.

Example of a Screen Forward of Stationary Main Body (Heavy Troop)

The 1st and 3d platoons deploy abreast along PL BOOT, the initial screen line, and establish OPs to maintain continuous surveillance of the high-speed avenues of approach into the troop sector. Route 220 in the 1st platoon sector is the most likely route of enemy approach. Scout platoons execute their approved patrol plans between OPs, and periodically check the areas that cannot be observed. The mortar section lays its guns in a centrally located position about 3 kilometers behind the initial screen line and orients on Route 220. The 2d platoon occupies a hide position behind the 1st platoon. The 4th platoon occupies a hide position north of checkpoint F. Both positions provide good lateral movement and several maneuver options. The troop CP positions south of TIRS A21. The 1SG positions the troop trains in the woods just north of contact point 8, which is near a good lateral road network. The troop commander and FIST collocate well forward with the 1st platoon leader, overlooking Route 220 (see Figure 4-11).

The 1st platoon leader reports two BMPs and one T-64 bounding across the open terrain, moving southwest astride Route 220 about 2,000 meters east of PL BOOT. He maintains contact with the patrol as it crosses PL BOOT and enters the troop sector. The troop commander orders 2d platoon to attack the patrol by fire. The 2d platoon moves into hull-down positions and ambushes the patrol. All three vehicles are hit and damaged. The 1st platoon leader sends his PSG to capture the surviving BMP crew members and to search the vehicles for vehicle unit symbols, maps, sketches, and any other items of intelligence value. The PSG links up with the 1SG, who has moved forward. He hands over the prisoners and material, and returns to his position on the screen line. The 1SG evacuates the prisoners and captured material to the squadron combat trains. The 3d platoon reports no contact in its sector. The XO reports the situation to the squadron.

The troop commander tells the troop to be alert for other divisional reconnaissance patrols in the sector, and to expect regimental reconnaissance within the next hour. The XO eavesdrops on the squadron OI net and monitors reports from Troop B that indicate it has encountered several divisional reconnaissance patrols. The XO passes this information to the troop commander.

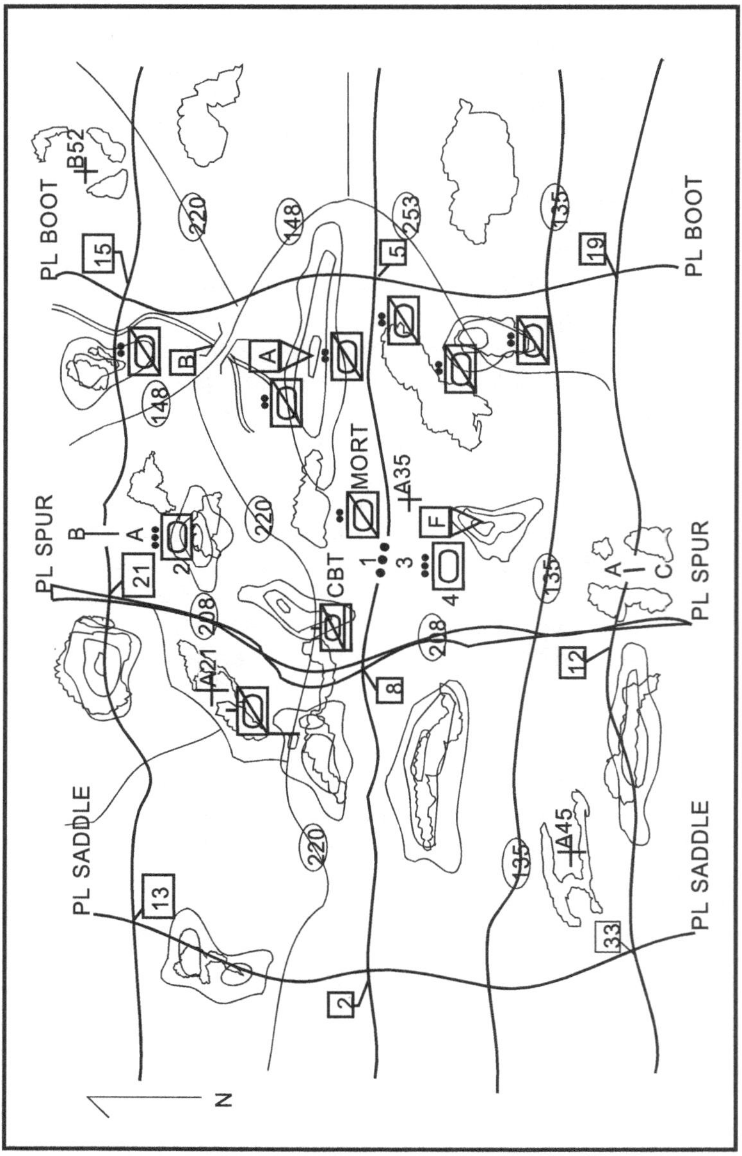

Figure 4-11. Troop screen.

Within the next 45 minutes, both scout platoons report reconnaissance patrols of two BMPs each in their sectors. The troop commander maneuvers a tank platoon into position to ambush each patrol. Three of the four BMPs are quickly destroyed. One BMP in the 1st platoon's sector is damaged, but is able to move into a covered and concealed position in the wooded area west of checkpoint A. Artillery begins to fall on the dominant terrain east of PL SPUR. The XO reports the situation to squadron.

The troop commander orders the 1st platoon to regain contact with the damaged BMP. The 1st platoon leader sends Bravo section to locate the damaged BMP. Bravo section moves into the wooded area and quickly finds and destroys the BMP.

The troop commander orders the scout platoons to move one of their scout sections back in sector in anticipation of the arrival of enemy reconnaissance elements. Both scout platoons reposition a scout section back towards PL SPUR while the troop commander shifts his tank platoons, mortar section, and trains.

The 1st platoon later reports a reinforced MRP moving southwest along Route 220. The 1st platoon continues to observe the enemy reconnaissance elements and reports their progress to the troop. The FSO engages the enemy with mortar high explosive (HE) and smoke rounds to disrupt their movement and keep them buttoned up. The troop commander tells the 4th platoon to move north and be prepared to counterattack the enemy by fire. He quickly outlines its route using TIRS. The enemy reconnaissance elements rapidly move in an attack formation toward the terrain previously occupied by the 2d platoon. It appears that the enemy elements are attempting to secure the dominant terrain in the 1st platoon sector and fix the tank platoon so the lead company of the enemy advance guard can maneuver around it.

The FSO continues to suppress the enemy with smoke and HE while it attempts to maneuver toward its objective. The troop commander directs 2d platoon into position to establish a base of fire and fix enemy elements while 4th platoon counterattacks by fire from the south. The FSO coordinates mortar and artillery fires to suppress and isolate enemy elements as the tank platoons execute their counterattack.

While the tank platoons complete the destruction of the enemy, heavy artillery begins to impact along and east of PL SPUR. The 1st platoon scouts report a tank platoon followed by six BTR-60s moving west along Route 220. The troop commander reports this information directly to the squadron commander. He orders the troops to fall back to PL SPUR, and to begin coordination for battle handover and passage of lines. The troop commander orders the 3d platoon to begin to bound back to PL SPUR. He tells the 1st platoon to maintain contact with the enemy lead company as it falls back to PL SPUR. The troop commander maneuvers the tank platoons west of PL SPUR.

The 1st platoon leader bounds his scout sections back to PL SPUR as he maintains contact with the lead company. The FSO engages the MRC with mortar and artillery fires to slow and disrupt its advance. The XO keeps squadron informed of the situation in the troop sector.

The troop XO moves back to the contact point to begin coordination for battle handover and passage of lines. The troop continues to maintain contact with the lead MRC as it maneuvers back in sector, while the FSO engages the MRC with mortar and artillery fires.

Section III. Guard

Cavalry squadrons conduct guard missions to provide another unit early warning of enemy approach, and to prevent enemy ground maneuver forces from coming within direct-fire range of the protected unit as it performs other missions or tasks. A guard is conducted to the front, flanks, or rear of a stationary or moving force. Many of the considerations for conducting a moving flank screen apply to conducting a moving flank guard.

A cavalry troop mission unique to moving flank guard operations is assigned to the lead troop of the squadron. This three-fold mission consists of the following tasks:

- Maintain contact with the protected force.
- Reconnoiter the zone between the protected force and the squadron route of advance.
- Reconnoiter the squadron route of advance.

The lead troop accomplishes these tasks by performing zone reconnaissance. The speed of the protected force determines how thoroughly the reconnaissance is performed. Assistance is required if the zone is too wide for the lead troop. An air cavalry troop may maintain contact with the protected force, or a following troop may perform route reconnaissance along the squadron's route of advance.

Note. For a detailed description of techniques for performing a squadron guard mission, see Chapter 4, FM 17-95.

The cavalry troop does not conduct guard operations independently. It normally performs reconnaissance, screen, defend, delay, and attack missions as part of a squadron guard operation. Review the appropriate chapter in this manual for specific mission details. See Chapter 3, Sections III and IV, Route and Zone Reconnaissance; Chapter 5, Section III, Hasty Attack; and Chapter 6, Sections II, III, and IV, Defend From a Troop Battle Position and Defend/Delay in Troop Sector.

Section IV. Cover

A covering force operates apart from the main body to develop the situation early and deceives, disorganizes, and destroys enemy forces. It accomplishes all the tasks of screening and guard forces. Unlike screening or guard forces, a covering force is a tactically self-contained force (that is, it is organized with sufficient combat support and combat service support assets to operate independently of the main body). Covering force operations can be either offensive or defensive in nature and can be conducted to the front, flanks, or rear of a stationary or moving force.

Cover operations are performed by the regiment (ACR or LACR) for the corps commander to provide the time and space he needs to make decisions and to achieve his operational objective.

Defensive cover operations, whether forward, flank, or rear, prevent the enemy from attacking the corps main body at the time and place and with the combat strength he chooses. Defensive cover operations are generally intended to destroy the enemy's initiative and set him up for defeat. Both heavy and light cavalry troops participate in defensive cover operations as part of their parent squadrons. Depending on the squadron commander's scheme of maneuver, they either screen, defend, delay, or attack.

Offensive cover operations are conducted by the regiment to seize and retain the initiative for the corps commander, and to allow him to attack decisively with the main body of the corps at the time and place he chooses. In offensive cover operations, the troop usually conducts movement to contact or zone reconnaissance missions. See the appropriate chapters in this manual for mission details.

Section V. Route Security

Cavalry squadrons and regiments conduct route security missions to prevent enemy ground maneuver forces from coming within direct-fire range of the protected route. A route security force operates on and to the flanks of a designated route. Route security operations are defensive in nature and, unlike guard operations, are terrain oriented. A route security force conducts reconnaissance, screens, attacks, defends, and occupies key locations along the route to prevent an enemy force from impeding, harassing, containing, seizing, or destroying traffic along the route.

Route security operations are not conducted independently by a cavalry troop. Both the heavy and light cavalry troops participate in route security operations as part of their parent squadrons and may conduct reconnaissance, screen, defend, delay, attack, or convoy security missions throughout the area of operations.

Section VI. Area Security

An area security force neutralizes or defeats enemy operations in a specified area. It operates in an area delineated by the headquarters assigning the area security mission. It screens, reconnoiters, attacks, defends, and delays as necessary to accomplish its mission. Area security operations may be offensive or defensive in nature and focus on the enemy, the force being protected, or a combination of the two.

Area security operations are conducted to deny the enemy the ability to influence friendly actions in a specific area or to deny the enemy use of an area for his own purposes. This may entail occupying and securing an area before the enemy can, or taking actions to destroy enemy forces already present.

The area to be secured may range from specific points (bridges, defiles) to areas such as terrain features (ridgelines, hills) to large population centers and adjacent areas. The factors of METT-T and unit capability will determine specific unit missions.

Area security missions are conducted by cavalry troops, squadrons, and regiments who employ the techniques of screen, guard, offense, and defense, depending on the nature and purpose of the mission.

Section VII. Convoy Security

Convoy security operations are performed as a minimum by a cavalry troop or a company team. Both heavy and light cavalry troops are suited to the requirements of protecting a convoy due to their organic reconnaissance capability and combat power. Both the cavalry troop and company team should be reinforced with engineers. METT-T considerations, such as restrictive terrain and limited time, may dictate a coordinated effort with air cavalry assets.

Convoy security operations are conducted when insufficient friendly forces are available to continuously secure lines of communication in an area of operations. They may also be conducted in conjunction with route security operations. A convoy security force operates to the front, flanks, and rear of a convoy element moving along a designated route. Convoy security operations are offensive in nature and orient on the force being protected.

A convoy security mission has certain critical tasks that guide planning and execution. To protect a convoy, the security force must accomplish the following critical tasks:

- Reconnoiter the route the convoy will travel.
- Clear the route of obstacles or positions from which the enemy could influence movement along the route.

- Provide early warning and prevent the enemy from impeding, harassing, containing, seizing, or destroying the convoy.

The convoy security force is organized into three or four elements to accomplish the following (see Figure 4-12):

- **Reconnaissance element.** The reconnaissance element performs tasks associated with zone and route reconnaissance forward of the convoy.
- **Screen element.** The screen element provides early warning and security to the convoy's flanks and rear.
- **Escort element.** The escort element provides close-in protection to the convoy. May also provide a reaction force to assist in repelling or destroying enemy contact.
- **Reaction force.** Provides firepower and support to the elements above in order to assist in developing the situation or conducting a hasty attack. May also perform duties of the escort element.

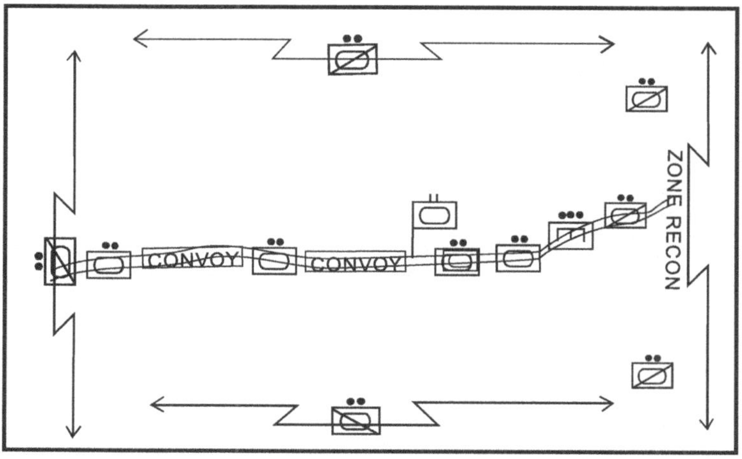

Figure 4-12. Cavalry troop conducts convoy security.

TECHNIQUES

The troop commander organizes and coordinates the efforts of his unit to fulfill the critical tasks associated with the convoy security mission.

Tasks of the reconnaissance element can usually be fulfilled by a single scout platoon. The troop commander ensures the reconnaissance element focuses on trafficability of the route and enemy forces that may influence movement along the route. METT-T may dictate the use of engineers to assist in reconnoitering and clearing the route. Convoy speed is determined by the pace of reconnaissance (METT-T). As a guide, the reconnaissance element should operate a minimum of 3 to 5 kilometers ahead of the main body of the convoy.

Tasks of the screen element can usually be fulfilled by a single scout platoon also. The troop commander ensures the screen element establishes OPs to provide early warning on critical portions of the route or key avenues of approach to the route. OPs have a limited ability to destroy enemy forces; therefore, their primary purpose is to acquire the enemy and direct reaction forces or indirect fire to destroy it.

Tasks of the escort element are best performed by tank platoons in a heavy troop. A light troop should consider a scramble of platoons (MK-19/M2/TOW—see Chapter 2, Figure 2-13) to gain the suppressive firepower necessary to protect the convoy. The troop commander ensures the escort element is positioned to provide security throughout the length of the convoy. This requires elements of the two tank (heavy troop)/mixed (light troop) platoons be dispersed throughout the convoy order of march. If there is no reaction force available or designated, a task of the escort element may be to provide reaction forces that respond to enemy forces identified by the reconnaissance or screening elements. Depending on the length of the convoy and METT-T considerations, the troop commander may keep one tank/mixed platoon consolidated or specifically designated as the reaction force.

Chapter 5

Offense

Offense is the decisive form of war. Brigades, battalion/task forces, and company teams are the principal offensive force for the corps or division. Cavalry units normally perform reconnaissance and security missions in support of corps and division offensive operations. Cavalry troops may perform certain offensive missions as part of a squadron, regiment, or other combined arms force. These offensive missions are normally performed during reconnaissance or security operations.

If required, cavalry troops may perform offensive operations within an economy-of-force role for a higher headquarters.

CONTENTS

Section I. Purpose and Fundamentals

PURPOSE

The main purpose of the offense is to defeat, destroy, or neutralize the enemy force. Offensive operations are also undertaken to secure decisive terrain, to deprive the enemy of resources, to gain information, to deceive and divert the enemy, to hold the enemy in position, to disrupt his attack, and to set up conditions for future successful operations.

FUNDAMENTALS

Successful offensive operations have four fundamentals in common.

- **Surprise.** Strike the enemy at the time and place or in a manner that he least expects.
- **Concentration.** Mass available forces, strive for overwhelming superiority in men, weapons, and firepower. With concentration, however, vulnerability becomes a factor. A force that is dispersed is much more survivable. The commander must maintain a high sense of situational awareness to anticipate the conditions of battle that will allow him to mass at the critical point, kill the enemy, and quickly disperse to survive.
- **Tempo.** Tempo is the rate of speed of military action. Controlling or altering the rate is essential for maintaining the initiative. Tempo can be fast or slow dependent on the capabilities of the troop relative to that of the enemy. Commanders must adjust tempo to ensure synchronization.
- **Audacity.** Boldness in the plan's execution is key to success in offensive operations. Commanders should understand when and where they are taking risks, but must not become tentative when executing their plan.

Section II. Movement to Contact

A movement to contact is a mission undertaken to gain or regain contact with the enemy. Unlike a zone reconnaissance, which is focused on reporting detailed information on the terrain and enemy within a given zone, movement to contact is focused on finding the enemy. The critical tasks are geared for achieving fast movement and rapid location of enemy forces.

The cavalry troop normally conducts a movement-to-contact mission as the lead element of a squadron or combined arms force hasty attack. The troop may also conduct the mission as part of a squadron movement to contact when the squadron is leading the advance of another combined arms force, such as a brigade or division.

A movement to contact is characterized by rapid aggressive action. The troop commander must rapidly develop the situation and may be permitted to bypass enemy forces to maintain momentum.

CRITICAL TASKS

During a movement to contact, certain critical tasks will be accomplished. Unless the squadron commander gives guidance otherwise, the troop will—

- Reconnoiter and determine the trafficability of all high-speed routes within the zone.
- Inspect and classify all bridges, culverts, overpasses, and underpasses along high-speed routes. Identify all bypasses and fords that cannot support rapid heavy-armor movement.
- Clear all high-speed routes of mines, obstacles, and barriers within its capability, or locate a bypass.
- Find and report all enemy forces within the zone and determine their size, composition, and activity.

TECHNIQUES

The cavalry troop can perform movement to contact in a zone up to 10 kilometers wide. The size of the zone depends on the terrain and enemy situation. Scout squads, and in some cases sections, must be mutually supporting during movement based on the enemy situation. Mutual support between squads and sections is gained by reducing the size of the zone based on the terrain. The more restrictive the terrain the less mutual support becomes available.

When considering techniques for conducting a movement to contact, remember the basics.

- Always retain the ability to maneuver.
- Make contact with the smallest force possible.
- Employ forces in depth; stay flexible.
- Develop the situation rapidly once contact is made with the enemy force. Operate at a tempo that forces him to react to you, not you to him.

When the troop receives a movement-to-contact mission, the zone of action is normally identified by lateral boundaries. A line of departure and an objective or limit of advance are specified to orient the troop's movement. The objective is usually a terrain feature placed deep enough to ensure contact with the enemy. Divide the troop zone into two platoon zones. Think about where to draw the boundary. Make sure it does not split enemy avenues of approach and that it is on easily identifiable terrain. Do not use a road for a boundary. To control progress through the zone, add phase lines every 5 to 8 kilometers on easily identifiable terrain. Place contact points near the intersection of the platoon boundary and all phase lines. If specific areas need to be checked, identify them using checkpoints for reference. Place TIRS on the map.

Use a troop vee or split vee formation when conducting a movement to contact. Deploy scout platoons abreast and have them accomplish all the related reconnaissance tasks. Move them out first across the line of departure. The mission of the scouts is to find the enemy, develop the situation, and recommend to the troop commander the best course of action for the employment of the tank/AT platoons.

The mortar section, under control of the troop fire support officer or mortar section sergeant, follows about 2 kilometers behind the scouts, centered in the zone, to provide continuous coverage forward of the scouts. The troop command post, under control of the XO, displaces in the zone, positioning on terrain that affords effective and continuous communication with troop elements and the squadron. The troop trains, under control of the 1SG, follow about 2 kilometers behind the tank/AT platoons, and bound from one covered and concealed position to another. The troop commander positions himself well forward to observe one scout platoon or the other. His position is usually located where he expects initial enemy contact or difficult situations.

Heavy Troop

Due to the survivability and firepower of scout platforms (M3) in the heavy troop, the tank platoons may be kept in considerable depth (1,000 to 3,000 meters) behind the scouts until enemy contact is made. Once contact is made and the scouts have developed the situation/recommended a course of action, the troop commander may then employ his tanks, using firepower and maneuverability to fix the enemy or destroy it by hasty attack.

Light Troop

In the light cavalry troop, the AT platoon(s) should be kept close (500 to 1,500 meters) to the scout platoons. Upon contact, the HMMWV scouts will require more time to develop the situation than Bradley-mounted scouts. The AT platoon(s) can assist the scouts in developing the situation by overwatching the scouts. Also the TOW takes time to employ, so reaction time is slower than with the tank. During movement to contact, the light troop commander must be able to execute his actions on contact very quickly to maintain the momentum of the operation. A technique the light troop might use is outlined below.

- Scouts make contact with an enemy force.
- The troop commander immediately moves one AT platoon forward to overwatch the scouts as they develop the situation.

- If the situation dictates, the AT platoon suppresses with direct fires to allow the scouts to maneuver.

- Once the situation is developed, the troop positions to either fix the enemy for the follow-on force or destroy it by a hasty attack.

EXAMPLE OF MOVEMENT TO CONTACT (HEAVY TROOP)

The troop is ordered to reestablish contact with elements of a withdrawing enemy mechanized infantry platoon. As the troop deploys in a split-vee formation, the 1st and 3d platoons cross the line of departure abreast, north and south respectively, and begin to reconnoiter in troop zone up to PL BOOT. The scout platoons determine the trafficability of high-speed routes within their zones, and search for the enemy in areas that dominate the routes. The bridge across the creek at checkpoint B will support only 20 tons, so the 1st platoon finds a solid-based fording site 100 meters south of the bridge. Thus far, the high-speed routes are not damaged or blocked. The 3d platoon reports finding tracks of what appears to be three BMPs heading northwest from Hill 244. The 1st platoon reports tracks of an enemy company-size force heading west along Route 220. Track patterns indicate BMPs and T-64s/T-72s. The mortar section eavesdrops on the troop command net, and moves with the scout platoons, staying back about 1.5 kilometers, centered in the troop zone. The 2d and 4th platoons are paired with their sister scout platoons and key their movement on the scout platoon's progress, staying back about 1 kilometer and using terrain that affords good cover and concealment. The troop CP and trains hold in place (see Figure 5-1).

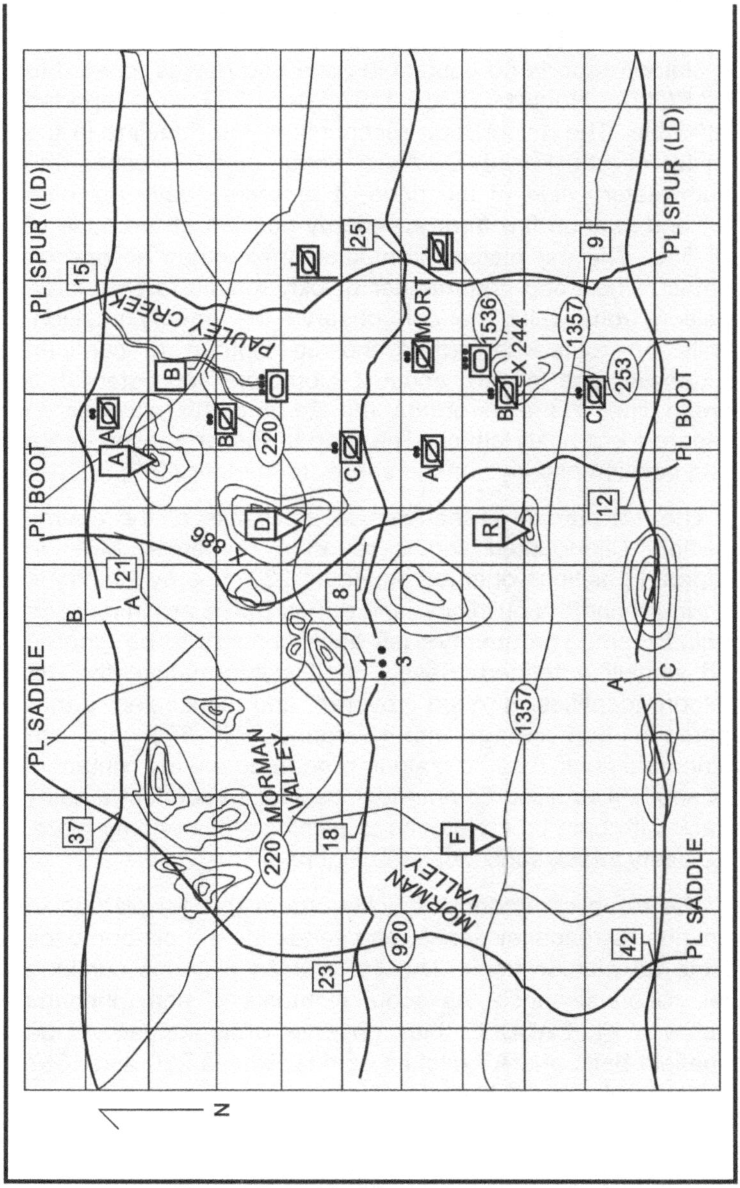

Figure 5-1. Movement to contact.

As the 1st platoon approaches PL BOOT, it reports two BRDM-2s withdrawing west at high speed on Route 220. The 3d platoon reports no contact in zone and moves forward to PL BOOT. Routes 1536, 1357, and 253 are reported trafficable. The 1st platoon reconnoiters the ridgeline in the vicinity of checkpoint D. As a scout squad crosses the southwestern side of the ridge, it receives heavy machine gun and cannon fire from an enemy position at the base of Hill 886. The 1st platoon immediately executes actions on contact. The troop commander quickly moves to a covered position from which he can observe the enemy position. While en route, he orders the 3d platoon to continue reconnaissance forward about 2 kilometers and establish a screen line. He also orders the 2d and 4th platoons to prepare for a hasty attack. The mortar section moves into a firing position and lays on the enemy (see Figure 5-2).

The 1st platoon maneuvers to the flanks of the enemy position. Dismounted scouts report a reinforced MRP in prepared positions oriented on Route 220. Enemy infantry is dug in around the buildings, and dug-in BMPs are in defilade behind them. The enemy's left flank appears to be ignored and weakly defended. Based on this information, the 1st platoon identifies a good covered and concealed armor approach that swings north around Hill 886 into the vulnerable flank. The 3d platoon reports no enemy contact in the south. The troop commander decides to conduct a hasty attack with both tank platoons to eliminate this enemy force. The hasty attack succeeds (see Figure 5-3).

The troop commander quickly orders the 1st platoon to continue its reconnaissance, and sends the 4th platoon back to the 3d platoon zone. The 1st and 3d platoons continue their reconnaissance. As scout elements of both platoons approach PL SADDLE, they observe what appears to be minefield belts and AT ditches astride Routes 220 and 1357 in the Morman Valley. Dismounted reconnaissance determines the extent of the obstacles, and reveals heavy concentrations of enemy forces in several company strongpoints along the dominating ridgeline by Highway 920.

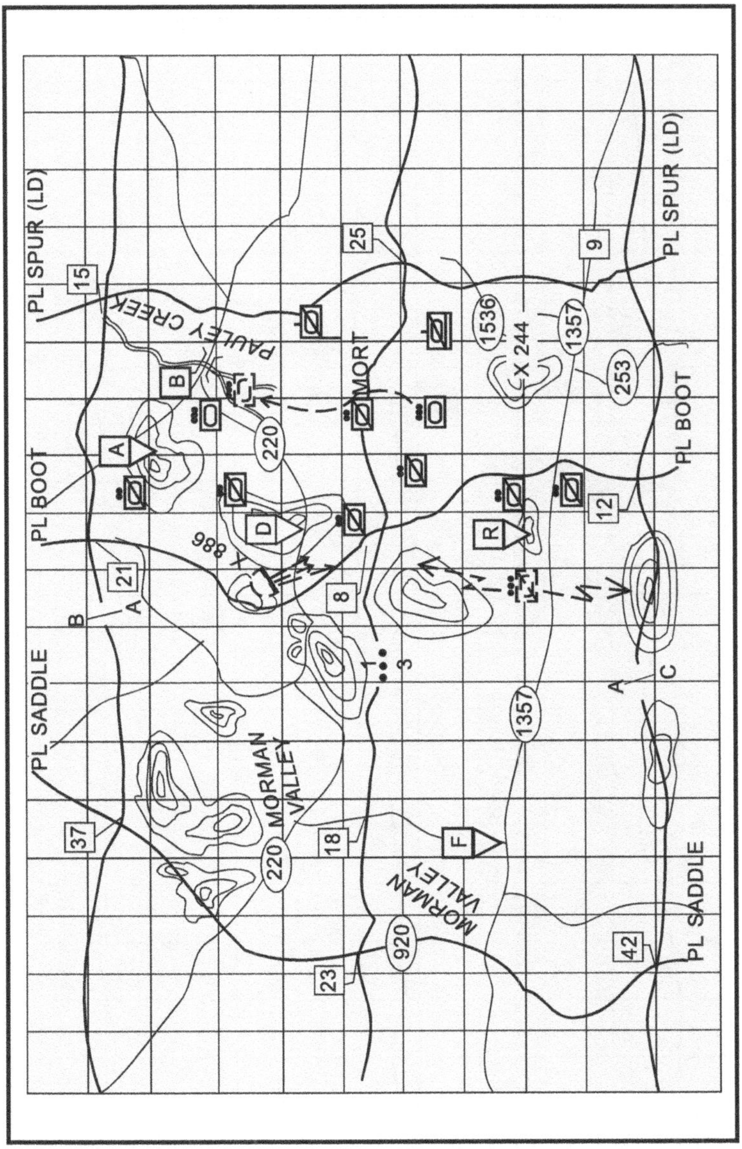

Figure 5-2. Actions on contact.

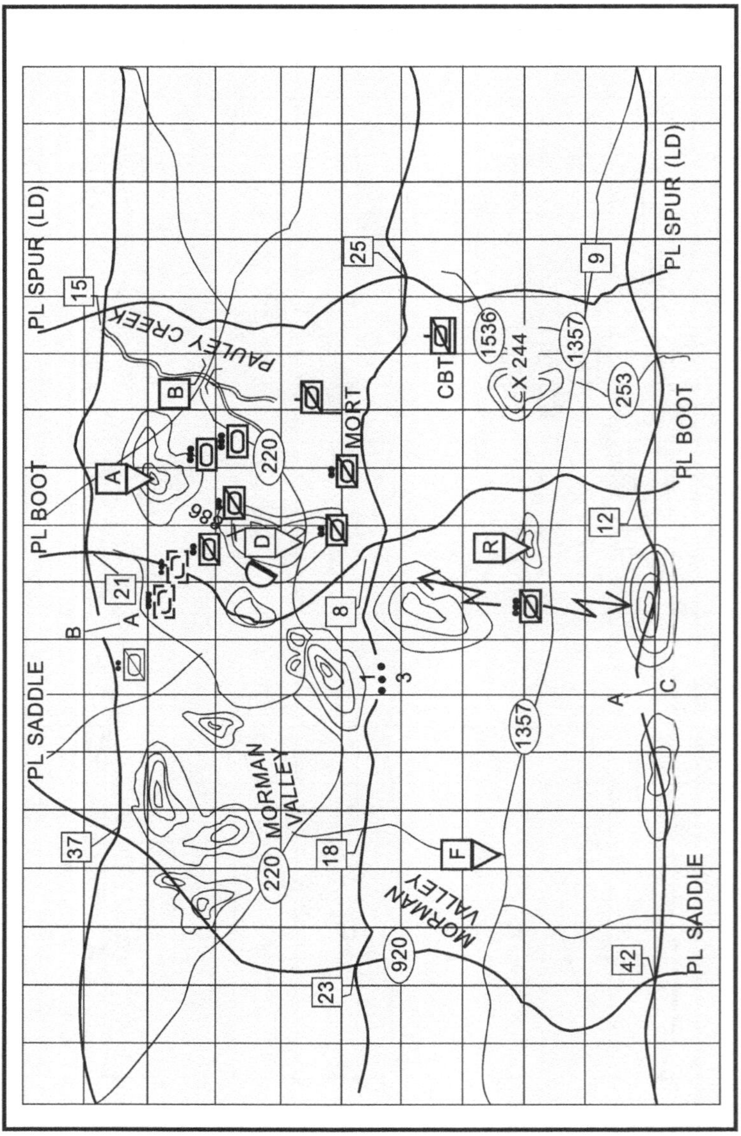

Figure 5-3. Hasty attack.

The troop commander takes a look. En route, he orders both scout platoons to determine if any gaps or boundaries exist between the enemy positions. After seeing the situation, the troop commander calls his squadron commander and recommends the troop establish a hasty defense and prepare to assist in battle handover and forward passage of following attack forces (see Figure 5-4).

Section III. Hasty Attack

A hasty attack is conducted with a minimum of preparation to defeat an enemy force that is not prepared or deployed to fight. It is a course of action routinely employed in cavalry operations to seize or retain the initiative, or to sustain the tempo of operations. A hasty attack can be executed while the troop is engaged with a zone reconnaissance mission or movement to contact. (See Chapter 3, Section II, and Section II of this chapter for other examples of hasty attack.)

CRITICAL TASKS

To execute a hasty attack, the following critical tasks must be accomplished:

- Reconnoiter and determine the size, composition, and orientation of the enemy force.
- Determine if the objective enemy force is supported by other units nearby.
- Find a high-speed covered and concealed approach into the enemy's flank(s).
- Establish a maneuver element (usually one or both tank platoons) to move to a position of advantage over the enemy and attack him by fire.

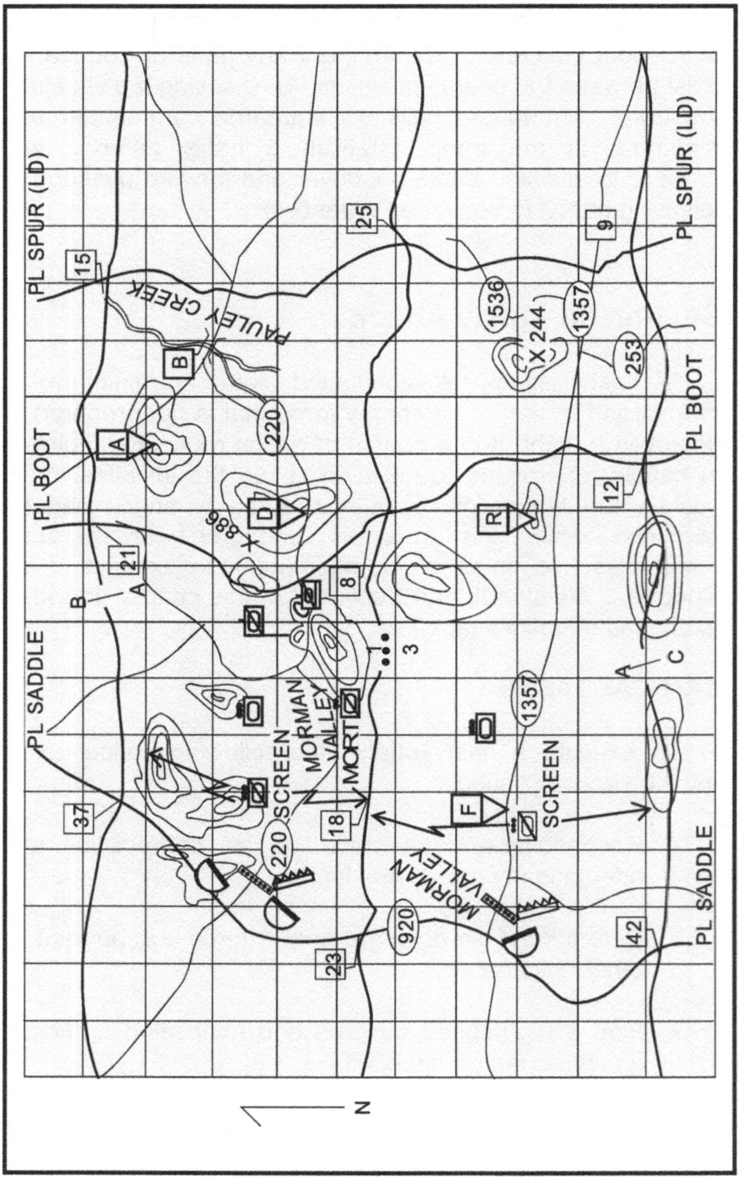

Figure 5-4. Hasty defense.

- Establish a base-of-fire element (usually one or both scout platoons) to defeat or suppress all observed enemy AT weapons with long-range direct and indirect fires before the maneuver force deploys into its attack.

- Isolate the objective enemy force from other mutually supporting units with indirect fires (usually with smoke and HE ammunition).

- Attack the enemy by fire or by fire and movement, and defeat him.

- Once the attack is completed, immediately establish hasty defensive positions and OPs on high-speed avenues of approach into the troop position.

TECHNIQUES

Each critical task has a time at which it will be accomplished in relation to all other critical tasks. A good hasty attack depends on the commander's sense of timing and on his ability to employ his forces to accomplish the tasks in the proper sequence. The commander has to synchronize—to concentrate and apply different forms of combat power against the enemy at the right times and places. The decision to conduct a hasty attack is usually made after an enemy force's reconnaissance and dispositions show that winning requires a quick strike, with little preparation. Tactics for conducting a hasty attack have three common features.

- Known or suspected enemy AT weapons are suppressed and destroyed with direct and/or indirect fires before the maneuver force is committed.

- The enemy is forced to fight in two directions.

- The enemy is suppressed and unable to react.

SETTING UP THE TROOP FOR A HASTY ATTACK

While conducting other missions, scouts will often make contact with an enemy force. In developing the situation, a scout platoon may recommend hasty attack as a course of action to the troop commander, who decides to execute the recommended course of action. The troop commander issues FRAGOs that will position forces to execute an attack simply and effectively.

The scout platoon in contact continues to reconnoiter the enemy's position and to accomplish its reconnaissance tasks. One section of the scout platoon remains in contact with the enemy. The other scout platoon continues its reconnaissance up to a limit of advance established by the troop commander. Both scout platoons continue to develop the situation further by looking for the presence of other enemy units, to the flanks or rear, supporting the enemy contact.

The FIST moves to a good position to see the battlefield and to control the indirect fires. The FSO places the mortar section on terrain where it establishes a firing position and prepares to suppress the enemy position.

The size and strength of the enemy may require the use of one or both tank/AT platoons. The commander may choose to do one of the following options:

- Move one tank/AT platoon into an overwatch position and attack by fire with one.
- Keep one in reserve behind the scout platoon not in contact and attack by fire with one.
- Attack by fire with both platoons.

The scouts determine a good attack position and attack-by-fire position for the tank/AT platoon(s). A scout squad from the scout platoon in contact moves to a link-up point with one or both tank/AT platoon(s) to guide them into the attack position.

The troop commander moves to collocate with the scout platoon in overwatch. The 1SG moves medics and recovery vehicles close to the battlefield. The XO assists the commander in control of the troop and keeps the higher commander informed.

INDIRECT FIRES IN SUPPORT OF A HASTY ATTACK

Indirect fires should complement the troop's scheme of maneuver. In the light troop, dependent on availability of supporting indirect-fire systems, indirect fires may prove to be the best weapon of destruction. The troop can acquire and engage the enemy with indirect fire from positions offering good protection from enemy direct/indirect fire. Because the light troop is in covered and concealed positions, the survivability of the troop is enhanced. Indirect fires must also be controlled to prevent fratricide. The troop commander must determine the following:

- Who will control the indirect fires during the hasty attack?
- Who will initiate indirect fires onto the objective?
- Who will lift and shift the indirect fires to subsequent targets?
- What will the signal be for lifting and shifting indirect fires?
- Are there any restrictive fire measures or restrictive fire areas?

There are many different answers to these questions. METT-T will determine which answer works best. Under most conditions the commander or the FSO will be in the best position to control the engagement of indirect-fire systems.

The troop commander should use available indirect fires from mortars and supporting artillery to—

- Suppress the enemy while scouts are maneuvering to develop the situation.
- Screen enemy observation of scouts or assault element during the conduct of the hasty attack.
- Isolate the enemy contact by firing HE and smoke between the enemy force and any possible supporting positions.
- Shift indirect fires off the objective to block enemy withdrawal routes.

DIRECT FIRES DURING HASTY ATTACK

When executing the hasty attack the troop must combine indirect fires with direct fires. The direct-fire systems available to the heavy troop commander are highly destructive. The high rate of fire from the M3 and M1 allows the commander to quickly place effective fires into the objective area while stationary or moving.

The systems available to the light troop commander can be just as destructive when used in combinations and against a comparably equipped enemy. The light troop commander should strive to engage the enemy with combinations of weapons. Using the bounding technique and move-set drills by platoons, the troop commander can place effective suppressive and destructive fires into the objective area. (The high rates of fire from the M2 caliber .50 and MK-19 should be used to suppress the enemy and cover the TOW systems as they engage.) See Figure 5-5.

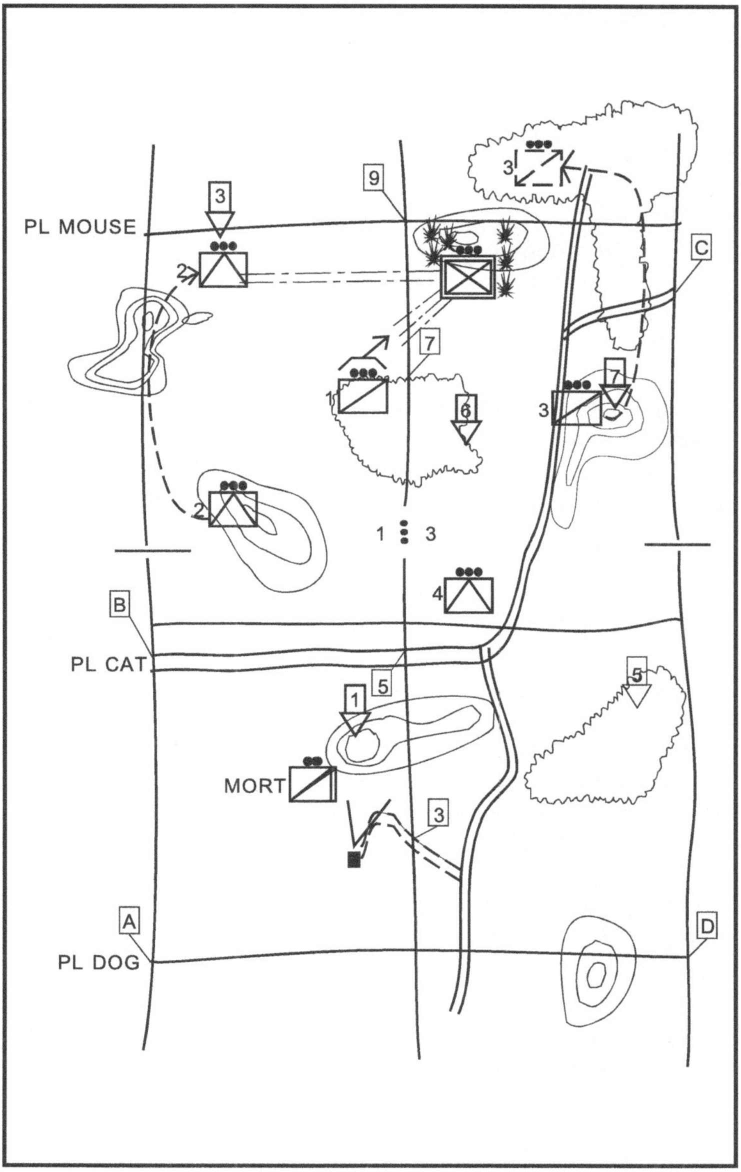

Figure 5-5. Light cavalry troop using fire and maneuver.

Like indirect fires, direct fires must be controlled. The commander must determine the following before executing a hasty attack:

- Who will initiate direct fires into the objective area and from where will they be initiated?
- What is the aim point for the support element? When do they shift fires and where do they shift fires to?
- What is the limit of advance for the assault force?

To determine the answers to these questions the commander must look at the enemy he is facing, the terrain he is operating in, and the forces he has to bring to bear against the enemy.

THE OBJECTIVE

Once the troop has initiated its attack and enemy resistance in the objective area has ceased, the troop may begin consolidating in preparation of continuing its mission. Have one or both scout platoons dismount scouts to sweep the area quickly for prisoners and other items of tactical value. Immediately establish a hasty defense with tank/AT platoons oriented on high-speed approaches into the troop position, and with scout platoons oriented on expected high-speed enemy approaches at OPs out from the defensive position. Position the mortar section and have the guns lay on the most likely enemy avenue of approach. Redistribute ammunition and balance crews as time allows. Replace key leaders quickly. If time is available and enemy counterattack unlikely, begin resupply operations.

EXAMPLE OF A HASTY ATTACK

A troop is conducting a zone reconnaissance moving toward its limit of advance (LOA). As the 3d platoon crosses the ridge (PL SABER) in its zone, Charlie scout section is hit by enemy tank cannon and heavy machine-gun fire from concealed positions near checkpoint 2. The platoon immediately takes cover and reports contact to the troop commander. The troop commander acknowledges and immediately moves to link up with the platoon leader. The platoon sergeant calls for indirect fire to suppress the enemy force. Under this protection, the platoon leader orders dismounted reconnaissance through the woods on the enemy's right to find his flank. He then sends his Bravo scout section forward on the enemy's left to find his other flank and to determine if the enemy is mutually supported by other enemy forces near checkpoint 2. The platoon leader and two scouts dismount, move in closer, and determine the enemy force consists of a reinforced MRP, two tanks, and three BMPs, in a hasty defensive position. Scouts report what appear to be minefields forward and to the flanks of the enemy position. The Alpha section identifies the right flank of the enemy position and discovers the flank is not protected by obstacles and not covered by direct-fire weapons (see Figure 5-6).

En route to the 3d platoon's zone, the troop commander tells the 1st platoon to continue reconnaissance forward to the LOA and find out if other enemy forces are providing mutual support to the enemy platoon in the 3d platoon's zone. He also tells the platoon leader to establish a screen along the LOA in his platoon zone. He then tells the 4th platoon to join the 2d platoon near checkpoint 12, using the ridgeline to cover its move, and prepare to conduct a hasty attack. Using TIRS, he outlines a route that runs from checkpoint 12 around the western edge of the village to checkpoint 7, then to a position from which it can engage the enemy on his most vulnerable flank (see Figure 5-7).

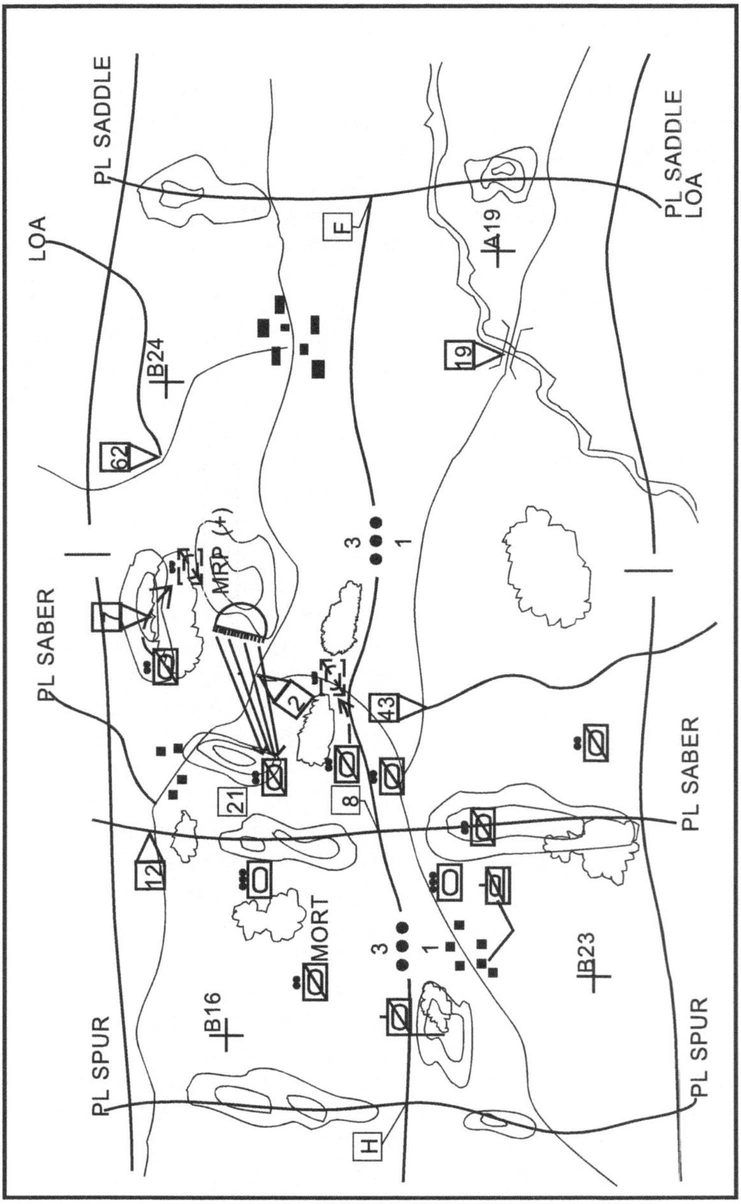

Figure 5-6. Develop the situation.

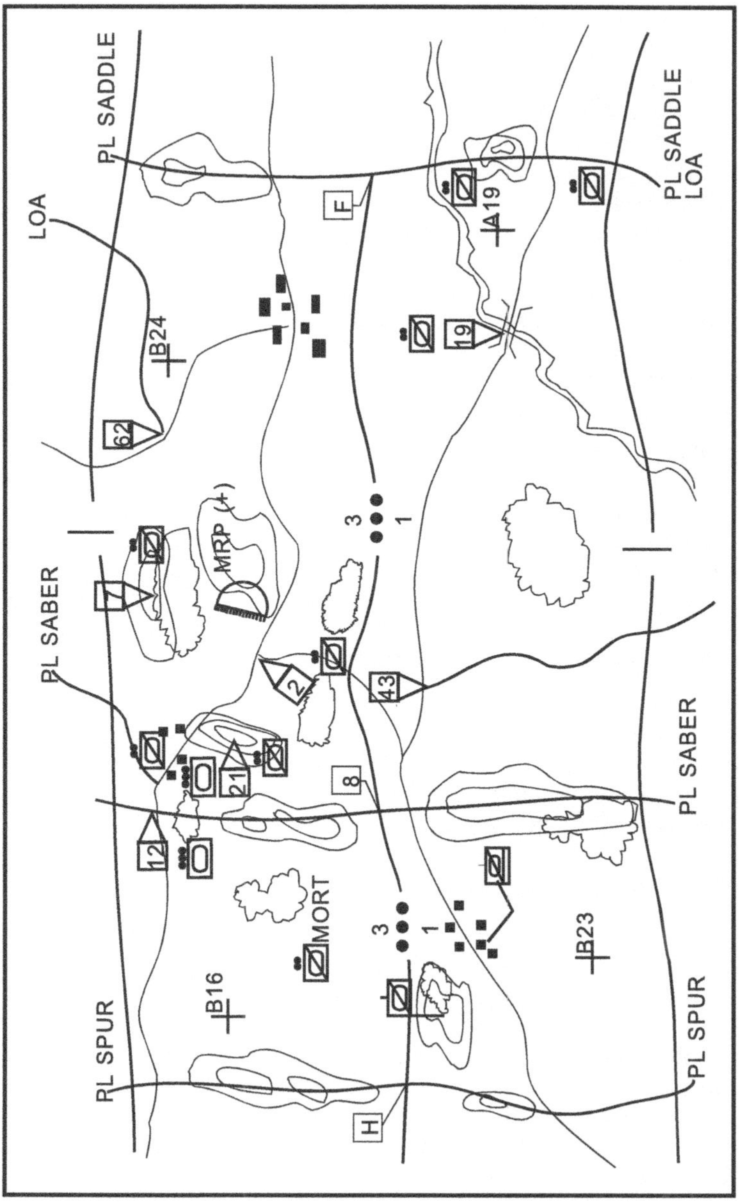

Figure 5-7. Set for hasty attack.

After a first-hand look at the situation in the 3d platoon's zone, the troop commander decides to attack. He tells the FSO to continue suppressing the enemy. Next, he tells the 3d platoon leader to establish a base of fire from covered positions near checkpoint 21. The 3d platoon is tasked to destroy or suppress all AT weapons observed. While heavy suppressive fires are raining on the enemy force, the commander joins the tank platoons at checkpoint 12 and leads them into the attack. After passing checkpoint 7, the tank platoons begin maneuvering. The troop commander tells the FSO to shift his fires on routes of withdrawal behind the enemy position. The tank platoons go around the woodline, deploy abreast, and attack by fire (see Figure 5-8).

Two tanks in the 4th platoon are hit and damaged; one soldier is killed and three are injured. The tank platoons consolidate in covered and concealed positions near checkpoint 7. Scouts from the 3d platoon move in quickly, capture several wounded prisoners, and search the vehicles and dead for anything of intelligence value. The 3d platoon leader orders one scout squad to stay and secure the prisoners, and then takes the rest of the platoon forward to finish zone reconnaissance up to the LOA.

The 1SG leads the medics and M88 to the 4th platoon's casualties and picks up the prisoners the 3d platoon secured. The XO coordinates with the S4 for immediate resupply, then collects and transmits final reconnaissance reports to the squadron.

Section IV. Deliberate Attack

A deliberate attack is an offensive mission conducted to defeat enemy forces in dug-in and prepared defensive positions or strongpoints. It is preceded by intensive intelligence gathering using sources such as patrols, aerial photographs, reconnaissance by fire, radio intercepts, radar surveillance, satellite reconnaissance, and prisoner interrogation. Large volumes of indirect preparatory fires are normally used. Detailed tactical planning and rehearsal of the attack are crucial to success.

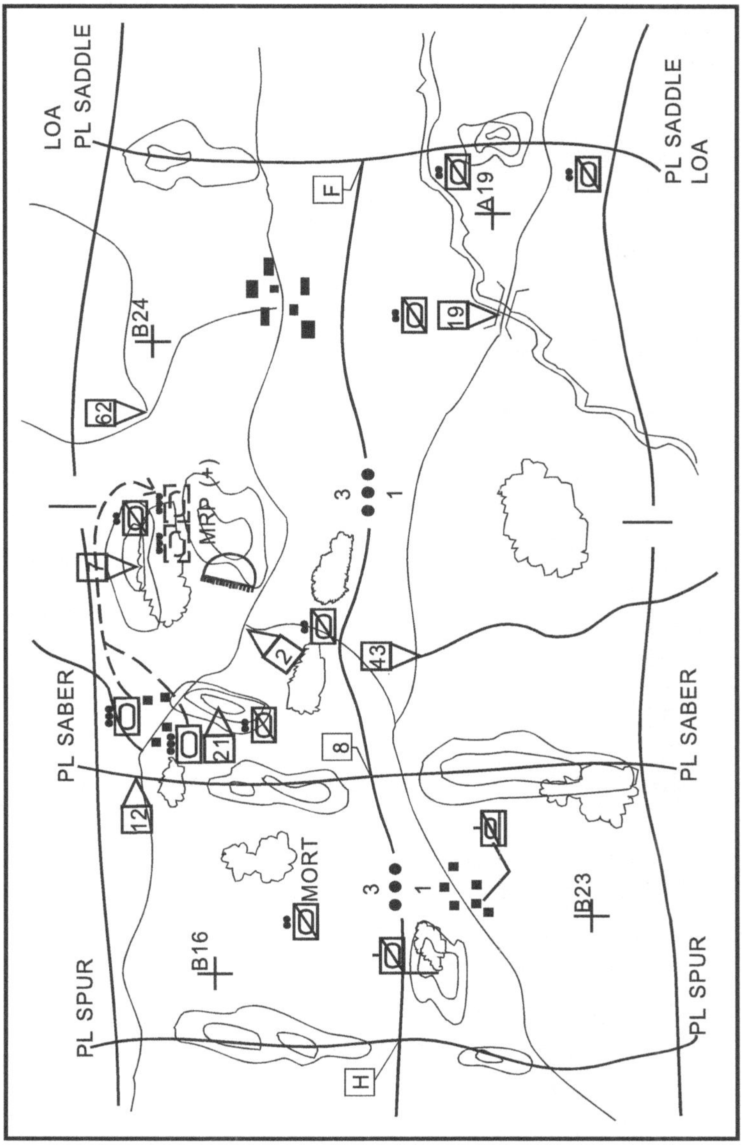

Figure 5-8. Tank platoons move into position for the hasty attack.

The cavalry troop alone cannot conduct a deliberate attack against prepared defensive positions. It usually participates with its parent squadron or other combined arms force as one of the three major elements in a breaching and assault operation. The troop can serve as the *overwatch/support element,* the *breaching element* (if reinforced with engineers), or the *assault element*. Doctrine for attacking a complex obstacle or a strongpoint is found in FM 71-1 and FM 17-95-10.

Section V. Raid

A raid is an attack into enemy territory for a specific purpose, with no intent to gain or hold terrain. The troop returns to friendly lines after the attack.

The cavalry troop normally conducts a raid against clear, well-defined targets whose destruction will have an adverse effect on an enemy force. Good targets for a troop raid are enemy artillery, rear service units, or command and control headquarters.

CRITICAL TASKS

During a raid, the following critical tasks must be accomplished:

- Coordinate and establish a restricted fire area around the raid area.
- Reconnoiter the troop's direction of attack from the point of departure to the objective.
- Establish visual contact with the enemy. Reconnoiter and determine the best place to initiate the attack.
- Establish a base of fire element to suppress the enemy force.

- Establish a maneuver element (usually one or both tank platoons) to move to a position of advantage over the enemy and attack them by fire.
- Suppress the enemy force with long-range direct fires while the maneuver element moves up to begin its attack.
- Establish surveillance on enemy avenues of approach to the objective, where follow-on enemy maneuver units would most likely appear.
- Attack the enemy by fire or by fire and movement to defeat them. Finish quickly.
- Establish security to protect the withdrawal of the remainder of the troop.
- Return rapidly to friendly lines and enter at the designated passage point.

TECHNIQUES

A raid achieves its purpose through speed, violent execution, and surprise. The troop's maneuver should produce that effect. There will probably not be a chance to reconnoiter before the attack; speed and surprise generate the security needed. Consequently, the troop might not always take the mortar section, combat trains, or command post. Vehicles disabled on a raid may be left behind in the interests of speed and security for the entire troop. When the troop receives a FRAGO to conduct a raid, the commander should receive the following graphic control measures from the S3 or squadron commander: a point of departure along the line of departure or line of contact, a direction of attack, an objective, a route of withdrawal, and a passage point to enter friendly lines. Analyze the terrain and the enemy situation, then issue the troop FRAGO.

The FSO should coordinate with the squadron FSO to establish a restricted fire area around the raid area. The 1SG should coordinate for immediate resupply of Class III and Class V when the troop returns. Have one scout platoon take the lead and reconnoiter the troop direction of attack from the point of departure on the line of departure to the objective. The lead platoon should establish visual contact with the raid objective to confirm the troop's scheme of maneuver and establish surveillance of the objective. The other scout platoon may act as the base-of-fire element, and follow about 1 kilometer behind the lead platoon. Have the lead platoon guide it into an overwatch position to deliver long-range suppressive fires on the enemy. Consolidate the tank platoons, move forward to join them, then follow about 1 kilometer behind the second scout platoon.

Once the base-of-fire element is set, have the platoon suppress the enemy with direct and indirect fires. Guide the maneuver element into a position to attack by fire. Finish quickly and move out.

Tell the scout platoon performing reconnaissance and surveillance tasks to screen and protect the withdrawal of the remainder of the troop back to friendly lines. Have the other scout platoon lead the way back on the designated route of withdrawal, and have the tank platoons follow. Stay back and supervise the conduct of the rear screen until all troop elements have entered friendly lines. Consolidate and resupply the troop with fuel and ammunition as soon as possible. Get ready to fight again.

EXAMPLE OF A RAID (HEAVY TROOP)

The squadron is defending in sector astride a likely Threat regimental zone of attack. The three armored cavalry troops are deployed abreast with the tank company in reserve. The squadron frontage is about 10 kilometers wide. Threat forces attack from the march. Squadron elements in the northern half of the sector are softened with long-range preparatory fires delivered by rocket and artillery forces. As artillery fires increase in intensity, the fire support element and the advance guard battalion of the regiment deploy from the march into prebattle formation and then attack. The advance guard battalion is quickly defeated by the concentrated fires of the northern cavalry troop and a surprise counterattack by the tank company. The squadron rapidly repositions to meet the assault of the regimental main body. Within one hour, the regimental main body approaches the line of contact with two MRBs abreast. Their axis of advance indicates most of their combat power will be concentrated in the northern half of the squadron's sector. Artillery suppressive fires increase as the Threat battalions deploy into attack formation. Aeroscouts report two 122-mm artillery battalions of the regimental artillery group moving into firing positions about 4 kilometers behind the attacking combat maneuver units. In this situation, with good defensible terrain, the squadron commander decides to concentrate the combat power of the troops, the tank company, and his available artillery in the north to defeat the advancing maneuver forces. He orders the southern cavalry troop to conduct a raid behind enemy lines to destroy the regimental artillery group simultaneously.

The troop commander quickly analyzes the terrain, and then issues a FRAGO to the troop. He tells the 1st platoon, which is currently on the screen line, to take the lead and reconnoiter the troop's direction of attack from the point of departure on the LD to the objective area. The platoon is tasked to establish visual contact with the Threat artillery units and to locate the best covered and concealed approach into their flanks or rear. The troop commander tells the 3d platoon to follow about 1 kilometer behind the 1st platoon and establish a base of fire to suppress and fix the artillery units. He orders the 2d and 4th platoons to consolidate rapidly near the point of departure. He tells the mortar section to remain in the troop sector. Priority of fires goes to the 1st platoon. As the 1st platoon approaches the point of departure, the troop FSO coordinates with the squadron FSO to establish a restricted fire area around the raid area. Once it is established and confirmed by the FSO, the troop commander orders the 1st platoon forward (see Figure 5-9).

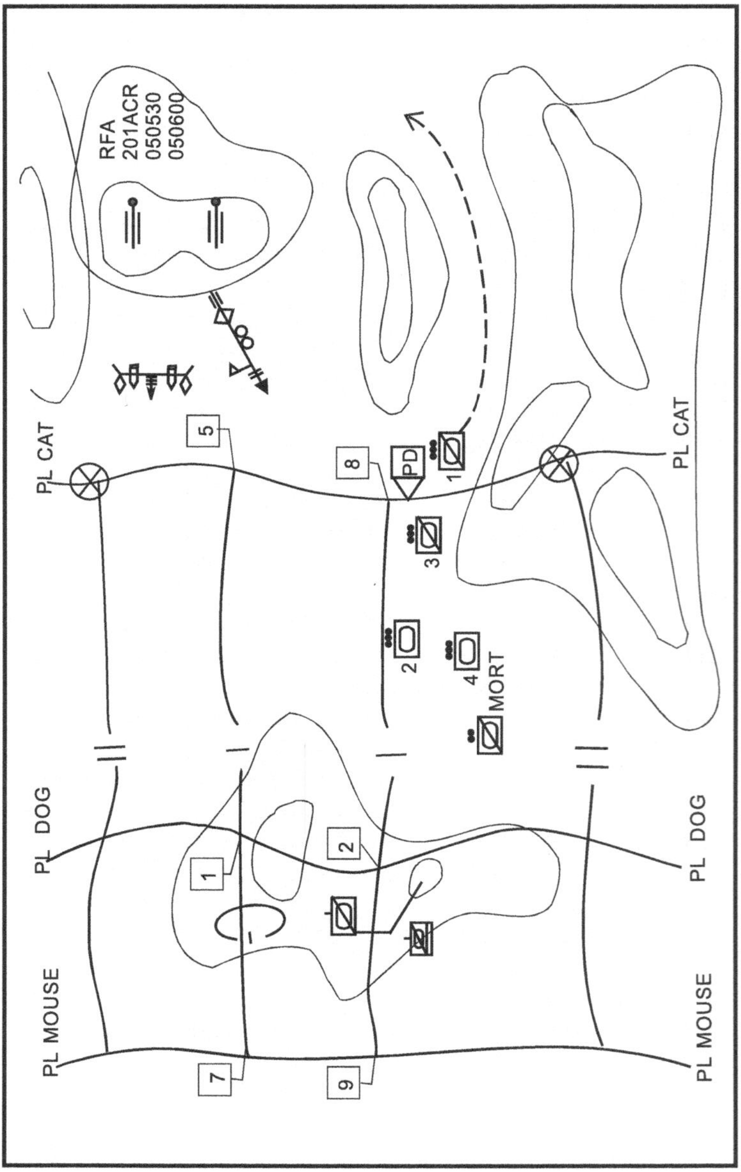

Figure 5-9. Heavy troop raid (part one).

The 1st platoon makes visual contact in about 10 minutes, and continues to reconnoiter for a good axis of attack. The 3d platoon moves into an overwatch position about 1,800 meters away and lays down a curtain of suppressive fire with chain guns, TOW missiles, and mortar fire. While suppressive fires are delivered, the tank platoons maneuver into the left rear of the artillery units, then deploy abreast and attack the Threat artillery battalions by fire (see Figure 5-10).

After silencing the battalions, the troop commander immediately orders the 3d platoon to lead the troop back to friendly lines along the designated route of withdrawal. The 2d and 4th platoons follow while the 1st platoon screens the withdrawal of the troop. The troop commander joins the 1st platoon and supervises the conduct of the rear screen until all troop elements have reentered friendly lines (see Figure 5-11). The troop returns to defensive positions, and the 1SG begins resupply actions.

EXAMPLE OF A RAID (LIGHT TROOP)

The light troop should consider the raid target when planning its operation. A combination of weapon systems may best suit the destruction of the raid objective rather than operating in pure platoons. A mix of weapon systems may also assist in security of the objective during the raid. The troop scramble should be considered as a possible task organization when planning a raid.

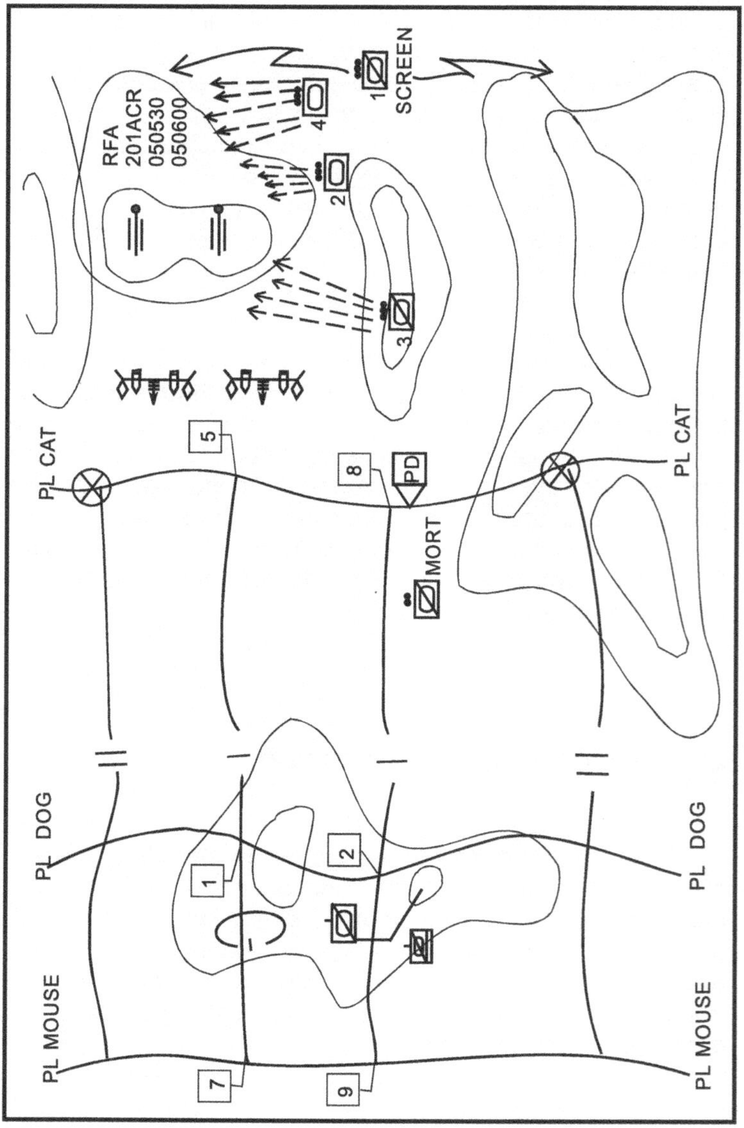

Figure 5-10. Heavy troop raid (part two).

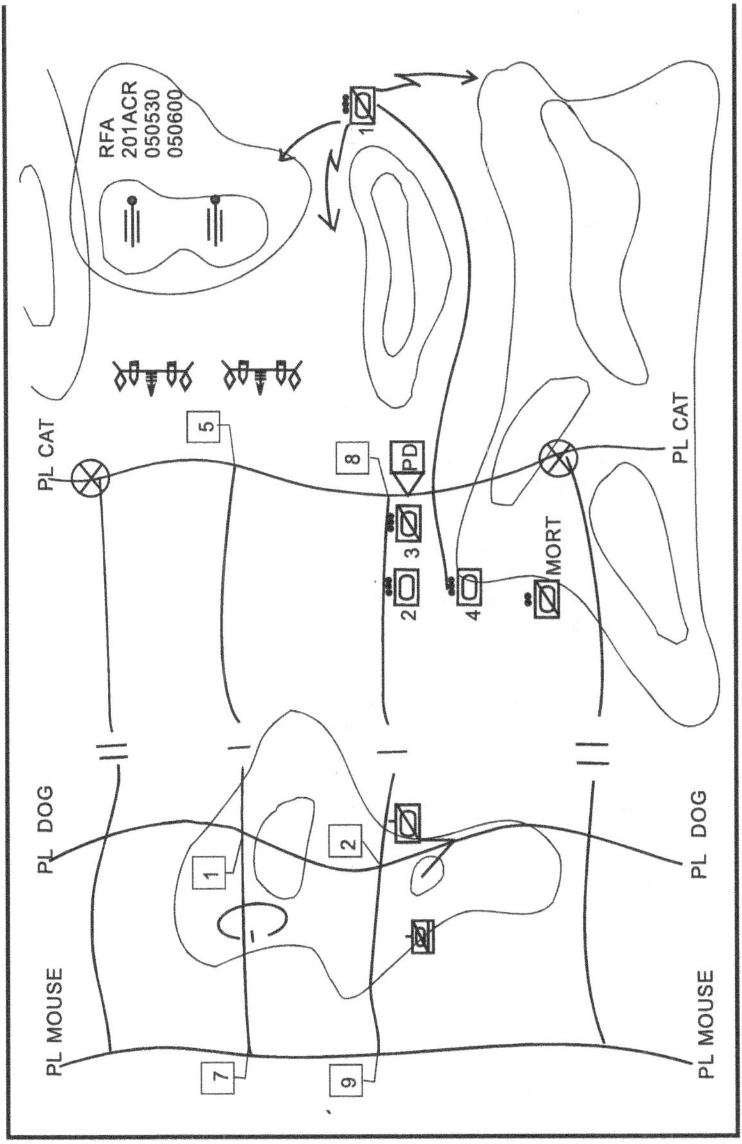

Figure 5-11. Heavy troop raid (part three).

Example steps of a light troop conducting a raid are as follows:

- Troop moves in column, platoons in a vee. Troop is a scramble organization with platoons integrating the AT vehicles into the scout sections. Each scout platoon is running two sections of three vehicles plus the platoon leader. See Figure 5-12.
- Lead platoon (1st platoon) conducts reconnaissance forward of the troop.
- 1st platoon establishes contact with the objective, then secures the flanks and conducts surveillance of the objective.
- 2d platoon, second in the order of movement, moves through 1st platoon, establishes security on the far side of the objective, orienting on the mounted avenues of approach into the objective area.
- 3d platoon establishes an attack position vicinity of BP1. 4th platoon establishes an attack position vicinity of BP2.
- Troop commander initiates the attack with indirect fires. 3d and 4th platoons move and attack by fire from BP1 and BP2.
- Troop engages enemy position using TOWs to destroy hard targets and fires from caliber .50 and MK-19s for soft targets and suppression of enemy defenses.
- Troop uses indirect fires to cover their withdrawal.
- 1st platoon covers the troop withdrawal.
- 4th platoon displaces first, followed by 3d platoon, then 2d, then 1st.
- Troop moves in column, platoons in vee with 1st platoon providing rear security. 4th platoon conducts a rapid zone reconnaissance forward of the troop as it moves toward friendly lines. See Figure 5-13.

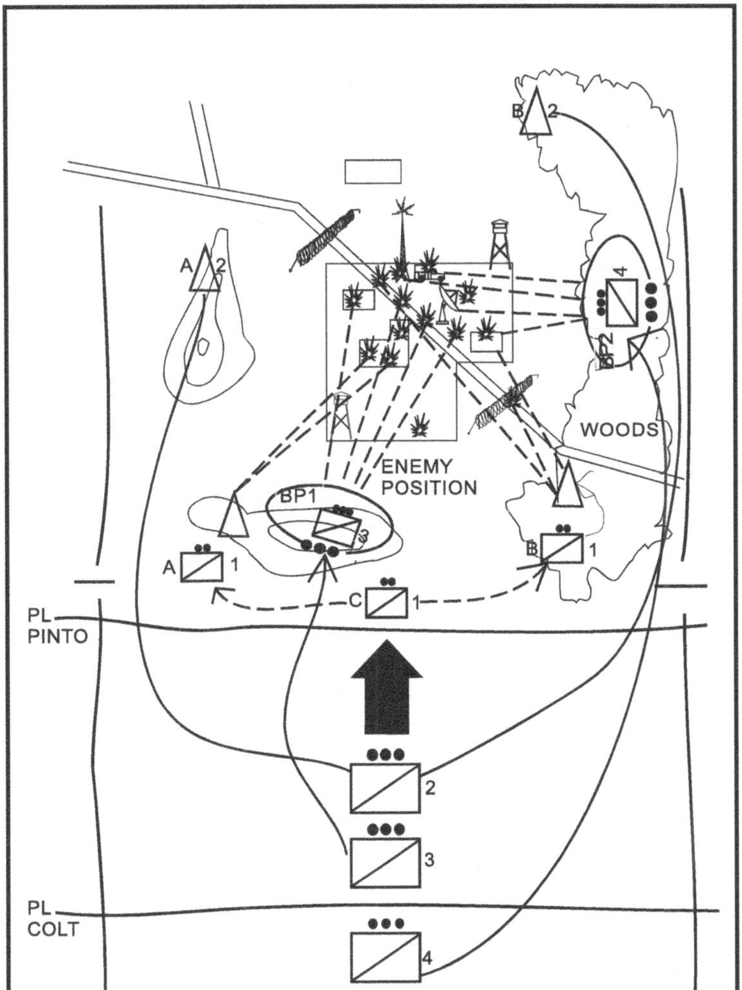

Figure 5-12. Light cavalry troop conducts a raid using a troop scramble.

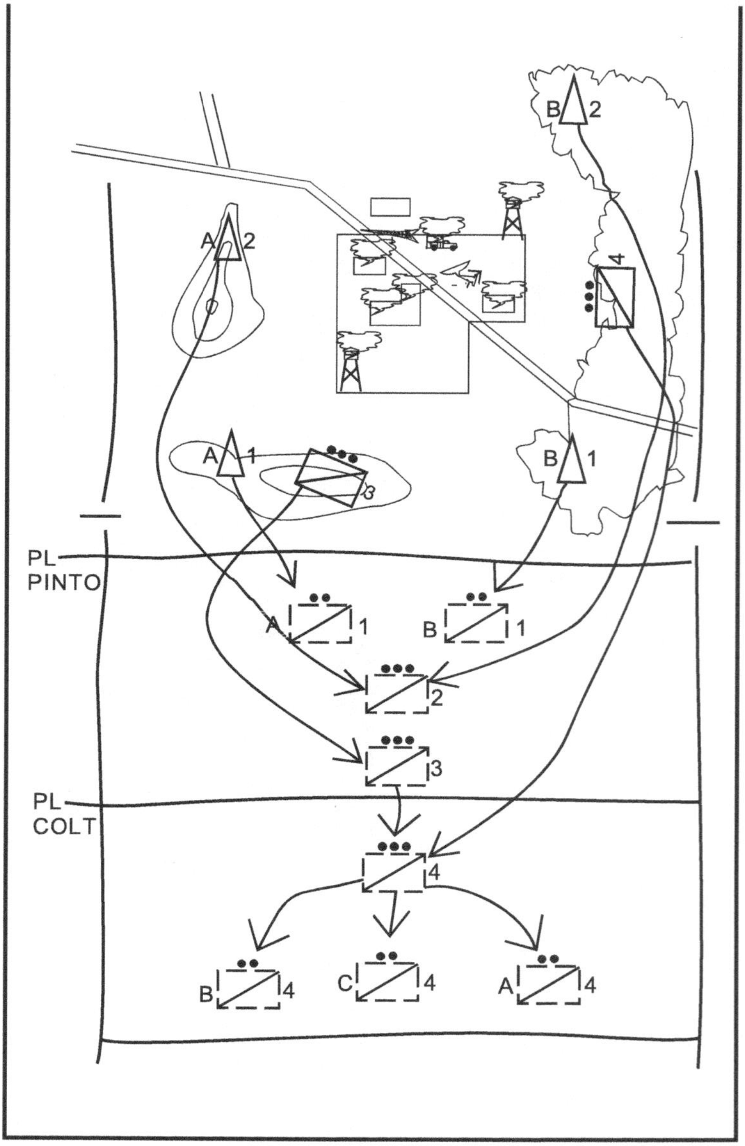

Figure 5-13. Light cavalry troop completes the raid and returns to friendly lines.

Chapter 6

Defense

Defense is not the decisive form of war. While defense can deny success to the enemy, it seldom assures victory. The defense is, however, the stronger form of war due to the inherent advantages of the defender. Army operations recognize the strength of the defense, but emphasize the necessity to transition to the offense quickly.

Brigades, battalion task forces, and company teams are the principal defensive forces for the corps or division. Cavalry units normally perform security missions for the defense or reconnaissance missions to support attacks. Cavalry units frequently perform defensive operations as a part of security missions, or when required to defend in an economy-of-force role for a higher headquarters.

Contents

Section I. Purpose, Fundamentals, and Schemes of Maneuver

PURPOSE

The immediate purpose of defensive operations is to defeat an enemy attack and create opportunities to go on the offense. For both the heavy and light cavalry troops, defensive operations may also achieve one or more of the following:

- Gain time.
- Concentrate forces elsewhere.
- Attrit enemy forces as a prelude to offensive operations.
- Control key or decisive terrain.
- Retain tactical, strategic, or political objectives.

In some cases, the troop may be required to defend because it cannot muster enough combat power to attack.

FUNDAMENTALS

Eight fundamentals are common to cavalry troop defensive operations.

Depth. Position platoons in depth, and place obstacles in depth. Depth allows the troop to—

- Gain enemy contact early.
- Perform counterreconnaissance tasks.
- Ascertain enemy direction of attack/intentions.
- Develop the situation, providing reaction time and maneuver space to concentrate combat power when and where it is needed.

Dispersion. Deploy subordinate elements as far apart as possible without losing their ability to concentrate (mass) firepower against the enemy. The more dispersed the troop, the harder it is for the enemy to mass fires against it as a whole. However, do not allow the enemy to concentrate its forces or fires against isolated elements.

Security. The troop may employ passive or active measures, or a combination of techniques. All must be considered in the defensive plan. Examples of passive and active security measures follow:

Passive	Active
Disperse vehicles and platoons.	Screen/Establish OPs.
Use camouflage/ cover and concealment.	Perform mounted/ dismounted patrols.
Impose radio listening silence.	Establish GSR posts.
Use hide positions.	Establish M8 chemical alarm net.
Enforce noise and light discipline.	
Minimize movement.	
Do not position in likely target areas.	

Maximize Terrain Advantages. Study the terrain. Reconnoiter it from both the troop commander's and the enemy's view, if possible, to determine—

- Avenues of approach.
- Reconnaissance avenues of approach.
- Restricted/severely restricted areas.
- Defiles (canalizing terrain).

- Engagement areas.
- Battle positions.
- Subsequent and alternate battle positions.
- Hide positions to support battle positions.
- OP positions forward of defensive positions.
- Subsequent and alternate OP positions.
- Positions where obstacles can be tied in with natural obstacles to turn, disrupt, or block the enemy.
- Positions that facilitate counterattacking by fire or by fire and maneuver into the flanks and throughout the depth of the enemy.
- Routes to and from each position.

Stop Enemy Rate of Advance. Offense is based on two principles—speed and mass. Develop a defensive plan that blunts the momentum of the enemy attack through the use of obstacles and fires (direct and indirect). If most of his combat power is killed, the enemy will be forced to dig in and establish a hasty defense or withdraw from combat.

Mass Combat Power at the Right Place and Time. In order to defeat a massed attack, the troop must mass fires against the enemy where and when he is least able to escape its effects. The ability of the troop to mass combat power when and where it is needed is a function of—

- Early warning/reaction time.
- Responsive/rapid maneuver by subordinate elements.

Force the Enemy to Fight in Two Directions. When engaging the attacking force, maneuver platoons into positions that force the enemy to turn and fight in two or more directions. This will force him to split his fires, preventing him from concentrating fires, and to expose his vulnerable flanks.

Counterattack. Draw the enemy into structured engagement areas and attack him *en masse* with overwhelming firepower to destroy him quickly and decisively. Wrest the initiative from him. Maneuver forces to exploit the situation. Attack by fire and by fire and maneuver into his vulnerable flanks and throughout the depth of his formation.

DEFENSIVE SCHEMES OF MANEUVER

There are three basic schemes of maneuver the commander can use in designating a course of action for a defensive mission. These schemes of maneuver center on the use of battle positions and sectors for subordinate platoons, or a combination of the two.

Defend from a Troop Battle Position. Both heavy and light cavalry troops may defend from a battle position. This method is be used when the enemy situation is clear, when there is only one avenue of approach, or when the troop commander must coordinate subordinate unit fires. In this scheme of maneuver the troop commander retains most of the authority for fighting the battle. The troop commander must understand his squadron commander's intent and concept to prevent holding the troop in place and risking its destruction.

Defend in Troop Sector. Both the heavy and light cavalry troops may defend in troop sector. This method may be used when the enemy situation is vague, when there is more than one avenue of approach, or when subordinate platoons require more freedom or action. In this scheme of maneuver, the troop commander delegates much of the responsibility for fighting the battle to his subordinate platoon leaders. He maintains control through the use of effective control measures and a clearly understood intent and concept.

Defend by Combination of Battle Positions and Sectors. Both heavy and light cavalry troops may defend using a combination of battle positions and sectors. This is the most common method of defense for a cavalry troop as it provides the troop commander maximum flexibility to maneuver and mass fires throughout the depth of the sector when changes in METT-T dictate.

Section II. Defend From a Troop Battle Position

Either the heavy or light cavalry troop may defend from a troop battle position. This mission is normally assigned when the squadron commander elects to concentrate the direct fires of the troop or squadron within an engagement area. The troop cannot maneuver outside the position without the squadron commander's permission. Within the battle position, the troop commander positions his platoons to concentrate all direct fires where the squadron has specified. The troop fights to retain the position unless ordered by the squadron commander to counterattack or withdraw. The troop may still retain the task to screen/maintain contact with the enemy forward of the battle position, depending on the squadron commander's intent. If so, the tank/AT platoons will initially be positioned within the troop battle position, and the scout platoons may assume adjacent positions within the battle position following execution of a security drill (see Security Drill paragraph in Chapter 4, page 4-15).

CRITICAL TASKS

To defend a troop battle position, the following critical tasks must be accomplished:

- Decide where the enemy will be killed and designate the engagement area.
- Establish OPs oriented forward and to the flanks of the BP to gain contact with the enemy force and provide early warning.
- Establish primary and alternate platoon BPs to concentrate direct fires within the engagement area as directed by the squadron commander.

- Designate supplementary platoon BPs to cover other routes of enemy approach.
- Establish sectors of fire for each platoon.
- Reconnoiter and establish platoon routes from hide positions to platoon BP, and for withdrawal to subsequent platoon/troop BPs.

TECHNIQUES

The squadron commander assigns troop battle positions when he can mass the fires of two or more troops in a squadron engagement area. The size of a troop battle position can vary, but it should provide enough depth and maneuver space for platoons to maneuver into alternate/supplementary positions and execute local counterattacks.

The troop commander conducts a thorough terrain study before the troop occupies the position, keeping in mind where the squadron commander wants the troop to concentrate its fires.

- Designate primary and alternate positions for each platoon.
- Position platoons to achieve flanking fires along the avenue of approach. Consider the effective range of each platoon's weapon systems.
- Position platoons to mass direct fires within the engagement area and to provide mutual support.
- Position platoons to cover any dead space in the engagement area. If not possible, plan indirect fires to cover the dead space.
- Use TRPs to control fires and orient weapon systems for each platoon.
- Occupy the BPs from the rear. Establish OPs to support the squadron plan. Allow platoon leaders time to reconnoiter and position vehicles to place effective direct fires within their sectors of fire. If better

defensive terrain lies outside the BP, call the S3 or commander for approval to adjust the boundaries.

- Once platoons are set, inspect them to make sure each platoon is properly oriented and has good fields of fire. When satisfied, have the platoons reconnoiter routes to subsequent positions and select firing positions for each vehicle. If time and engineer support are available, dig prepared positions for vehicles. Have the platoon leaders prepare platoon fire plans for approval.
- After preparing the BP, have all elements except OPs move to hide positions to reduce the risk of enemy observation and to decrease their vulnerability to enemy fires.
- Maintain security. Position the command post where it has FM communications with subordinate elements and the squadron. Position the troop trains behind good cover that is out of direct-fire range and allows quick access to each platoon position. The troop commander is positioned where he can observe the engagement area and control the troop. The FSO should be nearby to ensure coordinated fire support.
- Identify trigger points/lines.
  - Upon the enemy reaching what terrain feature does the troop initiate indirect fires? Direct fires?
  - Upon the enemy reaching what terrain feature does the troop displace to prevent decisive engagement?
- If the troop has to disengage and displace under fire to a subsequent position, bound the troop back by platoon(s), consistent with the squadron scheme of maneuver. If the troop disengagement and displacement are covered by another element, the troop may move as a whole to a subsequent position.

See Figure 6-1 for an illustration of the techniques described above.

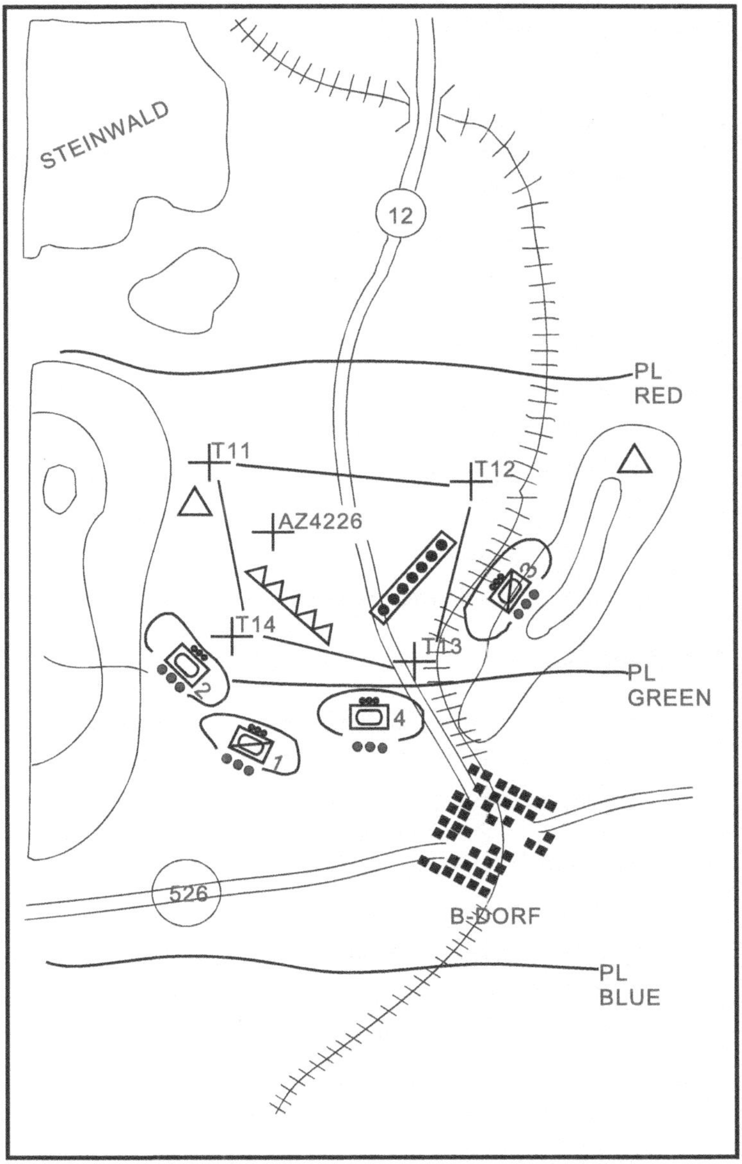

Figure 6-1. Troop engagement area.

Section III. Defend in Troop Sector

Both heavy and light cavalry troops may defend in sector. METT-T considerations determine optimal troop sector width; however, the troop is normally allocated a sector oriented on a single battalion-size avenue of approach.

Either troop may defend in sector when—

- The squadron cannot concentrate its fires due to—
  - extended frontages.
  - defending along a cross compartment.
  - multiple avenues of approach.
- Retention of specific terrain features is not necessary.
- The troop may use the depth of the sector to dissipate the enemy's attack.
- Maximum flexibility to maneuver is desired.

CRITICAL TASKS

- Maintain continuous surveillance of high-speed routes or avenues of approach into the troop sector (screen).
- Destroy or repel all enemy reconnaissance elements forward of the troop's initial defensive positions (counterreconnaissance).
- Structure engagement areas.
- Position platoon battle positions to support engagement areas.
- Engage the enemy from more than one direction.
- Determine criteria for initiating fires, counterattack, and disengagement.
- Prevent the enemy from penetrating the troop rear boundary or designated NPL (no penetration line).

TECHNIQUES

When given the order to defend in sector, the squadron will usually provide the following graphic control measures: *troop boundaries, an initial screen line, a rear boundary, phase lines, contact points between troops, TIRS, and TRPs that support any squadron engagement areas.*

Study the terrain in sector. Identify the terrain near the initial screen line from which OPs can maintain continuous long-range surveillance of enemy avenues of approach.

Determine where platoons can be positioned astride or on the enemy avenues of approach. Look for positions that provide good observation and fields of fire into the avenue of approach, and good cover and concealment for hide and defilade positions. Take a look at proposed platoon battle positions, and determine where troop fires can be massed on the avenue of approach. Use this portion of the avenue of approach to structure a troop engagement area(s).

The engagement area is where the troop will destroy an advancing enemy force. Establish a series of TRPs and use them to assign sectors of fire to each platoon. This allows the commander to control the fires of the troop and to achieve overlapping platoon fires.

With the assistance of supporting engineers, plan obstacles within the sector to support the defensive plan. Reinforce existing obstacles within engagement areas, and plan more obstacles to slow, canalize, or turn the enemy. Obstacles can buy the troop time to engage the enemy, and increase the effectiveness of indirect fires in the engagement area by compressing threat formations, slowing them down, and detaining them in the engagement area. Obstacles can give the commander time to maneuver platoons to counterattack or to move to subsequent positions. Plan obstacles in depth so the enemy gets bottled up in the engagement area and is confronted with a series of breaching operations. Make sure the troop can observe and place fires on all obstacles in the sector. Place the obstacles to achieve different effects. Obstacles emplaced on the reverse side of a hill or depression will cause the enemy to pile into them before he sees them. Strategically emplaced obstacles will cause the enemy to turn when he sees them, exposing his flanks to direct fires.

Give the FSO planning guidance so he can develop the troop indirect-fire support plan for the mission. Plan fires to support the scouts on the screen line forward of the troop. The scouts need indirect-fire support to engage enemy reconnaissance forces, to disrupt enemy lead echelon formations, and to attack follow-on forces. Plan indirect fires to engage enemy forces in the engagement area, when they are slowed by obstacles. These fires suppress, disrupt, and confuse the enemy and allow platoons to set up the direct-fire engagement. Plan indirect fires behind the engagement area to isolate the enemy. In addition, plan fires forward of the troop's positions to help disengage from the enemy in case the troop cannot stop him from initial positions. Identify rally points in sector behind the battle positions. Crews or troop elements that become separated or disorganized during battle move to these identified rally points to reassemble or reorganize.

Position platoons to maximize their weapons' effectiveness and crew/vehicle survivability based on the given terrain and the capabilities of the enemy. Platoons positioned at the base and along one or both flanks of the engagement area will force the enemy to fight in two or more directions.

Determine how to employ the scout platoons. Although the primary role of scout platoons is to conduct reconnaissance and screening in support of the troop, the troop commander may need to use their firepower to support troop defensive missions. Depending on sector width and number of avenues of approach, one scout platoon may be employed in a screen mission forward of the troop during a mission to defend in sector. It may fall back to a battle position after identifying the attacking enemy force (security drill), or it may stay forward of the troop, continuing to screen to identify follow-on forces. The other scout platoon may fight the attacking force from a battle position. The placement of the scout platoon's BP depends on the role the commander wants the platoon to play in the troop fight.

After making final adjustments to initial battle positions with the platoon leaders, plan alternate positions and subsequent positions in depth. Give platoon leaders time to reconnoiter covered and concealed withdrawal routes to their alternate subsequent positions.

Position the troop mortars where they can support the scouts on the screen line. Ensure they can engage targets from 3 to 3.5 kilometers beyond the screen line, or as far as the scouts can observe. Plan other mortar positions so that they can support the troop fight as the threat enters the engagement area. Also, plan positions through the sector.

The first sergeant positions the troop trains behind the initial troop battle positions, where they are responsive to troop needs but not vulnerable to direct fires. The XO positions the command post behind the initial platoon BPs on terrain that affords good FM radio communications with the troop elements and squadron headquarters. If possible, the XO positions the command post behind the subsequent troop positions; this reduces its vulnerability to fires, and allows it to remain stationary and maintain good FM communications while the troop displaces to other positions. The XO and first sergeant plan subsequent positions through the sector.

Position the FIST where it can maintain good digital FM communications with the supporting artillery unit. If possible, keep the FIST where it can use the laser designator to designate high-priority targets in the engagement area for Copperhead or other laser-guided munitions.

The troop commander must be in position to see the battlefield. To control the troop fires, he must understand which areas can and cannot be engaged by platoon battle positions.

The first critical task for a defend mission is to destroy or repel the threat reconnaissance. The scout platoon(s) on the screen line is responsible for identifying enemy reconnaissance forces, engaging them with indirect fire, and defeating them if possible. The other platoons may need to assist the scout platoon(s) in defeating the reconnaissance forces after the scouts identify them.

After destroying or repelling enemy reconnaissance forces, the troop is prepared to take on the lead echelon of the enemy force. Remember the scheme of maneuver. Let the enemy enter the engagement area and then mass the direct fires of the troop to strike a decisive blow. Depending on how the obstacles are set up, the troop commander may want to strike the enemy just before he reaches the obstacles; then, as the enemy deploys in reaction to troop fires, he hits the minefields and tank ditches. The commander may wish to wait until the enemy gets into the obstacles, and strike when he is confused and his formations are compressed.

Continue the fight by maneuvering platoons into alternate or supplementary positions while counterattacking by fire to complete the destruction of the lead echelon forces. If the troop is unable to defeat the enemy in the initial engagement area, it must be prepared to displace to subsequent positions. Do not allow the troop to become decisively engaged. Use the disengagement criteria from the scheme of maneuver to ensure adequate time to bound the troop back to subsequent positions by platoon(s). Keep one or two platoons in contact with the enemy, engaging him with direct and/or indirect fires. The other platoons move back in sector to subsequent positions. Once they are set, they engage the enemy with indirect fires or long-range TOW missile shots or tank main gun fire so the remainder of the troop can break contact and move to its subsequent positions.

Once the platoons are in their subsequent positions, make any needed adjustments to their positions or orientations based on what the enemy is attempting to do. Finish off the enemy from these positions, and then move forward and reoccupy initial or alternate positions, if possible.

EXAMPLE - DEFEND IN TROOP SECTOR (HEAVY TROOP)

The troop is given a mission to defend in sector and hold the threat forward of PL SABER. The troop commander completes his plan and issues the order. The 1st platoon moves forward and establishes a screen along PL SPUR. The 3d platoon prepares BP 20 at the base of the initial engagement area and occupies a hide position to the rear. BP 20 is about 2,500 meters away from the base of EA BEAR. The platoon's fires are concentrated in the direction of TRP 25. The 2d and 4th platoons occupy BP 30 and BP 40, respectively, along the eastern flank of EA BEAR. The 2d platoon orients on TRP 21, and the 4th platoon orients on TRP 27. The western flank is protected by a steep ridgeline. The troop command post is on high ground about 2 to 3 kilometers behind the initial defensive positions. The troop trains are just south of BP 30. The troop commander positions between BP 20 and BP 30, where he has a good view of all platoon positions and the engagement areas (see Figure 6-2).

A threat reconnaissance patrol of two BMPs approaches the screen line moving along the flanks of the high-speed route into the troop sector. As the patrol crosses PL SPUR, a 1st platoon scout element unmasks and quickly ambushes the patrol with cannon fire. One scout squad immediately searches the threat vehicles for intelligence information and captures two wounded prisoners. The troop commander tells the 1st platoon leader to expect a CRP within an hour. The rest of the troop stays in hide positions away from terrain most likely targeted by threat artillery units (see Figure 6-3).

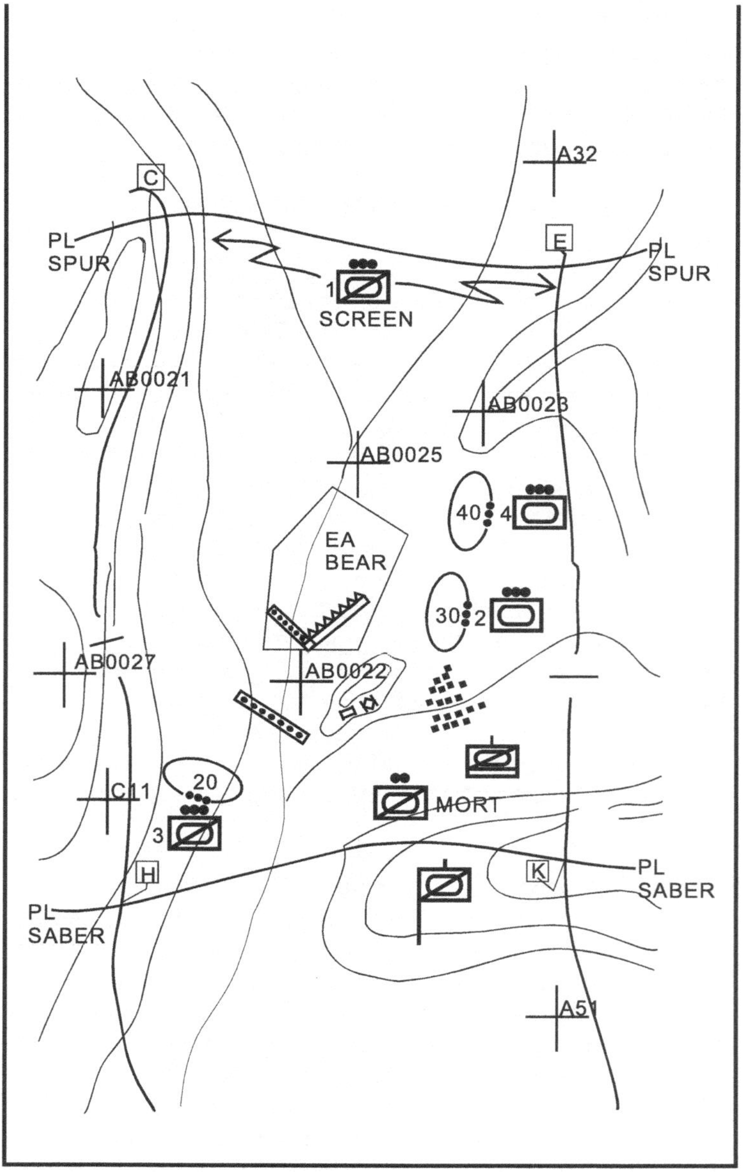

Figure 6-2. Defend in sector (part one).

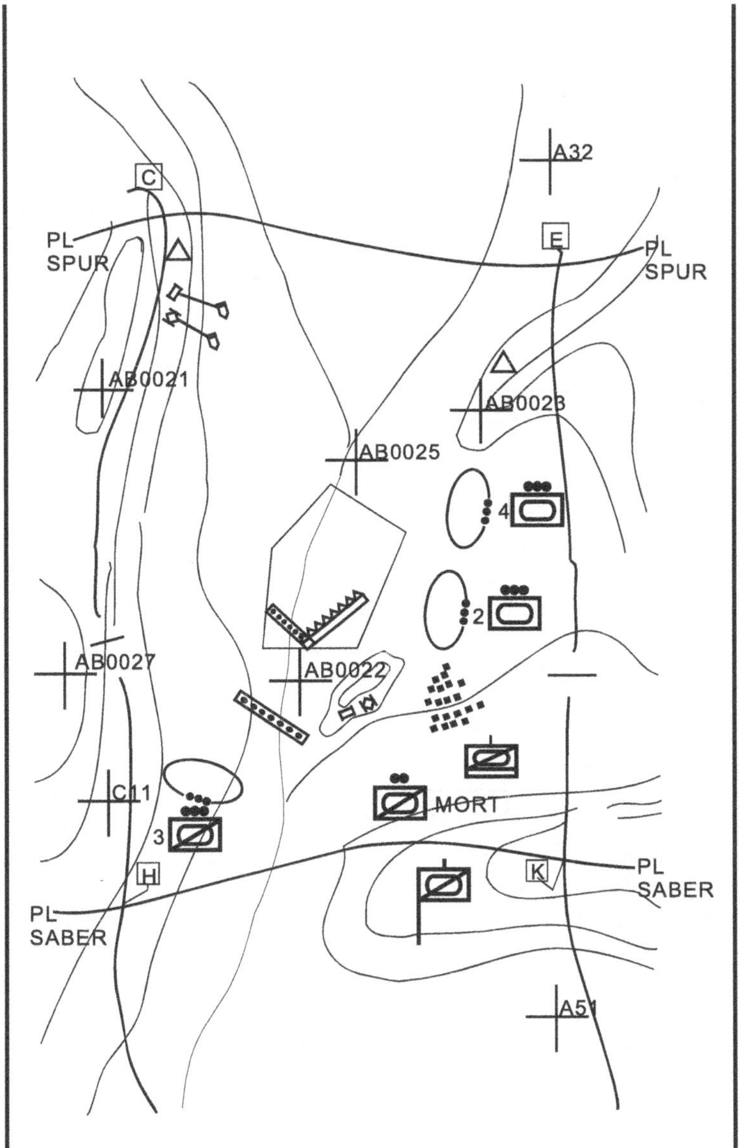

Figure 6-3. Defend in sector (part two).

About 45 minutes later, heavy concentrations of artillery and rocket fire begin falling along areas of high ground that dominate the avenue of approach near EA BEAR. Troop elements button up. When the suppression lifts, they immediately test for chemical agents. No chemicals are detected. The TCs unbutton. The 1st platoon reports two MRCs advancing abreast in prebattle formation about 2 kilometers forward of the screen line. The platoon sergeant contacts the FSO. Using the technique of fire "AT MY COMMAND," the platoon sergeant times the impact of artillery to coincide with the arrival of Threat formations at preplanned TRPs. Threat formations are disrupted, several vehicles sustain suspension damage, and the advance slows down. Threat leaders scramble to restore order, company formations are reformed, and the advance continues. The 1st platoon does not engage the advancing force, but maintains contact and reports the Threat's location and activity to the troop commander. The 1st platoon maintains its positions along PL SPUR to identify follow-on forces. It reports seeing a third MRC about 1,500 meters behind the lead companies. The troop commander now has a fairly clear picture of the Threat situation.

The platoon leaders and troop commander move into hide positions to observe the threat approach. The FSO continues to smoke and to suppress the threat lead companies with mortar fire, which keeps them buttoned up and slows their rate of advance. The troop commander orders the 3d platoon to move into firing positions and prepare to fire at his command. The 3d platoon moves into hull-down positions. The 2d and 4th platoons remain in hide positions. Their platoon leaders stay up and continue to observe. The lead threat MRCs appear about 2,800 meters away, with tanks leading platoon columns. The troop commander orders the 3d platoon to engage. The 3d platoon sends six TOW missiles down range, targeting mine-roller tanks and the threat platoon leader's BMPs (see Figure 6-4). The threat force detects the antitank guided missile (ATGM) fire and moves toward the 3d platoon, which quickly backs into defilade.

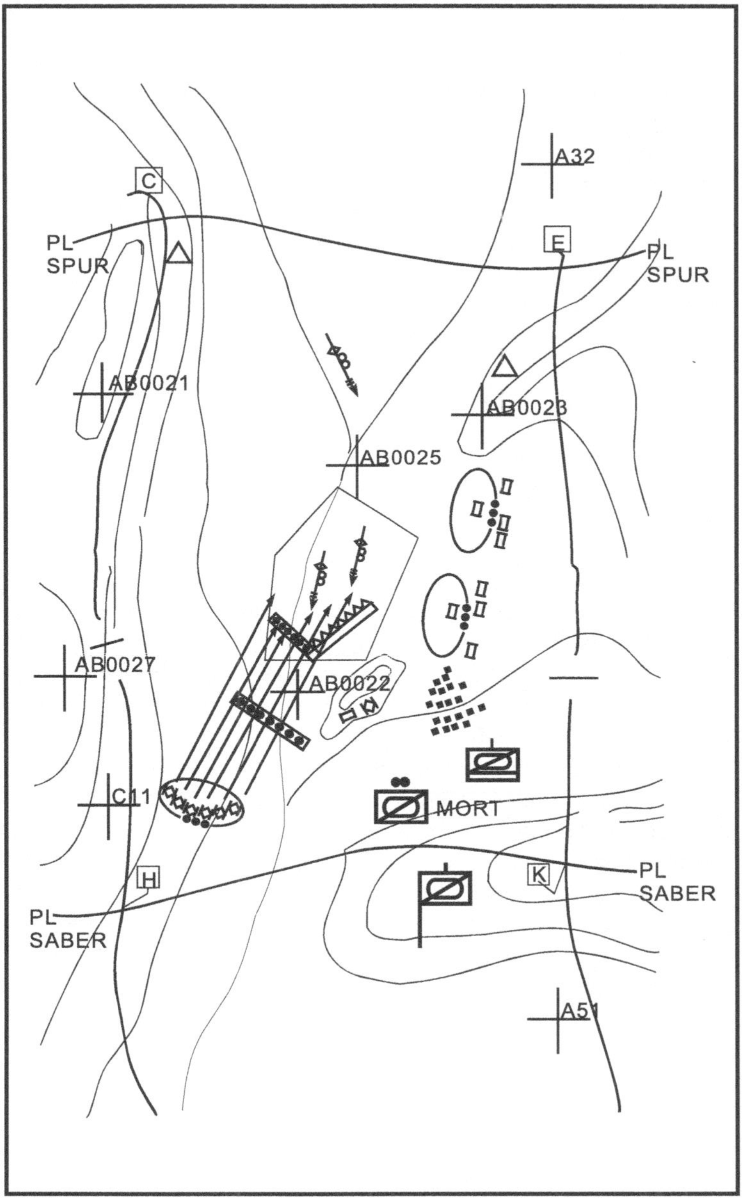

Figure 6-4. Defend in sector (part three).

The threat continues to advance through EA BEAR, piling into a minefield and tank ditch hidden on the reverse slope of a long, shallow draw. Several threat vehicles are caught in the obstacle. As the FSO observes this development, he concentrates all available fires on group targets that cover the obstacle. With the advance disrupted and stalled at the obstacle, the troop commander orders the 2d and 4th platoons to attack by fire into the flanks of the threat's lead echelon (see Figure 6-5). The concentrated cannon fire of both tank platoons quickly destroys most of the two lead companies.

The threat's third company comes into view in attack formation. The troop commander orders the 3d platoon to engage this force with long-range TOW missile fires. Survivors of this battalion begin to withdraw from the battlefield. The troop commander quickly orders the 2d and 4th platoons to counterattack and destroy the remnants of the battalion. The 3d platoon provides overwatch. Moving to alternate positions, the tank platoons counterattack by fire to finish off the remaining battalion vehicles. The troop commander then orders all platoons to reoccupy their initial hide positions and redistribute ammunition. He anticipates the second echelon will arrive within 30 minutes. The 1st platoon maintains its positions on the screen line. The first sergeant moves to each of the platoons to resupply. He collects the PWs from the 1st platoon. The XO collects all routine logistics reports and forwards them to squadron headquarters. The troop commander checks the status of leaders within the troop and designates replacements and cross-levels within the troop as necessary.

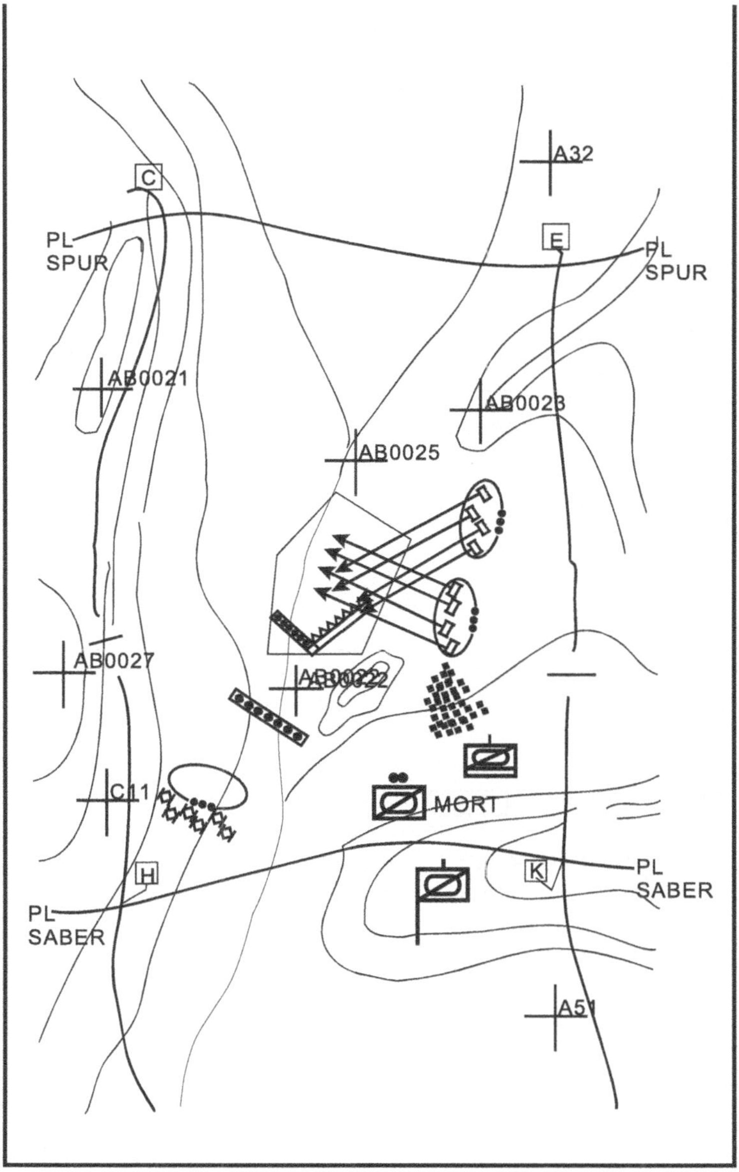

Figure 6-5. Defend in sector (part four).

Section IV. Delay in Troop Sector

Delay is a continuous series of defensive actions over successive positions in depth that trades the enemy space for time while retaining freedom of action. It is an economy-of-force operation that buys time to permit something else to happen at a more critical place on the battlefield.

CRITICAL TASKS

The critical tasks for delay include all the tasks associated with defend in sector as well as—

- Preserve freedom to maneuver.
- Cause the enemy to deploy from march or prebattle formation into attack formation as the troop moves to the rear.

TECHNIQUES

Planning and tactics for delay are identical to those for defend in sector, and vary only in their purpose. The flow of a delay resembles a "hit hard, then move" technique. The troop commander and subordinate platoon leaders must be very aware of disengagement criteria. The troop must mass the effects of fires to temporarily stop the enemy advance, then disengage and move to subsequent positions in depth. The troop cannot become decisively engaged. It must maintain a mobility advantage over the enemy. This means taking advantage of terrain, being familiar with high-speed routes of withdrawal, and rehearsing engagements and movements. The commander will probably have to use one or two platoons to assist disengagement of the others.

Chapter 7

Other Tactical Operations

Several combat operations are routinely associated with successfully accomplishing the missions described in Chapters 3 through 6. These operations require special planning and training considerations and techniques because of their complexity. At troop level, these operations are based on standing operating procedures (SOP) to ensure they can be conducted quickly and efficiently.

CONTENTS

Section I. Tactical Road Marches

Troops not in contact with the enemy often travel long distances to position themselves for future operations. A successful tactical road march depends on unit discipline and the ability to execute the plan strictly in accordance with the SOP.

CRITICAL TASKS

Accomplish the following critical tasks when planning, executing, and supervising a tactical road march:

- Establish the readiness condition (REDCON).
- Issue a warning order.
- Reconnoiter the route to the start point.
- Conduct quartering party operations.
- Issue a movement order.
- Execute a tactical road march in accordance with movement order and SOP.
  - Cross and report the start point, checkpoints, and release point on time.
  - Conduct actions at halts.
  - Conduct actions on contact.
- Maintain security throughout movement and during halts.
- Operate a trail party.
- Clear the release point.

TECHNIQUES

The march discipline necessary to execute a road march with routine precision is attained only by strict adherence to SOP. Ensure the tactical road march portion of the troop SOP addresses, as a minimum, the following:

- Order of march.
- March speed.

- Distance between vehicles.
- Actions on contact (air, ground, and indirect fire).
- Actions at halts.
- Security.
- Contingency plans for vehicle breakdowns, breaks in column, and lost vehicles.
- Quartering party.
- Trail party.

The basic considerations in planning any road march are listed below.

- Time available.
- Distance of the move.
- Current situation.
- Availability and condition of routes.
- Size of the unit.
- Types, numbers, and characteristics of vehicles that must move.

The troop will most often move as part of a squadron operation, and the move will be based on the squadron order. The troop commander must, however, be familiar with planning considerations so he can plan and execute an independent troop move, if required.

March Planning Sequence. If time permits, follow this sequence when preparing for a road march:

- Prepare and issue a warning order, giving the troop's movement and any follow-on missions. Give them time to prepare for the operation.
- Analyze the situation to determine if any of the movement factors (order of march, rate of march, or interval) specified in the troop SOP must be altered to meet mission requirements. If the troop must reconnoiter two assembly areas, the two scout platoons might have to be first in the order of march.
- Conduct a map reconnaissance of the route (if assigned) or determine the best available route.
- Organize and dispatch reconnaissance and quartering parties (if occupying an assembly area).
- Prepare detailed movement plans based on mission requirements and reconnaissance information.
- Prepare and issue the march order to the orders group.

Route Reconnaissance. Reconnaissance is essential to movement planning. It reveals accurate, up-to-date information about the route. A route reconnaissance determines travel times, and identifies capacities of underpasses and bridges, locations of fords and bypasses, and critical points and obstacles. This lets the commander plan his movement and avoid areas or situations that could disrupt it.

Conduct the route reconnaissance with one of the scout platoons. Before the reconnaissance, let members know what information is needed and when and where they should submit it. Get them out as early as possible; troop plans are based on what they see and report.

If no other traffic control assets are available or if the troop is moving independent of the squadron, use the scouts from the platoon conducting the reconnaissance as traffic control teams at critical points along the route. Determine through map reconnaissance any built-up areas or key intersections that could disrupt the troop's movement, and position traffic control teams there. Based on their reconnaissance, the scouts may need to refine these positions.

If the movement is being conducted as part of the squadron, one troop will usually reconnoiter the route. The troop must, however, reconnoiter the route from its present location to the start point to determine if the route is suitable and how long it will take to reach the start point.

Quartering Party. A quartering party is used to reconnoiter and prepare a position before the troop main body arrives. The troop quartering party will often move as part of the squadron quartering party. Organize it with a scout platoon leader or the first sergeant as the party leader, with guides from each platoon, and with any additional personnel needed to clear the area.

Before the quartering party leaves, it must know the troop's route, order of march, estimated time of arrival, and any specifics on establishing the assembly area. Before the main body arrives, the quartering party reconnoiters the area, marks routes, and prepares to guide the main body as it enters the new area.

Trail Party. The troop trail party is made up of personnel and equipment (normally the troop trains) to handle emergency vehicle repair and recovery, medical aid and evacuation, and emergency refueling. The trail party moves just forward of the last maneuver platoon in the main body. The troop motor sergeant or first sergeant is in charge.

Main Body. The troop normally moves as a single march unit in column formation when conducting a tactical road march. The organization of the troop and any attached elements should be standardized during movement. Alter this organization to meet specific mission requirements as needed.

The troop commander has no prescribed place in the column. He positions himself where he can best control the movement of the troop. He will usually be well forward in the column, behind the lead platoon, to respond to contingencies while on the move. As a security measure, the troop command post should be positioned farther back in the column to disperse troop command and control.

The troop's column organization must provide adequate security against air and ground threats, while on the move and during halts. See Figure 7-1 for one way to organize a troop march column.

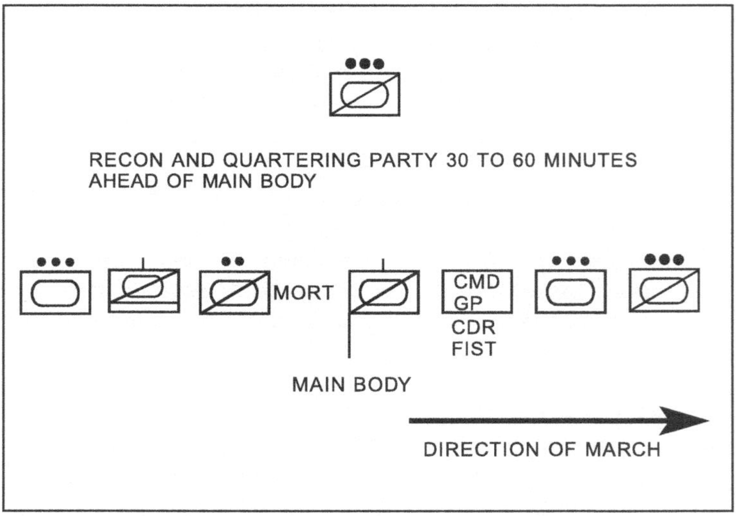

Figure 7-1. Troop column organization.

The 3d platoon, the first sergeant, and the quartering party move ahead of the troop. The troop order of march is 1st platoon, command group, 2d platoon, command post, mortars, trains, and 4th platoon. This order of march provides 360-degree security, disperses the command and control assets of the troop, and provides reconnaissance forward of the main body. Also, vehicle commanders assign sectors of observation to their crews, who search for air and ground threats (see Figure 7-2).

The troop performs the march in open or close column, depending on the situation. Use close column during limited visibility conditions. Vehicles are spaced 25 to 50 meters apart. This method takes advantage of the traffic capacity of the route, but provides little dispersion. Vehicle density is 15 to 30 vehicles per kilometer along the route of march. Use open column to provide greater dispersion and, thus, greater security. The distance between vehicles varies from 50 to 100 meters. Open column is normally used in daylight conditions. Base the troop's march speed on the slowest vehicle in the column. The troop trains will usually limit the rate of march.

Halts. Halts are used to rest personnel, provide personal comfort and relief, facilitate mess operations, refuel vehicles, maintain and inspect equipment, adjust the schedule, and allow other traffic to pass. The squadron march plan or troop SOP will specify the frequency and duration of halts, and will prioritize work to be done. For long movements, plan halts into the troop march table and ensure subordinate platoon leaders understand what actions must occur at the halt. When unscheduled halts occur, find the reason for the halt and let subordinates know how long it will last. Provide for security during halts, and establish OPs to provide early warning of enemy forces during long halts.

Vehicle commanders assign sectors of observation to their personnel so there is 360-degree observation. Each vehicle commander designates an air guard to provide air security, or specific vehicles may be designated air guard vehicles. In that case, the crews concentrate on only vehicle air observation rather than air and ground observation.

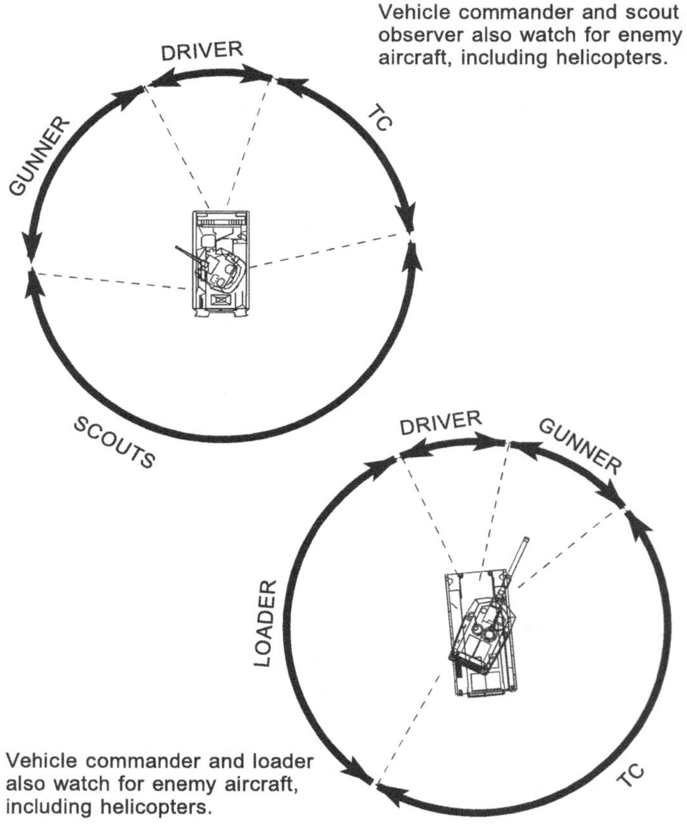

Figure 7-2. Maintain 360-degree security.

Vehicles that become disabled during movement must not obstruct traffic. The crew of the disabled vehicle must move the vehicle off the route, post guides to direct traffic, and find the problem. If the vehicle can be fixed, it rejoins the rear of the column. It does not return to its original position until the column has halted. If the vehicle cannot be readily repaired, the trail party recovers it.

Section II. Assembly Areas

An assembly area serves as a place where the troop gathers to prepare for future operations. In the assembly area, the troop prepares and issues orders, repairs and maintains vehicles, conducts resupply operations, and rests. As a minimum, assembly areas are positioned out of range of enemy light artillery.

TASKS

The following tasks are associated with assembly area operations:

- Plan the occupation of an assembly area.
- Conduct quartering party operations.
- Occupy an assembly area.
- Maintain security.
- Establish communications.
- Prepare for future combat operations.
- Depart an assembly area.

TECHNIQUES

The troop will normally be assigned a specific assembly area location. Within the area available to the troop, conduct a map reconnaissance, and if time is available, a ground reconnaissance. Look for an area that provides overhead concealment. This is extremely important if the troop will remain in the area for any length of time. Select an area that also provides—

- Cover from direct fire.
- Good drainage and a surface that will support troop vehicles.
- Good entrances and exits and an adequate internal road or trail network.
- Space for dispersion of vehicles, personnel, and equipment.

Quartering Party. The troop will often occupy an assembly area at the end of a road march. Before the quartering party leaves the troop's present location, tell them how to organize the assembly area. They must know any special requirements, such as a site for a logistics package (LOGPAC), so they can prepare the position. The quartering party must be organized to provide their own security during this operation. When the quartering party arrives at the forward assembly area, they must—

- Reconnoiter the area. If the area is not suitable, the leader of the quartering party must report immediately and provide a recommendation for another area.
- Organize the area. The leader of the quartering party selects locations for the platoons, mortars, command post, trains, and any attached elements, based on the commander's instructions. He may need to deviate from the commander's guidance to position an element in a suitable location.
- Improve and mark entrances, exits, and internal routes.

- Mark vehicle locations. Platoon representatives in the quartering party select general locations for vehicles in their platoon. Vehicle commanders and the chain of command refine these positions when they arrive.
- Perform guide duties. A platoon representative guides his platoon into position after the platoon clears the release point.

Occupation. Each platoon is guided from the release point into the assembly area by its quartering party member. Color-coded lights can be used to link up guides and lead vehicles (see FKSM 17-97-3). When the troop arrives at the assembly area, all elements move off the route without halting or slowing to keep the route of march clear. Keep this in mind when selecting routes, organizing the order of march for the road march, and allocating space in the assembly area. Once platoons have cleared the route and moved into their areas, they can adjust their positions without slowing the remainder of the troop.

Positioning troop elements in the assembly area is based on the size of the area, the terrain and avenues of approach, the length of time the troop will occupy the area, and any special requirements, such as resupply. Some considerations in allocating space for the troop are shown in Figure 7-3 and in the following list:

- If possible, space vehicles at least 100 meters apart to decrease exposure to enemy observation and fire.
- Keep vehicles in hide positions and establish OPs on terrain that provides good observation of approaches into the assembly areas for early warning of enemy movement. The vehicles can be moved up to a fighting position if necessary.
- Position platoons based on available terrain in the assembly area. Tank platoons may need more space to move vehicles in and out of position; scout platoons may occupy terrain that is more restrictive.

- Position the troop command post in the center of the assembly area. This makes establishing wire communications and issuing orders easier.
- Position troop trains where they have a good road network and space to conduct their maintenance and resupply operations. They may be used to secure the perimeter of the assembly area, but this will reduce their support ability.
- Emplace the mortars in a position from which they can support the troop with indirect fires. Overhead cover will reduce or eliminate their ability to fire.

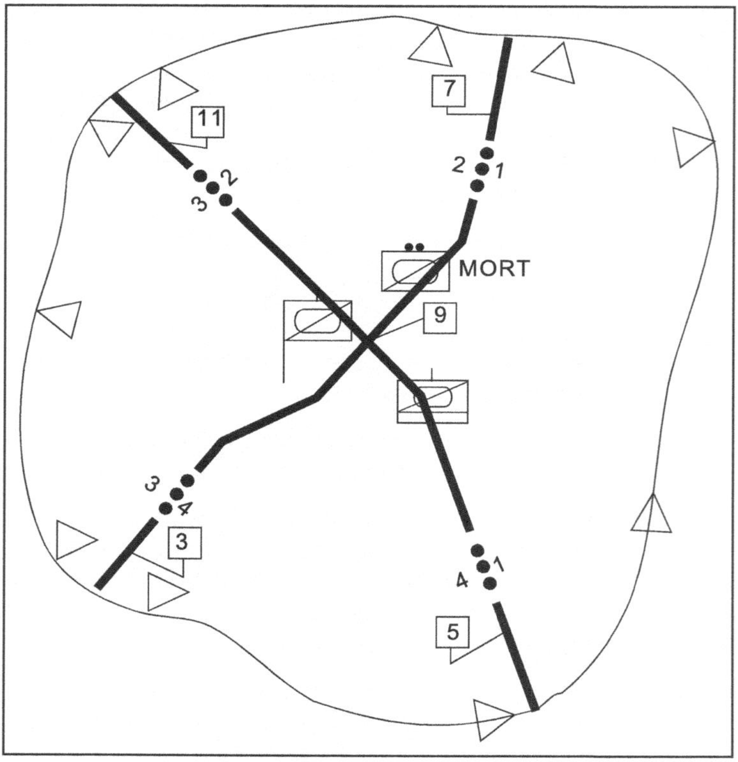

Figure 7-3. Troop assembly area.

Security. Although the assembly area is not a defensive position, the troop must be able to see and defeat enemy ground attacks. The best defense against air attacks is to remain hidden. Post guards at all entrances and exits to stop traffic that tries to enter the area. Establish OPs to observe key terrain features and likely avenues of approach for early warning of enemy approach. Each platoon must provide overlapping observation and fires within its platoon and with the platoons on its flanks. Establish a dismounted patrol plan so platoons make physical contact with their adjacent platoons. Ensure the platoons provide 360-degree coverage of the assembly area. Camouflage vehicles and equipment to prevent enemy detection from the ground and air. Place PEWS (platoon early warning system) in heavily vegetated areas or dead space to provide early warning of enemy movement in and around the position. Emplace NBC alarms upwind and about 150 meters from the troop's positions to provide early warning of an NBC attack. Assign the mortars an azimuth of fire on the most dangerous approach into the assembly area. Give the fire support officer guidance in preparing an indirect-fire plan.

Communications. Messenger and wire are the primary means of communications from the command post to each platoon and to each OP. If practicable, lay wire from the squadron to the troop. Plan on providing a messenger to the squadron command post. Use radio when no other means of communication can be used.

Preparation for Future Operations. Several tasks are routinely accomplished in an assembly area. These tasks are listed in the troop SOP (FKSM 17-97-3) under priority of tasks upon arrival in an assembly area, and include—

- Position vehicles.
- Establish local security.
- Establish OPs.
- Prepare fire plan.

- Establish wire communications.
- Maintain radio watch and man turret weapons.
- Camouflage positions.
- Prepare obstacles/mine plan.
- Select alternate and supplementary positions.
- Reconnoiter routes of withdrawal.
- Perform PMCS.
- Emplace NBC alarms and PEWS.
- Continue to improve positions.
- Conduct logistics resupply (Classes I, III, and V).
- Rest in accordance with REDCON status.

Modify this task listing to accomplish specific tasks (such as conduct rehearsals, test-fire weapons, and conduct inspections) in preparation for future operations. Ensure subordinates know how long the troop will remain in the assembly area and are told of any special requirements. Occupation of the assembly area will often be conducted and supervised by troop NCOs, while the commander and platoon leaders plan for upcoming operations.

Departure. Maintain the appropriate REDCON. Each REDCON level indicates critical tasks and time available to prepare for future operations.

- REDCON 1—be prepared to move immediately.
  - All personnel alert and ready for action.
  - Vehicles loaded and secured, and weapons manned.
  - Vehicle engines running and OPs not manned.

- REDCON 2—be prepared to move in 15 minutes.
  - All personnel alert.
  - OPs and wire pulled in.
- REDCON 3—be prepared to move in half an hour.
  - Fifty percent of each crew/squad stand down for rest, feeding, and maintenance.
  - Remaining 50 percent man vehicles, OPs, weapons, and monitor radios/phones.
- REDCON 4—be prepared to move in one hour.
  - Two men per platoon make dismounted checks of platoon area.
  - One man per vehicle monitors radios/phones and mans turret weapon.

All personnel remain at 100-percent alert until the prioritized work is complete after entering the assembly area. Initiate the appropriate REDCON when the work is finished. As the time for execution of a mission nears, increase the REDCON in accordance with guidance from squadron, assigning REDCON 1 just before the troop must move.

Section III. Relief in Place

A relief in place is an operation in which a unit in combat is replaced by another unit. Responsibilities for the combat mission and the assigned sector, battle position, or zone of the relieved unit are assumed by the relieving unit. A relief in place may be accomplished during offensive and defensive operations and may be conducted during any weather and light conditions.

The primary purpose of a relief in place is to maintain the combat effectiveness of committed units. A relief in place is conducted to replace a committed unit to give it the opportunity to reconstitute, rest, decontaminate, or perform a change in mission.

TASKS

The following tasks are associated with the relief in place:

- Plan and coordinate a relief in place.
- Establish communications.
- Establish liaison.
- Conduct reconnaissance.
- Initiate movement.
- Occupy positions.
- Maintain operations security.

TECHNIQUES

A troop will often perform a relief in place mission as part of a squadron operation to relieve or be relieved by a brigade. The troop will usually conduct a relief in place in conjunction with a battalion. The troop may relieve a battalion and operate in an economy-of-force role or be relieved by a battalion to conduct a change in mission.

To reduce confusion and maintain security, the following factors must be considered when planning a relief in place:

- The time that responsibility for the sector, battle position, or zone is to pass.
- Operations security.
- Deception plans.

- Time, method, and sequence of relief.
- Routes and critical control measures.
- Concept of subsequent missions.
- Plans for additional positions.
- Contingency plans.
- Location of obstacles and transfer of responsibility.
- Transfer of ammunition; wire lines; petroleum, oils, and lubricants (POL); and materiel to the relieving unit.

Communications. At a specified time or upon receipt of an order, the relieving unit minimizes radio traffic or begins to operate at radio-listening silence. The relieving unit command post monitors the command net of the unit to be relieved while continuing to operate on its internal command net.

Reconnaissance and Liaison. The orders group of the relieving unit moves to the command post of the unit being relieved for face-to-face coordination.

In preparation for the relief in place, the relieving unit moves to a forward assembly area under the command of the XO, if necessary.

If the situation permits, the relieving unit conducts a reconnaissance of the area when it completes coordination. The units establish a contact point behind the positions into which they will move; a guide from the relieved unit will meet them there to make any final coordination and guide them into position.

The commanders complete their plans and issue orders to their respective units after the reconnaissance and liaison are complete.

Methods. The relief in place can be conducted in several ways. Use the method appropriate to the troop's situation.

- One unit at a time. This is the slowest method, but the most secure.

- All units simultaneously. This is the quickest method, but the least secure.

- Center units first, followed by flank units.

- Flank units first, followed by relief of the center unit.

- Occupy adjacent or in-depth positions that cover the area of responsibility.

Passage of Command. The troop command post and battalion TOC collocate to facilitate command and control of the operation. The relieving unit departs its location in the prescribed sequence and lines up at the contact points with the guides from the relieved unit. The relieving unit moves into hide positions behind the relieved unit. The relieving unit moves into position after all pertinent information has been exchanged and the relieved unit moves out of position. The relieved unit moves to a rally point or assembly area to link up with the rest of the unit. When each element has completed the relief, the relieving unit assumes responsibility for the area and reports by messenger, wire, or radio to the command post. The relieved unit moves along its designated route to the new location. See Figures 7-4 and 7-5 for illustration of the relief in place.

The higher headquarters will often specify a time the relieving unit assumes responsibility for the sector, battle position, or zone. The commanders conducting the relief coordinate to ensure the relief is conducted on time. The relieving unit normally assumes responsibility when the relieved unit departs its positions. The commanders of the two units are usually collocated and responsibility passed face to face.

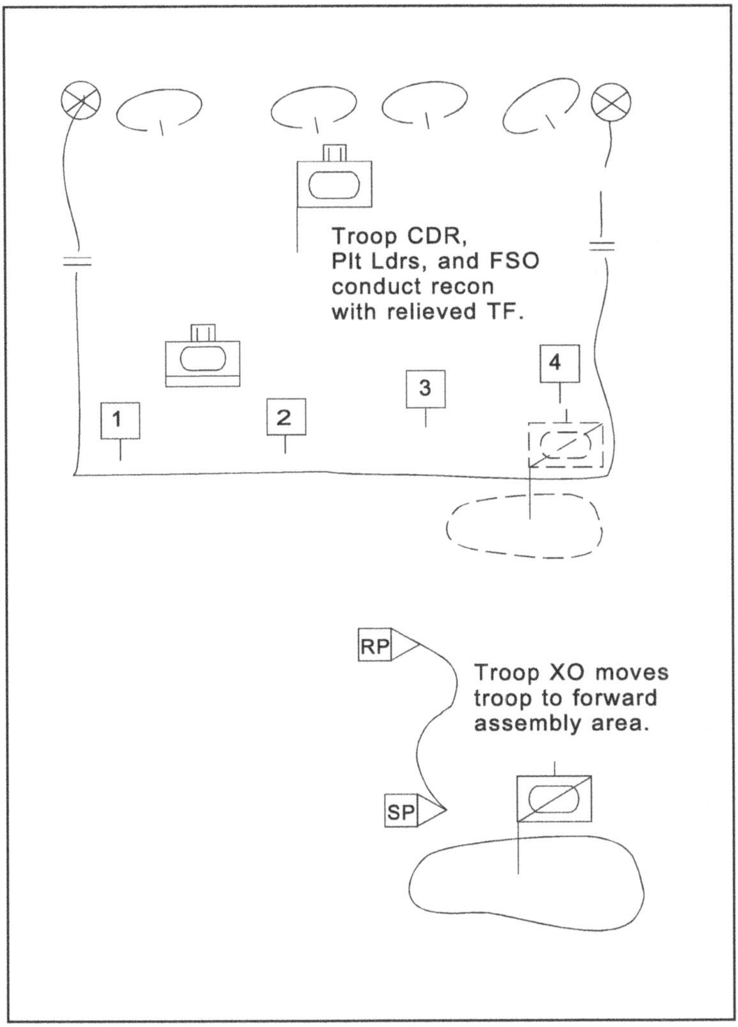

Figure 7-4. Troop prepares to conduct relief in place with task force.

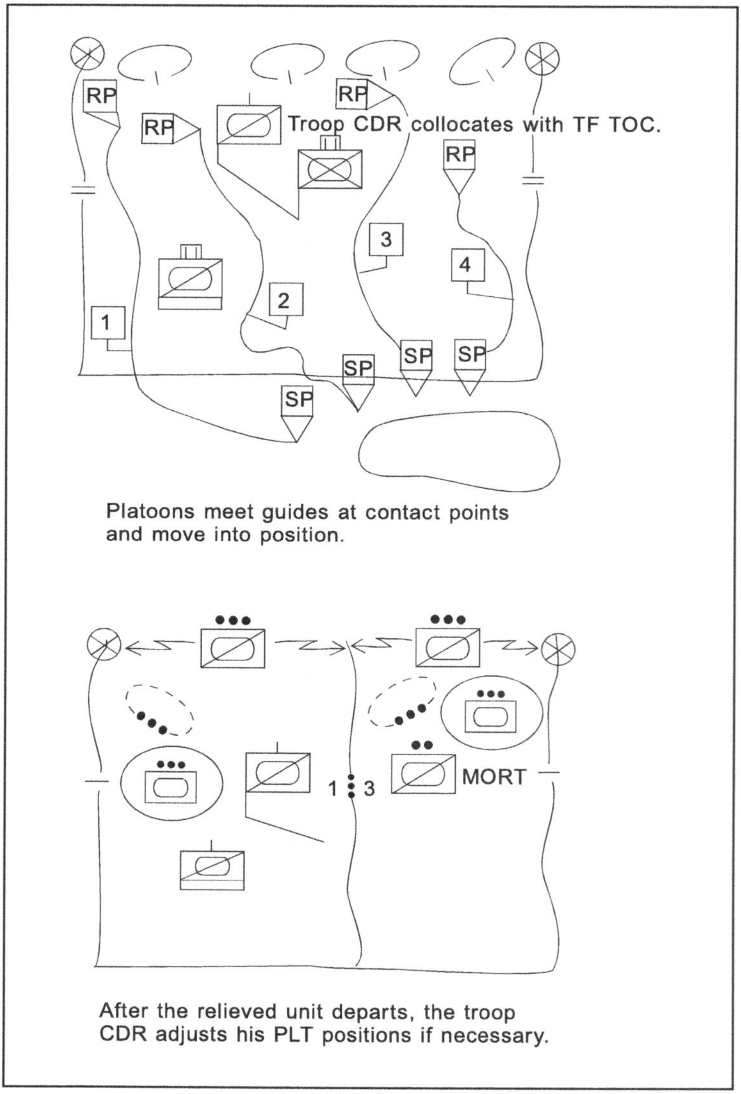

Figure 7-5. Troop completes relief in place.

Fire Support Assets. The troop fire support officer coordinates with the squadron fire support officer to ensure there is fire support throughout the operation. The fire support assets of the relieved unit remain in position throughout the relief of the maneuver forces, and are prepared to support both units. Fire support assets of the relieving unit position themselves as quickly as possible to provide additional support.

Enemy Contact. If either unit gains direct-fire contact with an enemy force, it immediately notifies the other unit and the higher headquarters directing the relief. If responsibility has not passed, the relieving unit becomes OPCON to the relieved unit. The relieving unit's mortars fire missions as directed by the relieved commander through his fire support officer. If responsibility has passed, the relieved unit becomes OPCON to the relieving unit. The collocation of the commanders and command posts facilitates rapid coordination and action in this situation.

Section IV. Battle Handover and Passage of Lines

Battle handover is an operation that transfers responsibility for fighting an enemy force from one unit to another in the close-in battle. Conducted by stationary and passing units, battle handover is designed to sustain continuity of the combined arms fight, and to prevent the enemy from getting a "free ride" anywhere on the battlefield as one force picks up the fight from another. It is also designed to preserve the fighting capabilities of both forces as they execute the operation. Battle handover is associated with almost all Army combat operations.

Passage of lines is a tactical event associated with battle handover. It is the controlled movement of one unit through the positions of another stationary unit that does not interfere with either unit's scheme of maneuver. A passage of lines is often used because the combat situation does not permit one unit to bypass another unit's positions. The passing unit must move through the positions of the stationary unit.

Passage of lines is often an integral part of the following operations:

- Deliberate attacks or counterattacks across the FLOT.
- Exploitation and pursuit.
- Route, zone, or area reconnaissance.
- Raids.
- Movement to contact.
- Defend, delay, or withdrawal.
- Defensive or offensive cover.
- Screen or guard.

Although a cavalry troop does not normally perform all of these operations, it may conduct them as part of a larger force.

CRITICAL TASKS

The three players in battle handover and passage of lines are: the *stationary unit*, the *passing unit*, and the *common commander* of both units. Each has critical tasks to perform to achieve smooth and efficient execution of the operation.

Critical Tasks of the Common Commander. The commander exercising command authority over both the stationary unit and the passing unit must accomplish three critical tasks.

- Establish where battle handover will occur by designating a phase line forward of the FEBA as the battle handover line (BHL). The line should be where combat maneuver forces of the stationary unit along the FEBA can effectively overwatch and protect the passing unit as it withdraws behind the FEBA or advances forward of the FEBA or FLOT. The distance forward of the FEBA or FLOT is limited to available fields of fire and the effective range of weapons of the stationary unit.

- Designate contact points just forward of the battle handover line at which stationary and passing units are required to conduct physical coordination (only in defensive operations).
- Ensure the passing unit is provided indirect-fire support while its artillery is displacing during battle handover and passage of lines.

The stationary unit normally recommends battle handover line and contact point locations to the common commander. It remains the responsibility of the common commander to establish the line and contact points. These control measures must be reflected graphically on an overlay and identified in the appropriate operation plan (OPLAN), OPORD, or FRAGO issued to subordinate units. The battle handover line, in effect, establishes a boundary between the ground owned and controlled by the stationary unit commander and the ground controlled by the passing unit commander. The stationary unit commander controls the ground forward of the FEBA up to the battle handover line. He can place security forces, obstacles, and direct and indirect fires into this area to support his scheme of maneuver.

While the battle handover line defines the point at which the battle handover should ideally occur, events normally preclude this from happening. The moving and stationary force commanders and their common commander should understand that the actual handover may occur in a "zone" centered on the battle handover line. Both the moving and stationary force must remain active and responsive enough to perform the battle handover anywhere in this zone, and should not dogmatically "stick to the plan."

Battle handover begins on the order of the common commander. To sustain unity of command in the passing operation, the passing unit is usually placed OPCON to the stationary unit to execute battle handover and passage of lines. OPCON by the stationary unit is limited to those actions necessary to get the passing unit through the stationary unit's area as quickly as possible. In defensive operations, battle handover from the passing unit to the stationary unit is not complete until the combat maneuver forces of the stationary unit along the FEBA have visual contact with the enemy, and the stationary unit commander indicates readiness to assume responsibility for the battle. In offensive operations, battle handover from the stationary unit to the passing unit is not complete until the combat maneuver elements of the passing unit have crossed the FEBA or FLOT, deployed for combat, and maneuvered across the battle handover line.

Passing Unit Critical Tasks. The passing unit must accomplish the following critical tasks to perform battle handover and passage of lines:

- Immediately establish communications with the stationary unit. Enter the command, operations and intelligence (OI), and fire support nets.
- Collocate a command post with the TAC CP or TOC of the stationary unit as soon as possible to enhance communications and unity of effort.
- Continuously report to the stationary unit the location, size, and composition of all enemy forces. Report the enemy's current activity. If the enemy is attacking, report his direction of movement, movement formation, and estimated rate of advance. If he is defending, report his locations, orientation, composition, fire sacks, reserves (if known), obstacle system, or flanks.
- Continuously report to the stationary unit the location, size, and activity of all subordinate elements to include combat support, combat service support, and command and control facilities.
- Given the current disposition of subordinate units, coordinate with the stationary unit to determine

contact points at which each subordinate company-size unit will physically coordinate handover and passage of lines with representatives of the stationary unit. Once contact points are determined, send a FRAGO to each subordinate unit specifying where they will physically coordinate passage with the stationary unit. Confirm recognition signals that must be displayed during passage (defensive operations).

- Once each subordinate unit acknowledges where it must physically coordinate passage, each unit will dispatch representatives to assigned contact points and coordinate passage for its unit. At the contact point, confirm recognition signals and exchange required information (defensive operations).
- Maintain visual contact with all enemy units and delay back to the battle handover line, avoiding decisive engagement (defensive operations).
- Display correct recognition signals and use correct challenge and password as specified in the CEOI during passage.
- Maintain proper weapons orientation.

Stationary Unit Critical Tasks. The stationary unit must accomplish the following critical tasks when ordered to conduct battle handover and passage of lines:

- Establish communications with the passing unit. Coordinate and direct the passing unit to contact points based on current dispositions of the subordinate units (defensive operations).
- Ensure contact points are manned and subordinate commanders have personal communications with their representatives (defensive operations).
- Ensure representatives at the contact points assign each passing unit a passage point into the area of operations and a route that extends from the passage points to the rear boundary or to an assembly area (defensive operations).

- Ensure representatives at the contact points exchange required information with the passing unit as outlined in FKSM 17-97-3.
- If security forces are employed, position them along the battle handover line to observe enemy avenues of approach. Adjust as needed for low visibility conditions (defensive operations).
- If obstacles are emplaced between the FEBA and the battle handover line, ensure routes through the obstacle system are clearly marked and physically controlled by guides, or provide an escort to the passing unit. Ensure that reserve targets on obligated routes are manned by soldiers in direct communication with their commanders.
- Ensure that all routes of withdrawal obligated to the passing unit are unobstructed and facilitate rapid movement to the release point (defensive operations).
- Ensure obligated routes of advance, attack positions, and routes to the battle handover line are unobstructed and permit rapid movement (offensive operations).

GRAPHIC CONTROL MEASURES

Figure 7-6 shows the graphic control measures that support battle handover and rearward passage of lines.

- Battle handover line. This line is established by the common commander in consultation with both commanders. The stationary commander has the major determination in locating the BHL, as his force must be able to overwatch the BHL with direct fires.
- Contact points. These are established on identifiable terrain and normally in the vicinity of the passage lanes. For rearward passage of lines, the contact points are established forward at the BHL. For forward passage of lines, the contact points are established in the stationary unit's rear area rearward of the passage lanes.

- Passage points. The passage point is that point on the passage lane where the moving unit moves through, and responsibility for the battle is passed to the stationary unit. It is usually placed where the passage lane begins.
- Passage lanes. The stationary unit establishes passage lanes to move the passing unit quickly through defending unit positions. This could include passing through gaps in friendly obstacles and moving near or through friendly engagement areas and battle positions. Lanes are restrictive; however, they should ideally be wide enough to allow the passing unit to move in a tactical formation. The passage lane begins at the passage point and ends at the rear of the stationary unit BPs. The passage is considered complete when the moving unit exits the lane.
- Routes. Routes are used to move the passing unit through the stationary unit rear area. The number of routes designated will vary based on METT-T, but as a general rule, multiple lanes/routes should be planned to facilitate rapid passage of moving units and to avoid unnecessary massing of units. The stationary unit may escort or guide the passing unit along the lane/route.
- Assembly area. An assembly area in the rear area of the stationary unit allows the passing unit to conduct hasty reorganization and emergency CSS actions. This assembly area is temporary in nature.
- Infiltration points. Units should plan infiltration points and lanes for personnel unable to complete the passage with the unit. Passing unit liaison officers may remain located with stationary unit CPs to serve as a point of contact for infiltrating personnel/equipment. Personnel who infiltrate must have some way of contacting the stationary unit before crossing into friendly territory.

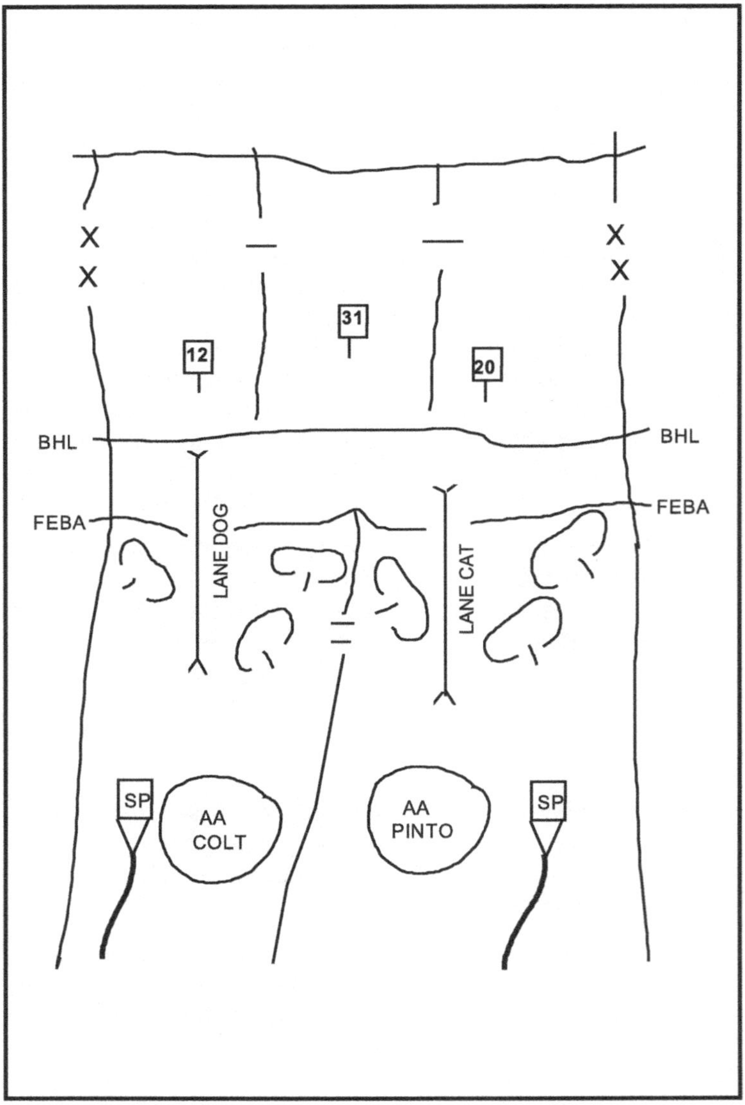

Figure 7-6. Rearward passage of lines.

TECHNIQUES

At troop level, the passage of lines will usually be performed as part of a squadron operation. The passage may be forward, such as to pass through a defending unit to conduct a counterattack, or rearward, such as when a covering force unit withdraws through units in the main battle area.

The troop is particularly vulnerable during a passage of lines. The unit may be concentrated and the fires of the stationary unit may be temporarily masked. Thorough reconnaissance and detailed coordination are critical to ensuring the operation is successful.

A troop will often perform a battle handover and passage of lines through a single battalion while the squadron conducts the operation through a brigade. Ideally, the boundaries will correspond to the battalion boundaries so that coordination for the battle handover is conducted through a single headquarters.

The troop commander has numerous considerations to ensure a successful passage of lines. During reconnaissance, he must confirm—

- The disposition of the stationary force through which the troop will pass.
- The location of contact points where both units are required to make physical contact at a predetermined time.
- The location of passage points on the battle handover line through which friendly forces will pass.
- The location of passage lanes that provide a clear route through a friendly position, and also facilitate a smooth and continuous passage. Areas selected for passage should be unoccupied or on the flanks of units in position. If possible, use multiple routes to reduce vulnerability during the operation.

- The location of an attack position (for forward passage) or assembly area. This position should provide cover and concealment and be located where the passing unit will not interfere with the stationary unit.
- The initial location for combat support and combat service support elements.

Based on the reconnaissance, the troop commander coordinates and plans for—

- Supporting fires. The stationary force supports the passing unit with direct and indirect fires up to the battle handover line. In forward passage, the stationary force supports the passing unit's move through the passage and until it crosses the battle handover line. In a rearward passage, the stationary unit supports the passing unit's move back across the battle handover line and through the passage of lines.

- Time of transfer of responsibility for control of the sector and handover of the enemy.

- Troop density. The passing troop commander should plan for multiple routes of passage to ensure rapid movement and to avoid congestion.

- Traffic control. Guides from the stationary unit pick up passing elements at each contact point and guide them through the position. The passing unit commander tells the stationary unit the type, number, and order of vehicles passing through each contact point.

- Communications. The leaders exchange CEOI information and mutually agreed upon recognition signals.

- CSS. The troop commander must coordinate the evacuation of casualties, PWs, vehicles, and resupply of fuel and ammunition. The stationary unit usually provides emergency service only. The passing unit supports itself.

- Liaison officers. The troop commander should designate a representative to perform the critical duties of a liaison officer. Commanders normally coordinate a forward passage of lines and the XO coordinates a rearward passage. Liaisons are normally located at critical points during the passage. If the commander or XO is not available, a scout platoon leader should perform liaison duties. Ensure he is thoroughly briefed on the situation and follows the checklist in the troop SOP.

Section V. Hasty Water Crossing

A hasty water crossing is the movement across an inland waterway using a crossing means at hand or readily available without significant delay once the waterway is reached. It is preplanned and conducted as a continuation of the operation underway. Although the crossing is termed hasty, detailed planning assures that fire support and crossing means are available on arrival at the water obstacle.

CRITICAL TASKS

The following critical tasks are associated with the successful conduct of a hasty water crossing:

- Plan the water crossing.
- Conduct reconnaissance.
- Establish security.
- Cross the obstacle.
- Provide continuous fire support.
- Continue the mission.

TECHNIQUES

Plan. A troop may perform a hasty water crossing independently or as part of a squadron. A hasty water crossing is performed as an extension of the ongoing operation. It gives the troop the ability to sustain the momentum of an operation, and is often associated with a movement to contact and a zone reconnaissance. The hasty water crossing is characterized by speed, surprise, minimum loss of momentum, and minimum concentration of forces.

The crossing should be preplanned to ensure fire support and crossing means are at the crossing site. The need for a water crossing should be determined in the mission analysis and added to the commander's concept of the operation. This ensures the troop's critical assets are positioned to support the water crossing. Seizing bridges intact before the enemy can destroy them is the quickest and most economical means of crossing, and is used when possible.

Reconnaissance. Scout platoons will be the first to encounter a water obstacle. When this occurs, they report it and reconnoiter it to obtain the following information, which is forwarded to the troop command post:

- Width and depth of the waterway.
- Water velocity.
- Possible entry and exit points and their conditions.
- Enemy situation on the far bank.

Scouts reconnoiter multiple crossing sites to prevent the enemy from discovering the true site and to reduce the concentration of vehicles at the site. If follow-on forces will cross at the same sites, scouts mark the entrances and exits for their use.

The troop commander, with the fire support officer, positions himself forward during the reconnaissance of the water obstacle so he can personally supervise the operation. The fire support officer will support the scouts with indirect and direct fires, if needed.

Security. A crossing can be performed with or without opposition. To avoid enemy fire while crossing, prevent him from knowing where the crossing will be.

Before the main body crosses, the scouts must secure the far bank by fording or swimming across the water obstacle and establishing positions. They should cross on both sides of the site and secure positions on the dominant terrain of the far side to provide early warning of threat activity. After securing the far side of the water obstacle, the scouts establish positions to the flanks of the crossing site to provide local security (see Figure 7-7).

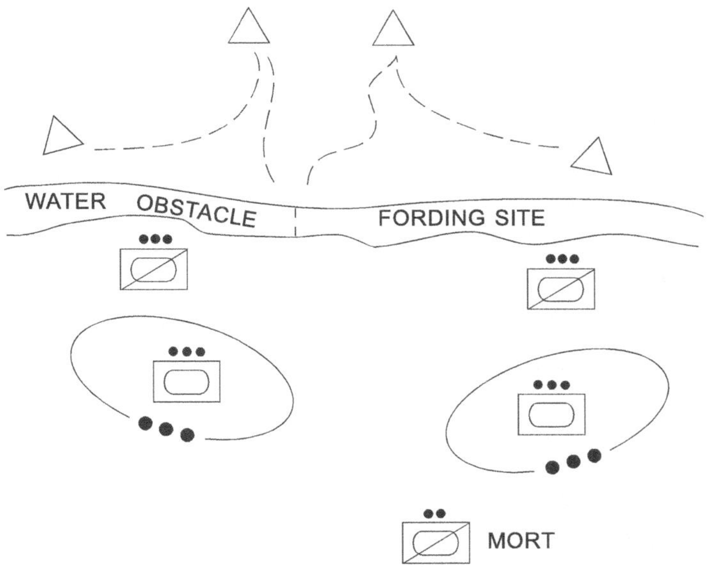

The scouts encounter a water obstacle, find fording sites, and secure the far side. Position the tank platoons and mortars to support with direct and indirect fires.

Figure 7-7. Secure the far bank.

The rest of the troop should not bunch up at the crossing site. Position them a terrain feature short of the water obstacle, so they have quick access to the crossing sites but are not exposed to enemy observation and fires.

Cross the Obstacle. Once the scouts have established positions on the far bank, the troop main body is ready to cross the obstacle. Time the movement to the crossing site so that no vehicles have to pause on the near side, but all can move directly through the crossing site and to their positions on the far side to continue the operation. Once the troop starts the crossing, complete it as quickly as possible (see Figure 7-8).

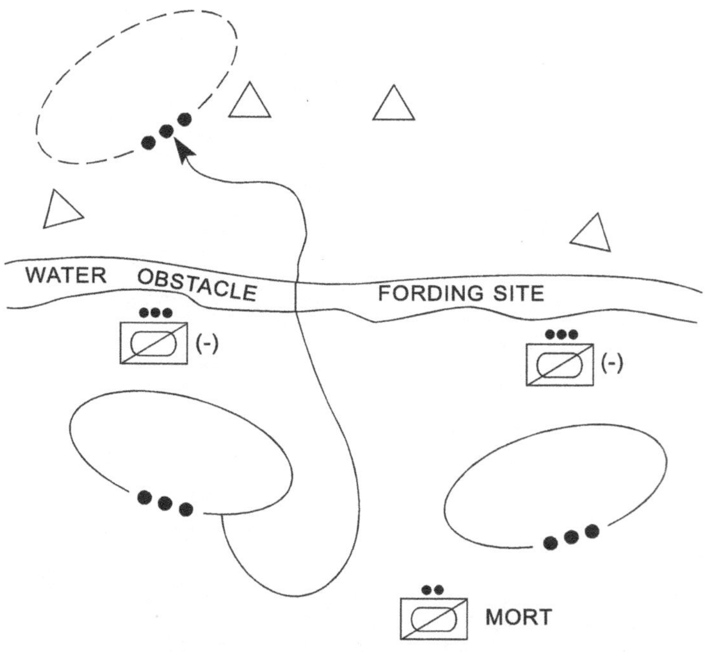

After the far side is secure, rapidly move platoons across the obstacle and continue the mission.

Figure 7-8. Cross the obstacle.

Fire Support. Fire support is used during the hasty water crossing to suppress known and suspected enemy positions at the crossing site. Smoke is employed to the front or flanks to screen the reconnaissance and the crossing. The fire support officer must time the movement of the mortars so they are in position to support the water crossing operation with continuous fire support.

Section VI. In-Stride Breach

When obstacles are encountered, an in-stride breach will maintain the tempo of offensive operations. Treat obstacles with caution and expect them to be covered by enemy observation and fires. The troop's ability to breach obstacles using available assets is important to ensure mission accomplishment. The size of the enemy force covering an obstacle with direct fire will, however, affect the troop's ability to conduct a hasty breach.

TASKS

The following combat tasks are associated with an in-stride breach:

- Detect the obstacle, reconnoiter it, and search for a bypass.
- Suppress all enemy positions with direct and indirect fires.
- Obscure enemy observation with smoke.
- Secure the near side of the obstacle.
- Breach or neutralize the obstacle.
- Move forces across the obstacle.
- Continue the mission.

TECHNIQUES

The enemy will reinforce natural obstacles with man-made ones to slow, disorganize, and canalize the troop. These obstacles may consist of—

- Minefields.
- Log obstacles, such as abatis, log cribs, stumps, and posts.
- Antitank ditches.
- Wire entanglements.

When scouts encounter an obstacle, they must follow the procedures for actions on enemy contact described in Chapter 3.

- Deploy and report.
- Develop the situation.
- Choose a course of action.
- Recommend a course of action.

As they develop the situation, the scouts should rapidly reconnoiter the obstacle to determine if bypass routes exist and whether or not they are covered by enemy fires, if breaches through the obstacle already exist, if the obstacle is part of an occupied defensive position, what the enemy strength is, and where he is located.

To conserve time and manpower, the troop should attempt to bypass all obstacles. Bypassing will not always be possible because available routes around the obstacle may lead into a fire sack. Forcing through an obstacle is the least desirable method of breaching because of the resulting loss of personnel and equipment. A hasty breach may be the only viable course of action.

Before deciding on a course of action, the troop commander should position himself to observe the situation and control the operation, if possible. Once the commander decides that a hasty breach is the only course of action, he must organize the troop into a breaching element and an overwatch element. When executing the hasty breach, the commander should avoid concentrating the troop in any one area, making it vulnerable to enemy fires.

During zone reconnaissance or movement to contact, the troop operates over a 6- to 10-kilometer-wide zone. During the hasty breach, the commander may not be able to use all the platoons in the troop. In this situation, until the breach operation is complete, give the scout platoon not involved in the breach operation a limit of advance so it does not move too far ahead of the rest of the troop. It can provide the troop with an early warning of enemy activity.

The Breaching Element. The breaching element is composed of scouts and engineers attached or in support of the troop. Their mission is to create lanes through the enemy obstacle system to allow the troop to pass.

The Overwatch Element. The overwatch element is the tank/AT platoons, scouts not in the breaching force, and any additional engineers. Its mission is to provide close, continuous overwatching fires in support of the breaching element. After the breaching element has cleared lanes through the obstacle, part of the overwatch element will assault through the lane to defeat any enemy in the area of the breach. The fire support officer coordinates the fires of the mortars and any supporting artillery to suppress enemy fires with HE and to obscure his observation with smoke.

Sequence of Events in a Hasty Breach. After encountering an obstacle and the commander chooses a hasty breach as the course of action, the sequence of events is as follows:

- Scouts.
  - Reconnoiter the near side of the obstacle to determine its front edge and lateral limits, enemy dispositions, if the obstacle is part of a defensive position, and if a partial or complete breach exists.
  - Secure flanks in support of the breaching operation.

- Overwatch element.

  – Occupy overwatch positions and provide direct and indirect suppressive fires on the enemy for elements moving to and through the obstacle.

  – Use smoke to obscure enemy observation of the obstacle (see Figure 7-9).

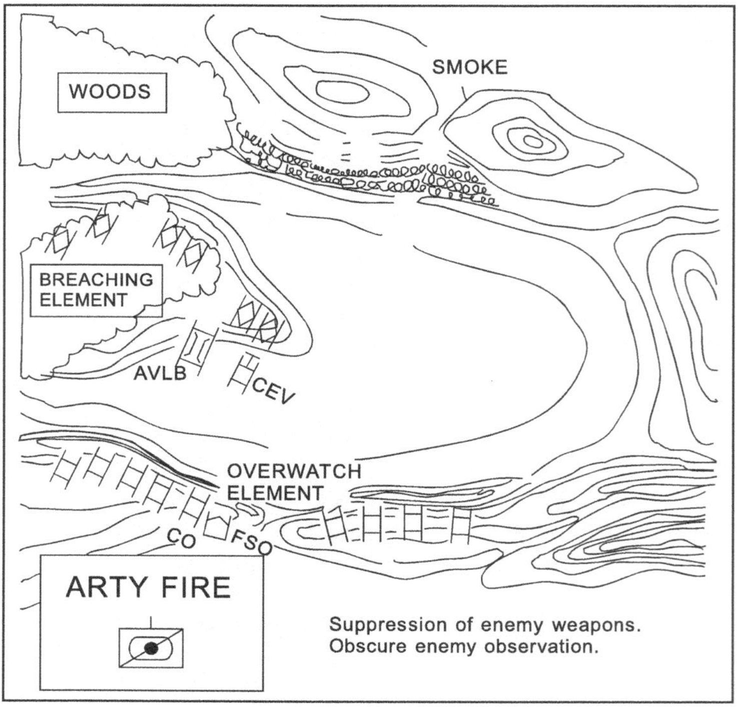

Figure 7-9. Obscure enemy observation.

- Breaching element.
  - Occupy covered and concealed positions to coordinate activities and prepare equipment, demolitions, and routes to the obstacles.
  - After enemy fire has been suppressed, breach the obstacle.
  - Secure fighting positions on the near and far sides of the obstacle (see Figure 7-10).

- Overwatch element.
  - As the breaching element breaches the obstacle, a tank platoon from the overwatch element prepares to attack through the obstacle.
  - Once the breaching element secures positions on the far side of the obstacle, the tank/AT or scout platoon attacks through the obstacle and destroys enemy elements that can fire directly on it. This assault element either continues to advance as the troop lead element or establishes hasty defensive positions while the remainder of the troop passes through the obstacle (see Figure 7-11).
- Troop. Continue the mission.

Neutralizing and Breaching Minefields. Minefields will differ in layout and composition, depending on the availability of mines and the nature of the avenue of approach. When possible, mines should be detonated in place. The objective in breaching the obstacle is to make a safe route to the far side. Multiple routes across the obstacle reduce vulnerability to enemy fires, but may be very time consuming and beyond the capabilities of the troop.

Use the following methods to neutralize and breach minefields:

- **Foot Lanes.** Establish foot lanes through the obstacle when a mounted assault breach is not at first practicable. This lane is normally two meters wide, and can later be widened into vehicle lanes. Foot lanes allow the troop to move forces through the obstacle to secure the far side.

- **The Bangalore Torpedo.** This device will clear foot lanes through mines and wire obstacles. It will clear a path three to four meters wide through wire entanglements and a narrow footpath through a minefield.

- **Vehicle Lanes.** After the first breach is made, foot lanes may be widened to one-way vehicle lanes at least eight meters wide. Vehicle lanes may also be breached separately from foot lanes. Use existing roads when possible. Vehicle-pushed rollers, if available, should be used to proof the lane cleared by line charges.

- **The M173 Rocket-Projected Line Charge.** This is an AT minefield clearing device. When the line charge explodes, it clears a vehicle lane about six meters wide.

Section VII. NBC Defensive Operations

OPERATIONS

NBC defense operations reduce casualties and damage to equipment and materiel, and minimize confusion and interruption of the troop's mission in the event of enemy NBC attacks. These operations are performed concurrently with all combat operations to preserve the fighting strength of the troop.

TASKS

The following combat tasks are associated with the conduct of troop NBC defense operations:

- Prepare for a nuclear attack.
- Respond to the initial effects of a nuclear attack.
- Respond to the residual effects of a nuclear attack.
- Cross a radiologically contaminated area.
- Conduct radiological reconnaissance.
- Perform radiological decontamination.
- Prepare for a chemical agent attack.
- Respond to a chemical agent attack.
- Cross a chemically contaminated area.
- Conduct a chemical reconnaissance.
- Perform hasty decontamination.
- Coordinate for deliberate decontamination of equipment.
- Exchange protective clothing.

ORGANIZATION

The troop's NBC defense system is composed of a trained NBC officer (usually the XO), an NBC NCO, and an enlisted alternate from troop headquarters. Troop soldiers are also designated and trained to operate all assigned NBC equipment and to assist decontamination teams.

The NBC officer supervises troop NBC defense activities and assists the commander in training NBC equipment operators and decontamination teams. The NBC NCO and his alternate directly supervise radiological monitoring, chemical detection, and decontamination operations. During combat operations, the NBC NCO is located in the command post where he—

- Receives, prepares, evaluates, and disseminates information and reports enemy and friendly NBC attacks.
- Supervises employment of detection, monitoring and survey, and decontamination teams.
- Maintains unit radiation exposure status records.
- Assists the troop commander in analyzing guidance from squadron, mission, threat, and weather as they affect NBC operations. Recommends appropriate MOPP level based on this information.

To facilitate operations in an NBC environment, designate the NBC teams from all platoons and sections by vehicle crews so vehicle commanders have responsibility for their crews and the execution of the NBC defense team mission. Each section in the troop has a chemical agent detection team and each platoon has a radiological monitoring and survey team.

Chemical Agent Detection Team. Each troop has twelve chemical agent detection teams. There are three in each scout platoon and one in each tank/AT platoon, one in the mortar section, one in the command post, and one in each combat trains and field trains. Each team has an M256 detector kit, a chemical agent alarm, detector paper, 400 meters of WD-1/TT for remoting the alarm, and enough batteries and supplies for seven days of continuous operations.

Radiological Monitoring and Survey Team. Each troop has seven teams, each equipped with an IM 174/PD dose rate instrument and an IM 93/147 dosimeter. One detection team from each platoon serves as a radiological monitoring and survey team. Also, the mortar section, maintenance section, and troop headquarters each have a radiological monitoring and survey team.

Deliberate Decontamination Team. Deliberate decontamination teams are organized by the vehicle crew. One crew from each scout and tank/AT platoon is designated as a decontamination team. This team is used to support the deliberate decontamination of the troop. One crew from each platoon should receive special training from the squadron chemical section to execute this task.

TECHNIQUES

Proficiency in NBC defense operations is attained only by strict adherence to standardized procedures as outlined in the troop SOP. The troop must be proficient in the following three fundamentals of NBC defense to survive and remain an effective fighting unit:

- Contamination avoidance (before, during, and after an attack).
- Protection (before, during, and after an attack).
- Decontamination (after an attack).

Actions Before the Attack. The best defense against a chemical or nuclear attack is to avoid being detected and targeted. To avoid becoming a lucrative target, implement and strictly enforce the following passive avoidance measures during all operations:

- **Dispersion.** Maintain dispersion between vehicles whether stationary or moving; vehicles bunched together are much easier to find. This is especially important when occupying assembly areas, conducting resupply operations, or when crossing obstacles and waterways.
- **Concealment.** Concealed positions prevent the enemy from observing vehicles and personnel from the ground or air.
- **Camouflage.** When concealment is not available or adequate, use camouflage to hide vehicles and positions.
- **Communications security (COMSEC)/signals security (SIGSEC).** The enemy can easily locate positions with direction-finding equipment if the troop does not adhere to proper communications procedures. Ensure all troop elements properly encode messages sent over unsecured nets.
- **OPSEC.** Take all possible actions to deceive the enemy about troop movement, positioning, intentions, and size. Mask movement by using folds in the terrain. Use hide positions when not moving or engaging the enemy. Enforce noise and light discipline. To deceive the enemy about the number of troop vehicles, start all vehicles simultaneously, and cover all track marks that lead into positions.

Nuclear/Chemical Vulnerability Analysis. Based on the NBC threat and the troop mission, determine the troop's vulnerability and incorporate those factors into the commander's planning.

Take the following actions to reduce the effects of a chemical or nuclear attack:

- **Improve positions.** Use available natural and man-made cover, such as folds in the earth, buildings, and dug-in positions, to protect personnel, equipment, materials, and supplies.
- **Alert personnel.** Inform soldiers to assume the appropriate MOPP level (see Table 7-1), based on the possibility of enemy attack and the MOPP analysis. Inspect unit and individual NBC equipment and prepare for operations.

Table 7-1. Mission-oriented protection posture levels.

MOPP LEVEL	OVERGARMENT	BOOTIES	MASK/HOOD**	GLOVES
0	Stored Nearby*	Stored Nearby*	Carried	Stored Nearby*
MASK ONLY	Carried	Carried	Worn	Carried
1	Worn, open or closed based on temperature	Carried	Carried	Carried
2	Same as MOPP 1	Worn	Carried	Carried
3	Same as MOPP 1	Worn	Worn, hood open or closed based on temperature	Carried
4	Worn, closed	Worn	Worn	Worn

*On combat vehicle or in fighting position.
**Includes M258A1 kit and detector paper.

- **Protect equipment and supplies.** Protect against chemical contamination inside a vehicle since it is very hard to decontaminate. Keep all hatches closed, if possible, and turn on vehicle overpressure system. Cover equipment and supplies stored outdoors or in tank bustles. Nonporous plastic sheets make good covers. Use tarpaulins if plastic is not available.
- **Position alarms.** Activate and emplace automatic alarms to provide the earliest possible warning of an attack. Position the alarms upwind and at least 150 meters away from stationary positions. Keep adjacent alarms about 300 meters apart.
- **Alert NBC personnel and teams.** Ensure NBC personnel and teams are available and prepared to perform their assigned missions.

Actions in Response to an Attack. Every soldier must be proficient at individual NBC basic skills. Masking is conducted immediately in response to these conditions:

- Chemical alarm sounds.
- Positive reading on detector paper.
- Individuals exhibiting symptoms of chemical or biological agent poisoning.

All soldiers mask automatically when an attack is possible, after the initiation of chemical/biological warfare, in response to artillery shells exploding in the area, and when aircraft drop bomblets or spray a mist or fog. In response to these conditions, all soldiers assume MOPP 4 and remain at maximum protection until the absence of contamination is confirmed or the commander has completed the analysis and determined a lower level of MOPP is appropriate.

Actions in response to a chemical or nuclear attack are standardized and included in the troop SOP.

- Initiate immediate masking and sound alarm (visual, audible, or voice).
- Continue mission.

- Assume MOPP 4 as rapidly as possible.
- Treat casualties.
- Send NBC-1 report and update it as more information becomes available.
- Identify the agent.
- Initiate decontamination operations.

These procedures must be rehearsed, and all individuals must adhere to them.

Actions After an Attack. Warn all subordinate elements immediately so they can take appropriate actions. In addition, alert squadron and all adjacent units, giving priority to the unit downwind from the attack. The troop and squadron are extremely vulnerable at the beginning of a chemical attack. Continue the troop mission while preparing for the enemy; he will take advantage of the chemical attack by conducting tactical operations.

In accordance with the troop SOP, prepare and send an NBC-1 report immediately. The personnel observing the attack send the report to the troop XO and NBC NCO in the command post. The XO forwards the report to the squadron TOC. Send the report immediately, and update it as more information is available. Treat any casualties of the attack using self-aid and buddy-aid, and report their status through the troop command post to the squadron S1. Operating in MOPP 4 will, over time, reduce the troop's effectiveness in combat operations. Soldiers become fatigued and lose mental and physical dexterity. Maintaining command and control becomes very difficult during extended MOPP 4 operations.

The troop must determine the type of agent used in the attack and the extent of contamination. Initiate tests with the M256 detector kit to identify the type of agent. The M256 kit detects and identifies field concentrations of nerve, blister, or blood agents. This test will help determine when unmasking may be safe after a chemical attack. Update the NBC-1 report with any information provided by the test. Decontaminate if required, and unmask as soon as possible.

When the troop is contaminated with persistent or semipersistent chemical agents, decontamination is necessary before reducing the MOPP level. Implement the following decontamination techniques:

- Start the skin decontamination within one minute of contamination.
- Apply decontaminating agents to equipment and vehicles to limit the spread of contamination and shorten the duration of the hazard. Use the M258A1 kit for personnel wipe down, and the M11 or M13 decontamination apparatus for vehicle/equipment spray down.

Decontaminate as far forward as possible. After leaving the contaminated area, perform hasty decontamination as soon as the situation permits. Hasty decontamination consists of exchanging MOPP gear and washing vehicles with hot, soapy water or high-pressure water to remove gross contamination. This operation is conducted at platoon or troop level, assisted by the decontamination team and other assets provided by the squadron. Select a site that is uncontaminated, near a water source, and concealed. Hasty decontamination may allow temporary relief from MOPP 4 for eating, drinking, and personal hygiene. Deliberate decontamination is required later when time and additional decontamination assets are available.

Initiate unmasking procedures as soon as possible after a chemical attack, to reduce the performance degradation that occurs while soldiers are in MOPP 4. Unmasking procedures differ based on the availability of detection equipment. When nonpersistent agents are used in the attack and an M256 detector kit is available, use the kit to determine when the agent is no longer present. Instruct two or three individuals to unmask for 5 minutes, remask, and sit in a shady area for 10 minutes. Check the individuals for chemical agent symptoms. If no symptoms appear, it is safe to unmask.

When no detector kit is available, direct two or three soldiers to perform the following procedures. Have them keep their eyes open, take a deep breath, hold it, and break the seal on their mask for 15 seconds. Reseal and clear the mask, and wait 10 minutes in a shady area. Check them for chemical agent symptoms. Have them break the seal and take three or four breaths and reseal the mask. Wait 10 minutes and check them for symptoms. Have them unmask for 5 minutes and remask for 10. Check for symptoms. If no symptoms appear, assume it is safe for all to unmask.

Following a persistent or semipersistent attack, begin unmasking procedures after hasty decontamination is complete and the M256 kit indicates no agent is present.

OPERATIONS IN A NUCLEAR ENVIRONMENT

Radiological Monitoring. This is the detection of radiation and the measurement of dose rate with radiac instruments. Monitoring finds a hazard that would otherwise go unmeasured or undetected. It is conducted by the troop radiological monitoring and survey team, which monitors—

- Total dose using the IM 93/147.
- Dose rate using the IM 174/PD.
- Food, water, and personnel contamination using the AN/PDR-27.

Conduct periodic monitoring when directed, or after nuclear war breaks out. Monitor a specific point in the area at least once an hour. Start continuous monitoring after receiving a fallout warning; when conducting a movement; when a nuclear burst is reported, seen, or heard; after detecting radiation above 1 cGy/hr during periodic monitoring; or when the squadron commander orders. Stop continuous monitoring when the squadron commander orders or, except while on the move, when the dose rate falls below 1 cGy/hr.

Based on NBC-4 monitoring reports, the troop may conduct a radiological survey to determine the degree and extent of contamination. This information is forwarded to squadron in an NBC-4 report and on DA Form 1971-R.

Exposure Control. While operating in a nuclear environment, the troop must control exposure to nuclear radiation. An operational exposure guide is a method of determining the maximum radiation dosage to which platoons can be exposed and still accomplish a mission. Determine radiation exposure by the cumulative dose or the platoon's radiation history.

Platoons record and report their total dose radiation daily, per SOP, to the troop command post where the NBC NCO compiles the data. The dose is determined by averaging the readings from both platoon dosimeters and rounding to the nearest 10. The NBC NCO totals each platoon's average dosage, and coverts it to a radiation status (see Table 7-2). He maintains a record of the radiation exposure status of each platoon and submits a daily radiation exposure status report to the squadron (see Table 7-3).

Table 7-2. Radiation dose categories.

RADIATION DOSE CATEGORIES

THE TABLE INDICATES THAT THE NUMERICAL VALUE OF cGys RECEIVED BY EACH PLATOON IS CONVERTED TO A RADIATION EXPOSURE STATUS (RES) FOR EACH PLATOON. THE CONVERSION FACTORS ARE SHOWN BELOW.

RADIATION STATUS CATEGORY	TOTAL CUMULATIVE DOSE (cGy)
RES-0	0 (NO EXPOSURE)
RES-1	GREATER THAN 0 BUT LESS THAN OR EQUAL TO 70
RES-2	GREATER THAN 70 BUT LESS THAN OR EQUAL TO 150
RES-3	GREATER THAN 150

The regimental and squadron commanders may move units on the battlefield, or relieve a particular unit that has been exposed to a high level of radiation. The squadron commander specifies the degree of exposure that units will not exceed for each operation. There are three degrees of radiation exposure risk—negligible, moderate, and emergency. Each level of risk is correlated to unit radiation status (see Table 7-4).

Table 7-4. Unit radiation status.

RADIATION STATUS (RS) CATEGORY	TOTAL PAST CUMULATIVE	EXPOSURE CRITERIA FOR A SINGLE OPERATION (cGy)
RS-0 UNITS	NO EXPOSURE	NEG RISK: 50 MOD RISK: 70 EMERG RISK: 150
RS-1 UNITS	GREATER THAN 0 BUT NOT MORE THAN 70	NEG RISK: 50-PAST DOSE MOD RISK: 70-PAST DOSE EMERG RISK: 150-PAST DOSE
RS-2 UNITS	GREATER THAN 70, NOT MORE THAN 150	ANY FURTHER EXPOSURE IS CONSIDERED EMERG RISK
RS-3 UNITS	GREATER THAN 150	ALL FURTHER EXPOSURE WILL EXCEED EMERG RISK

1. RADIATION EXPOSURE STATUS CATEGORIES ARE BASED ON PREVIOUS EXPOSURE TO RADIATION.

2. RECLASSIFICATION OF UNITS FROM ONE RADIATION EXPOSURE STATUS CATEGORY TO A LESS SERIOUS ONE IS MADE BY THE COMMANDER UPON ADVICE OF THE MEDICAL PLATOON LEADER AFTER AMPLE OBSERVATION OF ACTUAL STATE OF HEALTH OF THE EXPOSED PERSONNEL.

3. ALL EXPOSURES TO RADIATION ARE CONSIDERED TOTAL BODY AND SIMPLY ADDITIVE. NO ALLOWANCE IS MADE FOR BODY RECOVERY FROM RADIATION INJURY.

4. NEGLIGIBLE RISK WILL CAUSE NO CASUALTIES AND A LESS THAN 2.5 PERCENT INCIDENCE OF VOMITING. MODERATE RISK WILL RESULT IN A LESS THAN 5 PERCENT INCIDENCE OF VOMITING AND HEADACHES, BUT WILL CAUSE NO CASUALTIES. EMERGENCY RISK WILL RESULT IN LESS THAN 5 PERCENT CASUALTIES AND A HIGHER PERCENTAGE OF NAUSEA, VOMITING, AND HEADACHES.

Minimize the Effects of Friendly Nuclear and Chemical Attacks. A NUCWARN/CHEMWARN message is normally not sent below squadron level. Squadron alerts the troops using a code word prescribed in the OPORD. Squadron will also issue a FRAGO containing relocation instructions, the level of protection required, and the time of the friendly strike. To assure maximum reaction time, each soldier in the troop must be informed. Each crew immediately prepares its vehicles, equipment, and position for the attack as is appropriate to its location within the attack area. Crews prepare their vehicles by taking the following actions:

- Move if necessary.
- Position vehicle behind best available cover with front of vehicle toward the blast.
- Point gun away from blast.
- Lock brakes.
- Secure loose equipment in vehicle.
- Secure inside the vehicle all exterior components that could be damaged by the blast.
- Close and lock all hatches, to include ballistic shields.
- Wear helmets and protect eyes.
- Turn off and disconnect all radios but one, which is needed to retain communications with squadron.

Complete these tasks within five minutes of receiving notification. After the friendly strike, continue radiological monitoring and report if necessary. Quickly prepare to continue operations.

If the troop is contaminated with fallout, the level of contamination is monitored by the monitor and survey teams. Decontamination is accomplished by brushing or wiping the dust off the equipment. Brush off or shake out MOPP gear and wash down the equipment and vehicles to reduce radiological contamination. Use the AN/PDR-27 to determine decontamination requirements and to determine if supplies of food and water must be destroyed.

Chapter 8

Combat Support

The effective integration of combat support (CS) will spell the difference between success and failure on the battlefield. The commander must know the capabilities and limitations of CS assets, and how to employ them properly.

CONTENTS

Section I. The Fire Support Team

CAPABILITIES AND LIMITATIONS

The FIST is organized, equipped, and trained to provide—

- A fire support advisor/coordinator—the FSO.
- An observation/laser designation capability.
- A communications link to all available fire support.

The FIST has secure FM communications on four radios, and digital communications on two nets through the use of digital message devices (AN/PSG-5 and AN/PSG-2). These capabilities enable the FIST to maintain communications on the following nets (see Figure 8-1):

- Troop command net.
- Troop fire support net.
- Supporting artillery fire net (digital).
- Squadron fire support net.

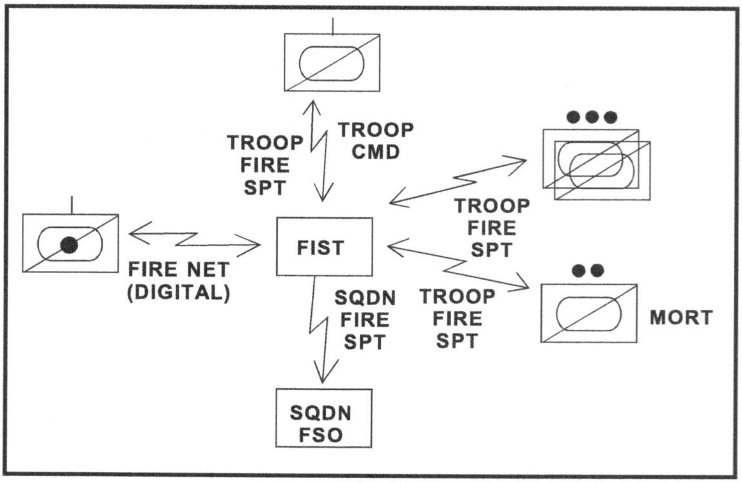

Figure 8-1. Fire support team communication nets.

The FIST is also equipped with a target laser locator designator set to designate targets for terminally guided munitions and to determine target locations accurately at ranges up to 10 kilometers (limited by line of sight). The target designator set is equipped with a thermal sight to use under limited visibility conditions.

The digital message devices allow the FIST to transmit and receive high-speed digital messages from the howitzer battery computer system, the direct support artillery battalion TACFIRE, OH-58D aircraft, and the squadron FSO's variable format message entry devices, among other digital message devices. Digital message devices can quickly transmit standard fire requests, fire adjustments, and plain text messages. The FIST can store nine fire requests on the digital message devices for rapid transmission in support of troop operations.

The FIST's abilities may be constrained by its communications with the supporting field artillery (FA). To coordinate any fire support for the troop, the FIST must be positioned where it can maintain at least a voice communications link to the supporting artillery. The FIST may not be able to see the battlefield and use the laser designator capability in all situations. When the troop is operating in a large sector, the FIST will be able to see only a small portion of the area. The scouts, therefore, must be proficient in requesting indirect-fire support. The scouts' calls for fire will be relayed through the FIST to the appropriate fire support unit.

EMPLOYMENT TECHNIQUES

Planning and Coordination. Fire support planning and coordination begin on receipt of a mission, and continue throughout its planning and execution. The troop FSO should always accompany the commander to mission briefings at squadron, so that he is fully aware of the squadron scheme of maneuver and the availability of fire support assets. He and the squadron FSO can also begin coordinating any specific requirements for fire support, such as preparations, groups, and series.

The troop commander's guidance should include the following to allow the troop FSO to plan the operation:

- What is the purpose of FA and mortar fires?
- How will FA fires support the troop's maneuver?
- What are the attack criteria? Will the commander use FA for immediate suppression, screening, or destruction?
- Are the attack and engagement criteria for mortars different from FA?
- On what and where does the commander want to mass FA and mortar fires?

Plan indirect fires on known or suspected enemy positions; at choke points; on the troop's objective or flanks; or on any other areas that would support the scheme of maneuver. These targets can be fired at a specific time, on call, or when a particular event takes place.

The FSO incorporates his approved fire plan into a target list and forwards it to the squadron FSO. The squadron FSO compiles the lists from all the troop FSOs, eliminates any duplicate targets, forwards the list to the supporting artillery, and informs the troop FSOs of any changes to their target lists.

Once he has completed his plan, the FSO will brief the commander on planned targets, scheduled fires, and fire support coordination measures that will be in effect. The commander will make the final decision on the fire support plan. Ensure it complements the scheme of maneuver and maximizes the available firepower for the troop. Once approved, the fire plan is continuously modified and updated to meet changing situations.

Silence is Consent. Once the FSO has briefed the commander and the commander approves the troop fire support plan, calls for fire are executed during the conduct of the operation by the silence is consent method. This method allows subordinate leaders within the troop to request indirect fires in support of their maneuver within the framework of the commander's guidance and the troop fire support plan. All initial requests for fires will go over the troop command net and the commander will simply acknowledge the request with "ROGER—OUT." The troop FSO will process the call for fire through the appropriate indirect-fire system based on the commander's guidance. All adjustments and corrections for the fires will be communicated on the troop fire direction net directly between the observer and the troop FSO. If the request for fire that is submitted by the subordinate leader is not within the commander's fire support guidance, the troop commander will deny fires or modify his guidance based on the situation.

Execution of the Fire Support Plan. During the operation, the FSO is responsible for—

- Keeping the commander advised of all fire support matters.
- Resolving any fire support conflicts that may arise during the planning and execution of the operation.
- Attacking targets with the most suitable fire support means available to support the scheme of maneuver.

Also, the commander may make the FSO responsible for positioning the troop mortars to best support the operation.

The FSO operates on the four radio nets mentioned previously. The scouts in the troop normally serve as the observers calling for indirect-fire support. The FIST monitors all spot reports on the troop command net and, as necessary, prepares calls for fire. The FSO will allocate fires from mortars and artillery, based on the commander's guidance. The commander may give the FSO additional guidance concerning a fire mission as it is requested (see *Silence is Consent* paragraph). Further information about the fire mission and all other calls for fire are passed on the troop fire support net. Keep the troop command net free for other traffic.

Both the FIST and the mortars monitor the troop fire support net. When scouts forward adjustments or corrections for indirect-fires, they use the troop fire support net. If the FSO forwards the mission to supporting artillery, he sends the message "LIGHTNING UP" on the troop fire support net. This tells the scouts calling for fire and the mortars that the mission is being fired by supporting artillery and that all adjustments and the "end of mission" report must be relayed through the FSO. If the mortars will fire the mission, the FSO sends the message "THUNDER DOWN" on the troop fire support net. This tells the mortar squad to fire the mission, and tells the scouts to send all adjustments and the "end of mission" report to the mortar squad (see Figures 8-2 and 8-3).

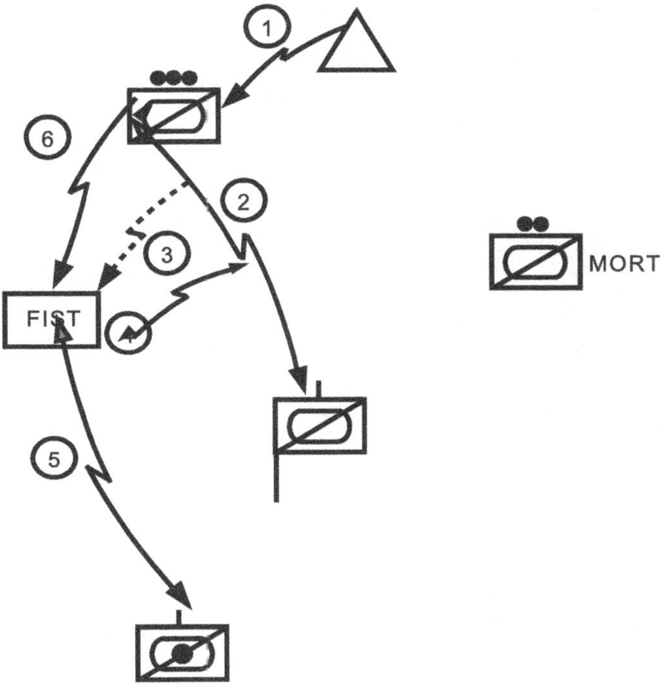

(1) SCOUTS SEND SPOT REPORT TO PLATOON LEADER (PLATOON NET).

(2) PLATOON LEADER FORWARDS THE SPOT REPORT TO THE CP (TROOP COMMAND NET) WITH REQUEST FOR FIRE.

(3) FIST MONITORS SPOT REPORT, RECEIVES PERMISSION TO FIRE THE MISSION, AND DETERMINES WHAT ARTILLERY TO USE.

(4) FIST SENDS MESSAGE "LIGHTNING UP" ON TROOP COMMAND NET. SCOUTS AND MORTARS MONITOR.

(5) FIST SENDS DIGITAL CALL FOR FIRE TO DS ARTILLERY. FIST SENDS SHOT ON TROOP COMMAND NET.

(6) SCOUTS RELAY ALL ADJUSTMENTS TO FIST ON TROOP FSN.

Figure 8-2. Fire request channels (artillery).

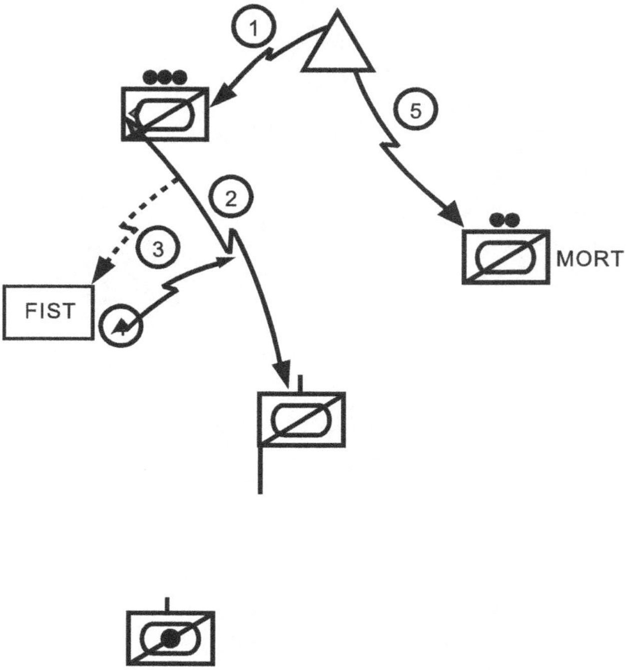

① SCOUTS SEND SPOT REPORT TO PLATOON LEADER (PLATOON NET).

② PLATOON LEADER FORWARDS THE SPOT REPORT TO THE CP (TROOP COMMAND NET) WITH REQUEST FOR FIRE.

③ FIST MONITORS SPOT REPORT AND RECEIVES PERMISSION FROM THE TROOP CDR TO SUPPORT WITH INDIRECT FIRES.

④ FIST SENDS MESSAGE "THUNDER DOWN" ON TROOP COMMAND NET. SCOUTS AND MORTARS MONITOR. FIST SENDS SHOT ON TROOP COMMAND NET.

⑤ SCOUTS CALL THE MORTARS ON TROOP FSN TO REQUEST INDIRECT FIRES.

Figure 8-3. Fire request channels (mortar).

Section II. Mortar Support

CAPABILITIES AND LIMITATIONS

The mortar section provides a heavy volume of accurate, sustained indirect fires in response to the needs of the troop. They are ideal weapons for attacking targets on reverse slopes, in narrow ravines, built-up areas and in other areas that are difficult to strike with low-angle fires. Mortars are most effective for—

- **Suppression**. Suppression forces the enemy to button up or move to less advantageous positions. High explosive (HE) ammunition with a variable-time fuze or quick fuze does this best.

- **Screening.** Obscuration smoke is placed directly on or just in front of enemy positions to obscure vision. Screening smoke is placed between troop positions and the enemy to conceal the troop's movements. Smoke will degrade the effectiveness of laser range finders, so the tank platoons must adjust their gunnery techniques when engaging the enemy.

- **Illumination.** The use of illumination permits daytime engagements during some limited visibility conditions by improving the effectiveness of image intensification devices. Illumination rounds are used to mark targets for close air support and attack helicopters, and to allow observers to adjust artillery onto a target. Illumination rounds are usually adjusted so they illuminate above or behind the enemy location to silhouette him.

The height of the illumination should be adjusted to burn out about 200 meters above the ground. This gives maximum illumination time, about 90 seconds for one round, and will not degrade the effectiveness of image intensification devices. Because illumination rounds light the battlefield for both sides, it should be used when night-vision devices are not available in sufficient quantities or when ambient light levels are very low.

EMPLOYMENT TECHNIQUES

The troop mortar section is employed as a separate element during operations. It moves independently of the platoons, and provides its own security. The troop commander normally designates positions for the section sergeant and provides guidance for support forward of the scouts. The commander may find that having the FIST position the mortars is easiest and most effective. This technique allows the mortar section sergeant the ability to concentrate on providing fires and computing fire data and the commander to focus on maneuvering the troop, not one section.

The mortars can operate on two nets. If the FIST is responsible for moving the section, both nets may be on the troop fire support net. The FSO will keep them updated on the situation. However, since many spot reports over the troop net become fire missions, the mortars can eavesdrop and provide more responsive fires if they stay abreast of the situation themselves. If the commander gives the mortar section sergeant authority to select positions and to move the section, he must operate on the troop command net to stay abreast of the situation and on the troop fire support net.

Offense. During offensive operations the movement of the mortars will be based on the progress of the scout platoons. While the section is on the move, it must be prepared to provide immediate fires using direct lay, direct alignment, or hip shoot. Plan the moves of the section so it is in position to support the troop at critical times, such as during river crossings and counterattacks, and when clearing defiles.

Defense. Considerations for using mortars in the defense are similar to those for using them in the offense. Plan on prestocking ammunition at subsequent positions to reduce resupply problems during a defensive operation.

Reconnaissance and Security. During reconnaissance and security missions, the troop will often be operating over a large frontage that cannot be completely covered by the section. The commander must decide whether to position the mortars to cover the most dangerous area, or to move them to a position where they can cover a part of different areas and adjust as necessary. Knowing what other artillery assets are available will help in making that decision.

Commander's Guidance. A list of considerations the commander can use to help plan mortar section employment follows:

- Ammunition constraints by type and quantity.
- Priority of fire to a designated platoon.
- Anticipated changes in mortar employment.
- Communications constraints.
- General designation of positions.
- Movement guidance.
- Coordination requirements.
- Resupply.

Section III. Artillery Support

ORGANIZATION AND SUPPORT RELATIONSHIP

The howitzer battery organic to the regimental cavalry squadron consists of eight 155-mm self-propelled howitzers. The squadron will often be supported by an artillery battalion. The artillery battalion usually has three batteries of 155-mm self-propelled howitzers, and provides direct support to the squadron.

CAPABILITIES AND LIMITATIONS

Field artillery provides the troop its main fire support. It gives accurate fires with a wide variety of munitions. Field artillery adds a powerful dimension to troop direct-fire and maneuver capabilities.

- Artillery can—
  - Provide fire support under all weather conditions and terrain types.
  - Shift and mass fires rapidly without the requirement to displace.
  - Support the battle in depth with long-range fires.
  - Provide a variety of conventional shell/fuze combinations.
  - Provide continuous fire support by careful displacement.
  - Be as mobile as the supported unit.

- Artillery has—
  - Limited capability against moving targets.
  - Limited self-defense capability against air and ground attack.
  - Limited capability to destroy point targets without considerable ammunition expenditure. (Copperhead is the exception. It can be used to engage point targets.)
  - Vulnerability to detection by enemy target acquisition systems because of firing signatures.

Field artillery has a wide variety of munitions tailored for the engagement of different types of targets. The ammunition types include—

AMMUNITION	TARGET
High explosive	Personnel, field fortifications, vehicles
HC smoke	Obscuration and screening
White phosphorus	Obscuration
Illumination	
Cannon-launched guided projectiles (Copperhead)	Armored vehicles or high-payoff targets requiring precision target engagement

In addition, the troop must become acquainted with the characteristics of improved conventional munitions (ICM) and scatterable mines. ICMs include antipersonnel (APERS) and dual-purpose ICM. The commander must consider the danger to friendly troops in areas in which APERS is fired. Also, the high ICM dud rate makes maneuver hazardous in the area of an ICM field. Scatterable mines are area denial munitions for use against personnel and remote antiarmor mines for use against armored vehicles.

EMPLOYMENT TECHNIQUES

Artillery units normally provide direct support to the squadron, and are positioned by the artillery commander in close coordination with the squadron S3 and FSO. The troop FSO is the troop's link to the supporting artillery (see Section I, The Fire Support Team, and Chapters 1 and 2).

Section IV. Engineer Support

ORGANIZATION AND SUPPORT RELATIONSHIP

A combat engineer company is organic to the heavy and light armored cavalry regiment. Additional engineer assets are often provided from divisional and corps engineer units to support the regiment. Up to a platoon of engineers is frequently attached to or placed in direct support of a cavalry troop for combat operations. A combat engineer platoon from the heavy regimental engineer company is organized for combat as shown in Figure 8-4.

HEAVY ENGINEER CAPABILITIES AND LIMITATIONS

A combat engineer platoon is uniquely equipped and trained to conduct mobility, countermobility, and survivability operations in support of troop operations. The platoon headquarters has two M9 armored combat earthmovers, which are highly mobile, armored, and amphibious (see Figure 8-5).

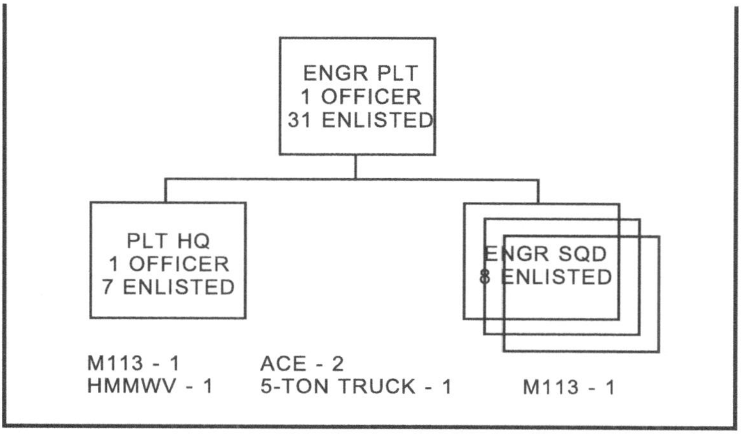

Figure 8-4. Regimental engineer company (heavy) organization.

Figure 8-5. Armored combat earthmover, M9.

In mobility operations, the engineer platoon can provide—

- Obstacle reduction. The engineers can reduce or negate the effects of obstacles to improve the troop's ability to maneuver.
- Route construction. The engineers can construct, improve, and maintain roads, bridges, and fords.

In a countermobility role, the engineers can assist with obstacle construction. The engineers can reinforce terrain by constructing obstacles to delay, canalize, disrupt, and destroy the enemy force to support the scheme of maneuver.

In survivability operations, the engineers can improve positions by constructing berms, dug-in positions, and overhead protection to reduce the effectiveness of enemy weapons.

A cavalry troop is frequently supported by additional engineer assets, such as an armored vehicle launched bridge (AVLB) and combat engineer vehicles (CEV) from the engineer company. The AVLB section of headquarters and headquarters troop of the armored cavalry squadron consists of three AVLBs that could be available during combat operations.

An AVLB is a tank chassis modified to transport, launch, and retrieve a 60-foot class-60 bridge. The bridge is capable of carrying military load class-60 track vehicles across a 17-meter gap and military load class-70 track loads across a 15-meter gap (see Figure 8-6).

Figure 8-6. Armored vehicle launched bridge.

The CEV is an M60A1 tank mounted with a hydraulically operated dozer blade, a 165-mm turret-mounted demolition gun, a retractable boom, and a winch (see Figure 8-7).

Figure 8-7. Combat engineer vehicle.

The small emplacement excavator (SEE) is a small highly mobile truck with a front-end loader and a backhoe (see Figure 8-8).

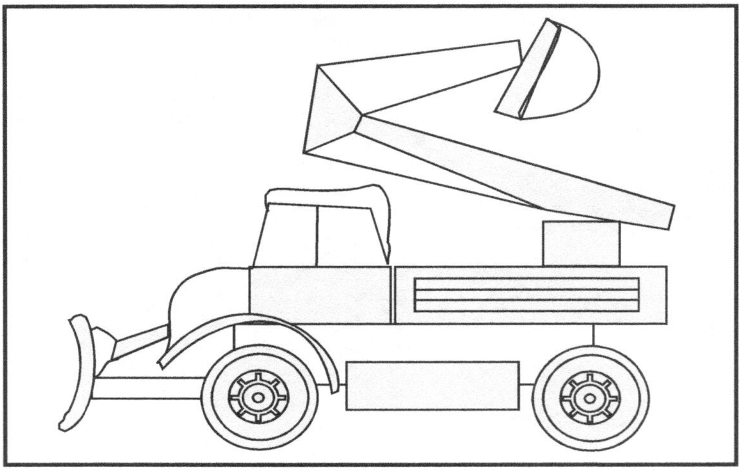

Figure 8-8. Small emplacement excavator.

LIGHT ENGINEER CAPABILITIES AND LIMITATIONS

The engineer platoon from the light regimental engineer company is equipped with five HMMWVs. The platoon can accomplish most of the same tasks that a heavy platoon can, but the time to complete most jobs is greater. The platoon can provide the troop with the following:

- In mobility operations.
  - Obstacle reduction. Engineers can reduce or negate the effects of obstacles to improve the troop's maneuver ability. To improve the platoon's ability, they may be reinforced with an M9 ACE, SEE, MICLIC, and Bangalor torpedo. The engineer company from the light regiment does not provide the regiment (troop) with any bridging capability.
  - Route reconnaissance. Engineers can conduct route reconnaissance.

- In a countermobility role, the engineers are usually left under squadron control to complete priorities of work in accordance with the squadron obstacle plan. The engineers can reinforce terrain by constructing obstacles to delay, canalize, disrupt, and destroy the enemy force to support the scheme of maneuver.

- In survivability operations, the engineers can improve positions by constructing berms, dug-in positions, and overhead protection to reduce the effectiveness of enemy weapons (see Figure 8-9).

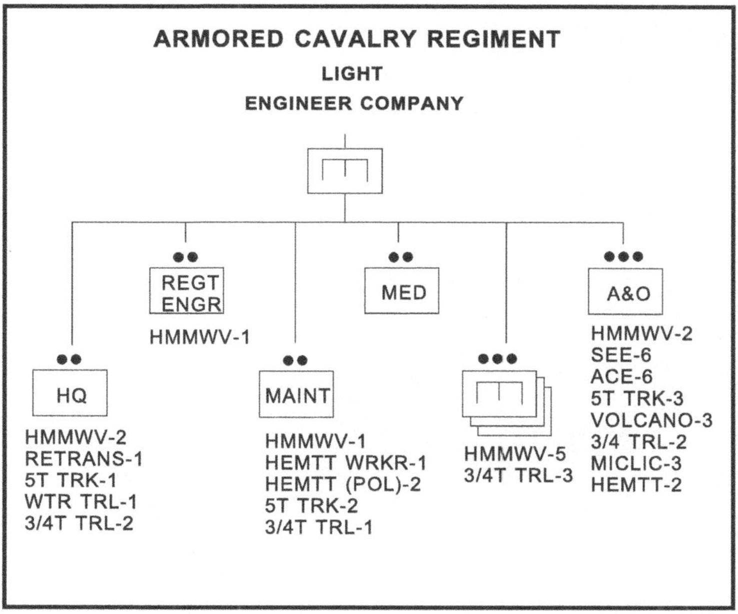

Figure 8-9. Regimental engineer company (light) organization.

EMPLOYMENT TECHNIQUES

The troop has several options in the task organization and employment of engineer assets to support its operation. Normally, the troop will keep the entire platoon under troop control and assign missions just as with a maneuver platoon. The commander can also task organize the engineer platoon, and assign squads to scout platoons to support the operation. Whatever method is chosen, the engineer platoon leader needs to be involved in the troop planning process.

Section V. Ground Surveillance Radar Support

ORGANIZATION AND SUPPORT RELATIONSHIP

A GSR team is organized for combat as shown in Figure 8-10. The team(s) is normally attached or OPCON to a troop during combat operations. GSR is found only in heavy regiments and divisions.

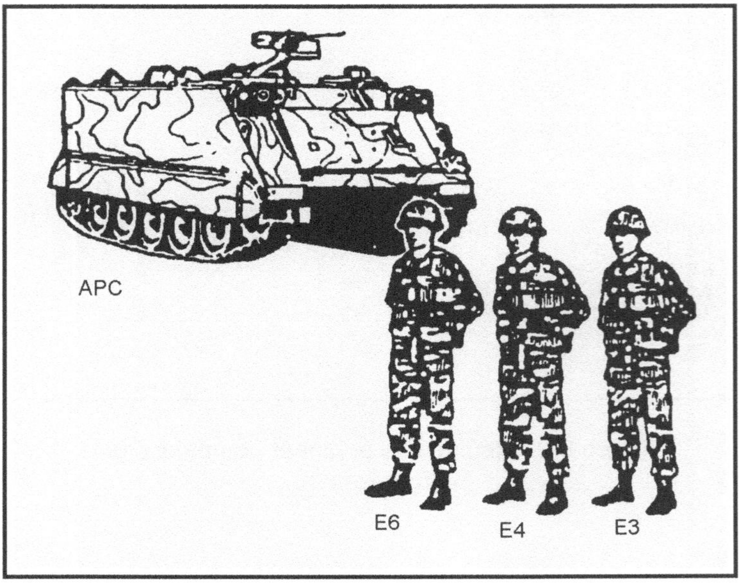

Figure 8-10. Ground surveillance radar team.

CAPABILITIES AND LIMITATIONS

GSRs provide mobile, all-weather battlefield surveillance. GSRs employed in pairs can provide observation from a given vantage point 24 hours a day, detecting targets and providing accurate range and azimuth readings.

The AN/PPS-5 has a line-of-sight range of 10,000 meters against vehicles and 6,000 meters against personnel. A GSR can detect targets through light camouflage, smoke, haze, light snow and rain, darkness, and light foliage. Heavy rain and snow seriously restrict its radar detection capability.

The GSR is generally ineffective against air targets unless the aircraft is flying close to the ground. The GSR is designed to detect targets moving in the presence of a background. It is vulnerable to enemy direction finding and jamming.

EMPLOYMENT TECHNIQUES

The troop will normally employ supporting GSR teams to augment the surveillance capabilities of its scout platoons. The commander should assign each GSR team a specific sector of surveillance and a frequency of coverage, or give the scout platoon leader guidance for employing the GSR team. The enemy can detect radar signals, so the GSR cannot conduct continuous surveillance. The tasks assigned GSR surveillance teams include—

- Searching avenues of approach or possible enemy positions on a time-schedule basis or randomly to determine location, size, composition, and nature of enemy activity.
- Monitoring point targets such as bridges, defiles, and road junctions and reporting quantity, type, and direction of enemy movement through the target.

- Extending the observation capabilities of scouts by enabling them to survey distant points and areas of special interest.
- Vectoring patrols to keep them oriented during times of limited visibility.

Position the GSRs in an area that is free of ground clutter and in one that affords them long-range observation and a wide field of view. Normally, they should be assigned a general area by the troop commander or scout platoon leader and the GSR team leader will select the specific location. The team should be prepared for rapid displacement, and have several alternate positions selected and reconnoitered to avoid enemy suppressive fires.

Offensive Operations. Employ GSRs in the offense to augment reconnaissance and security efforts. Fast-moving operations may preclude the continuous, effective use of GSRs.

In a movement to contact or a zone reconnaissance, employ GSRs on high-speed avenues of approach to the front or flanks of the troop to provide information on enemy movement. A GSR team can provide long-range observation while the scouts conduct a detailed reconnaissance of the local area. Employ GSR teams on the flanks of the troop's movement in pairs so they can bound forward. In this manner, the GSR can provide continuous early warning and keep pace with the operation.

Defensive Operations. In defensive operations, employ GSR teams to augment scouts. Position them to provide long-range observation on expected avenues of approach or to maintain surveillance of flanks.

Position the GSR teams behind the screen line if the terrain affords them long-range observation. Use GSR to assist scouts in maintaining contact with the enemy as the scouts deploy to subsequent screen lines. This also reduces their vulnerability to enemy fires.

Section VI. Air Defense

SUPPORT RELATIONSHIP

An air defense artillery (ADA) battery is organic to the regiment. For combat operations, these assets are normally attached or direct support to the squadron. The squadron commander may provide a slice of ADA to the troop.

Air defense assets are scarce, so the troop cannot plan on dedicated air defense protection. It is the troop's responsibility to protect itself from enemy air attack. The commander must take measures to avoid enemy air attack, to limit the damage if attacked and, if necessary, to fight back.

AIR DEFENSE EMPLOYMENT

Include air defense planning in all tactical operations. Air defense includes passive and active air defense and the employment of ADA assets.

Passive Air Defense

Passive air defense measures are the troop's first line of defense against enemy air attack. They include all measures, other than active defense, taken to minimize the effects of hostile air action. There are two types of passive air defense—attack avoidance and damage limiting measures.

Attack Avoidance. If the enemy pilot cannot find the troop, he cannot attack it. Use concealment, camouflage, obscuration, and deception to hide from the enemy.

Select positions that provide good concealment. When adequate concealment is not available, camouflage the vehicles to blend with their surroundings. Cover all track marks leading into the troop's position. Cover all shiny objects that could reflect light and attract attention. Take all actions necessary to reduce the enemy's ability to find the troop.

Damage Limiting Measures. Damage limiting measures are taken to reduce the effects of enemy air attack. Disperse vehicles while occupying static positions such as assembly areas or when preparing to cross a water obstacle or a breached obstacle. Dispersion reduces the effects of munitions. While on the move, air guards must be alert for enemy air attacks. When an enemy air attack is identified, quickly disperse, go to a concealed position, if possible, and then stop moving. A stationary vehicle is more difficult to see than a moving vehicle, and dispersion will reduce the effect of the air attack.

Use natural or man-made cover to reduce the effects of enemy air attack. Use folds in the earth, depressions, buildings, and sandbagged positions as damage limiting cover. These methods to reduce the effects of air attack must be used scrupulously by troop trains and command posts to reduce their vulnerability on the battlefield.

Active Air Defense

Although passive measures are the first line of defense against air attack, the troop must be prepared to fight back if necessary. The troop gains two things by fighting back. It may kill or drive off the attacker, and the morale and spirit of the troops improve by fighting back.

The decision to fight an air threat is based on the situation and the capabilities of the troop's weapon systems. Generally the troop may defend itself against direct attack, but unless attacked, it does not engage aircraft unless directed to do so by the squadron commander.

Stinger/Avenger Employment

Stingers/Avengers are used most effectively in point defense. However, mobility during tactical operations requires that Stinger crews move with the troop during maneuvering to provide the best possible air defense.

The troop commander must determine the air defense priorities for each phase of the operation and brief the Stinger/Avenger section chief accordingly. In order for the section chief to design an effective troop air defense plan for the commander, the commander must furnish the section chief the mission, objective, routes of march, intended scheme of maneuver, battle formations, and intended response to hostile air activity.

Unit air defense falls under two major categories: air defense of a fixed asset (assembly area or defensive position), and air defense of a mobile asset (convoy or maneuvering troop).

Fixed Asset. The air defense section leader places his crews from 2 to 4 kilometers apart near the troop's position, which provides overlapping and mutually supporting fires for troop defense. He provides command and control to his section, and air defense early warning to the troop unit and his deployed crews. The commander must provide logistical support for the section, indirect protective fires for their defense, and security within the troop defensive perimeter during hours of darkness.

Mobile Asset. While the unit moves, either in a tactical road march or during movement to contact, the air defense crews must move with each element to provide uninterrupted air defense protection.

Convoy operations require the air defense section leader to weight his crews near the front and rear of the troop's column(s), and to spread the remainder of his crews evenly throughout the column(s). While traveling, he provides command and control to his crews and air defense early warning to the troop and his crews.

Section VII. Army Aviation Support

SUPPORT RELATIONSHIP

Army aviation units are seldom OPCON to the troop; however, the troop frequently conducts joint operations with air cavalry or attack helicopter units within the troop area of operations. Air and ground units work together to make an effective team in conducting cavalry operations.

AVIATION TROOP COMMANDER

Air operations are controlled and coordinated by one man, the aviation troop commander. He coordinates directly with the ground troop commander, on the command net, when air operations extend into the troop's area of operations. When time is available, coordinate face-to-face to prepare for an operation. Let him know—

- Situation.
  - Enemy. Ground and ADA units by type and location.
  - Friendly. Locations and FLOT to include supporting artillery.
- Mission.
- Execution.
  - Scheme of maneuver.
  - Fire support.

The aviation troop commander tells the ground troop commander—

- His capabilities.
  - Number and type of aircraft.
  - Armament.
  - Time available.

- Execution. Scheme of maneuver to include approach direction into area.
- Command and signal.
  - Aviation troop commander, frequency, and call sign.
  - Succession of command and designated platoon leader in the troop commander's absence.

The aviation troop commander reports all spot reports and fire requests through the troop command post or FIST while operating in the ground troop's area of operations.

Reconnaissance. The capabilities of air and ground scouts complement each other. Air scouts can quickly reconnoiter a large area while ground scouts are conducting a detailed terrain reconnaissance. Employ the air scouts forward to provide information about areas where the ground scouts need to concentrate their efforts, such as at possible fording sites or route restrictions. The ground scouts can focus their efforts where the air scouts have directed them, instead of conducting a complete and time-consuming reconnaissance. The air scouts can also provide early warning of enemy movement to the front or flanks of the troop.

Security. During screen missions, employ the air assets to the front or flanks of the troop to observe areas between OPs and those areas that are difficult to observe from the ground. The air scouts can place effective indirect fires on the enemy and in maintaining contact with the enemy when the ground scouts move to subsequent screen lines. In offensive and defensive operations, Army aviation will normally work for the squadron to complement the squadron scheme of maneuver.

Section VIII. Close Air Support

CAS sorties are usually allocated to the regiment by the corps commander to meet preplanned requests submitted by the regimental fire support element. CAS for the troop is obtained by submitting an immediate CAS request.

CAS is requested by submitting an immediate request to the squadron air liaison officer. When CAS is on station, the tactical air control party may talk directly to the troop commander to determine enemy strengths, descriptions, and locations, as well as friendly unit locations. A forward air controller (FAC) normally directs the air strike. In the absence of a FAC, the troop FSO is trained to direct air strikes.

The ground troop may be required to mark the target for CAS aircraft. Use the mortar section for this. Fire white phosphorus or illumination rounds set for a ground burst. The FIST can designate targets for aircraft equipped with laser-spot trackers, such as Pave Penny or laser-guided bombs.

CAS capabilities are listed below.

- High speed and long range.
- Versatile weapon and ammunition mix.
- Accurate delivery.
- Excellent air-ground communications in A-10 and A-16 aircraft.
- Can locate and strike moving targets.

CAS limitations include—

- Limited resources.
- Delivery restrictions caused by limited visibility and weather or the proximity of friendly forces.
- Flight restrictions imposed by enemy air defense.
- Delayed response and short time on station.

Chapter 9

Combat Service Support

CONTENTS

Mission accomplishment depends on the troop's ability to obtain, transport, and distribute fighting resources such as fuel, ammunition, replacement personnel, and rations. Dependable ways to treat and evacuate wounded soldiers are crucial to good morale. Evacuation and repair of combat equipment sustain troop combat power and readiness to fight.

Section I. Organization

Troop trains are organized as combat trains and field trains. The combat trains are made up of immediate battlefield service support. The rest of the troop CSS elements are collocated with squadron CSS assets in the squadron field trains and at the unit maintenance collection point (UMCP). The troop CSS team consists of the troop XO, first sergeant, supply sergeant, communications sergeant, maintenance sergeant, and senior troop aidman.

There is no dedicated service support radio net at troop level. All logistical reports and initial requests for logistical support are conducted on the troop command net. Additional coordination is conducted on the platoon radio nets. Routine reports are sent before and after combat operations, and are delivered by messenger when possible. All service support coordination with squadron is conducted on the squadron administration and logistics (A/L) net. The first sergeant operates on this net. The command post does not routinely operate on the A/L net, but uses it when necessary to forward reports and to conduct logistical coordination.

COMBAT TRAINS

The troop combat trains provide CSS for the troop during combat operations. They are organized for combat as shown in Figures 9-1 and 9-2.

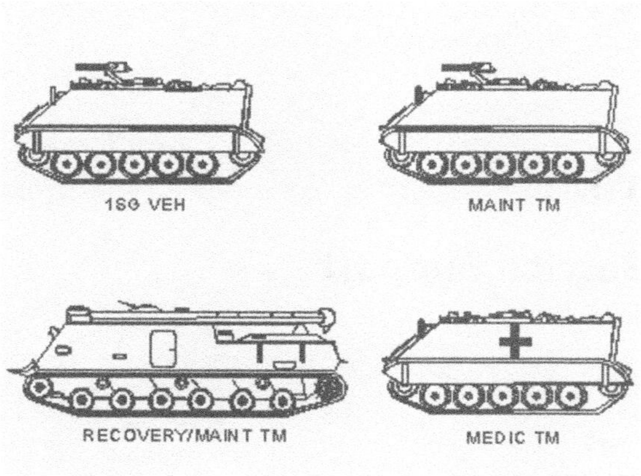

Figure 9-1. Heavy troop combat trains.

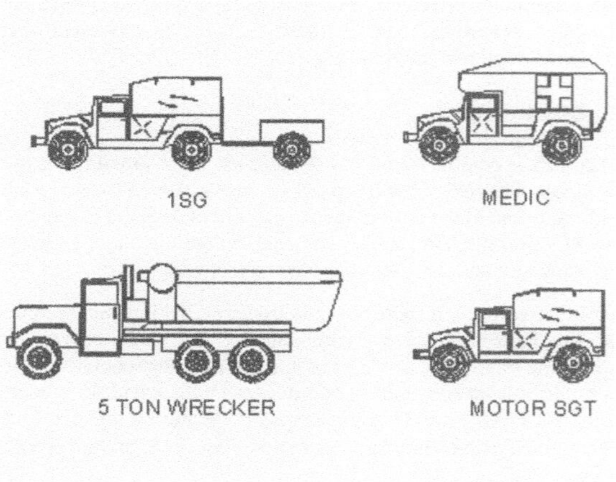

Figure 9-2. Light troop combat trains.

The first sergeant, assisted by the maintenance sergeant, directs movement and employment of the combat trains. He receives logistical reports directly from the platoon sergeants or through the troop command post, coordinates logistical matters with the XO, and assists the XO in ensuring the troop is logistically ready for battle. During operations, the first sergeant directs aidman teams and recovery teams forward as needed, leads the troop LOGPAC forward from the logistics release point (LRP) to resupply the troop and then leads it back to the LRP, and organizes the LOGPAC to conduct troop resupply operations. He also directs the medical evacuation of soldiers who are wounded in action. Additionally, he directs the evacuation of soldiers killed in action, enemy prisoners of war, and disabled vehicles.

The troop combat trains can be consolidated and moved under the control of the first sergeant, or they can be split and moved. If the combat trains are split, the motor sergeant and first sergeant will divide the medics, mechanics, special tools, and high-use PLL items into equal portions and move parallel and behind troop combat elements to provide responsive medical evacuation and battle damage assessment across the troop zone. Recovery assets will continue to move center of zone to facilitate recovery. This technique may be used if the troop zone is wide and split operations are necessary to provide proper support.

Position the combat trains where they can support the troop, but will not be exposed to enemy direct fires. This location is usually from 2 to 5 kilometers to the rear.

Offensive and Reconnaissance Operations.During such operations as a movement to contact or a zone reconnaissance, position the trains in the center of the troop zone and about 2 kilometers, or one terrain feature, behind the tank platoons. Use the established road network, if possible, to move the trains. Bound the trains forward to successive concealed positions, based on the movement of the troop, to keep them in position to support the troop.

Defensive and Security Operations.During defensive and security operations, position the combat trains about 3 kilometers behind the tank platoon positions in the center of the troop sector. If possible, hide the trains in a small built-up area, which provides cover and concealment and a road network to facilitate the movement of the trains. The trains must be ready to move to support the troop or displace to a new location. The XO and first sergeant plan subsequent locations, and as the situation requires, the first sergeant or maintenance sergeant moves the trains to the next location.

Security.Trains must provide their own local security. They do not have the manpower or equipment to clear large areas, so they should move into areas that have already been cleared by scouts. The trains' best defense is to avoid detection. The trains should disperse their vehicles and camouflage the positions by using natural concealment like a wooded area or a small built-up area. A built-up area is best, because there are a limited number of approaches and the thermal signature of the trains is concealed. Enforce strict noise and light discipline. Hide vehicles in barns or garages, if available. The buildings in a built-up area and the trees in a wooded area provide some protection against the effects of indirect fires and air-delivered munitions. Dispersion also limits the damaging effects of these weapons.

Establish OPs around the trains to provide early warning of enemy movement toward the position. Put them where they can cover major avenues of approach. An OP has at least two soldiers, a crew-served weapon (if available), a map, binoculars, a night observation device, and a radio or field phone. The OP must immediately report, by radio or phone, enemy ground and air attacks so the trains can take appropriate actions.

Passive defense is the trains' best defense against air attack. The steps taken to locate them properly, disperse vehicles, camouflage positions, and use the cover and concealment of built-up or wooded areas help prevent the enemy from detecting and attacking, and limit the effects of enemy munitions if attacked.

Air attacks cannot always be avoided. If attacked, take active air defense measures. The first sergeant or senior soldier present must coordinate the small arms fires of the trains for them to be effective against an air threat. Refer to Chapter 8, Section VI, Air Defense, for a complete discussion of passive and active air defense.

FIELD TRAINS

Troop CSS assets not in the combat trains are collocated with the squadron field trains or the UMCP, and sustain the troop's fighting capability by moving rations, ammunition, fuel, repair parts, and replacements from the squadron rear area to the troop on the battlefield. Troop field trains are controlled primarily by the supply sergeant, and consist of the personnel and equipment shown in Figures 9-3 and 9-4. Troop maintenance assets in the field trains may be consolidated under the control of the squadron maintenance officer.

The troop supply sergeant is the troop representative in the squadron field trains. The supply sergeant is responsible for the following tasks:

- Maintaining records of troop clothing and equipment.
- Processing or forwarding requests for all Class I, II, III, IV, V, VI, and VII. He also assists the maintenance section in ordering Class IX repair parts and the medics in resupplying Class VIII (medical).
- Assembling and leading all vehicles in the troop LOGPAC from the field trains to the LRP. If the squadron is conducting resupply operations, the supply sergeant may move under the control and direction of one of the following squadron personnel: squadron support platoon leader or platoon sergeant, S4, HHT commander, or XO.
- Assisting the first sergeant in organizing the LOGPAC site for troop resupply and in evacuating soldiers who are killed in action, enemy prisoners of war, and disabled vehicles; and in medically evacuating wounded and injured soldiers.
- Delivering mail, when available, via the troop LOGPAC.

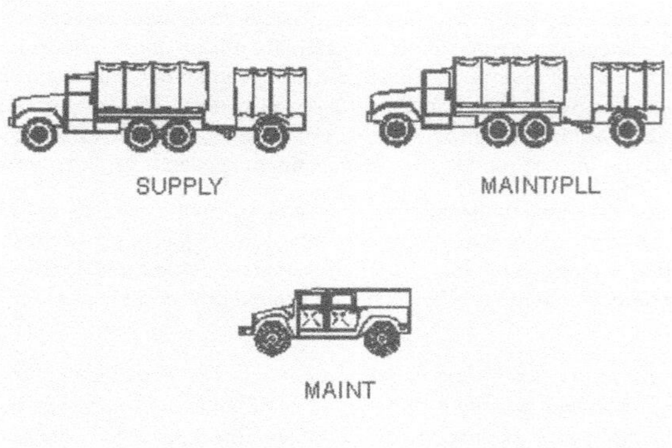

Figure 9-3. Heavy troop field trains.

Figure 9-4. Light troop field trains.

At the squadron field trains, the supply sergeant coordinates with the S4 and support platoon leader to ensure the LOGPAC meets the troop's CSS requirements.

The supply sergeant will coordinate with the troop PLL clerk for parts and requisitions and with the squadron personnel and administration center (PAC) section for mail, promotions, awards, Standardization Installation/Division Personnel System (SIDPERS), and other personnel actions. He will coordinate with squadron maintenance and communication for repaired vehicles and equipment and with the squadron food service sergeant for rations. The supply sergeant must frequently check with the HHT command post in the field trains for any additional troop requests reported on the A/L net.

The squadron field trains are responsible for their own local security. The supply sergeant will be tasked to assist the HHT commander in maintaining security of the field trains. He must execute and supervise security operations in accordance with the HHT commander's plan. During the movement through nonsecure areas, the supply sergeant should ensure that all crew-served and individual weapons are manned and ready. He should also ensure that all members of the troop field trains are briefed on immediate action drills for enemy contact or vehicle breakdowns.

If available, the troop should provide the supply sergeant with a radio to enhance communications between the supply sergeant and the first sergeant.

SQUADRON COMBAT SERVICE SUPPORT

In addition to organic troop CSS, one or two medical/evacuation teams are attached to the troop for combat operations from the squadron medical platoon. During resupply operations, the troop normally receives two fuel and two ammunition carriers from the transportation section of the support platoon. These carriers are not usually attached, but are provided to the troop as needed or as part of a standardized LOGPAC. The LOGPAC is discussed in Section II, Logistics.

To ensure responsive CSS for the squadron, the S4 and squadron maintenance officer organize the squadron support into echeloned trains and establish a main supply route (MSR), a UMCP, and an LRP. These trains are made up of combat trains (which provide immediate recovery, maintenance, medical, and emergency resupply support), and field trains (which provide the remainder of the squadron service support and limited direct-support maintenance). The UMCP is established to provide maintenance support for combat operations. It is positioned near or collocated with the squadron combat trains, and is the closest point to which damaged or failed equipment and systems are recovered. The UMCP is the focal point of the squadron maintenance effort. The MSR links troop combat trains to the squadron combat trains or UMCP and the field trains. LRPs are established along the MSR to facilitate the handover of LOGPACs and equipment between troops and squadron. An example of a squadron support plan is shown in Figure 9-5.

Section II. Logistics

SUPPLY OPERATIONS

The supply sergeant is responsible for getting supplies and delivering them to the troop. He delivers small items, but is largely dependent on support platoon assets to deliver bulky or high expenditure items. The commander establishes priorities for delivery, but the demands of combat normally dictate Classes I, III, V, VIII, and IX as most critical.

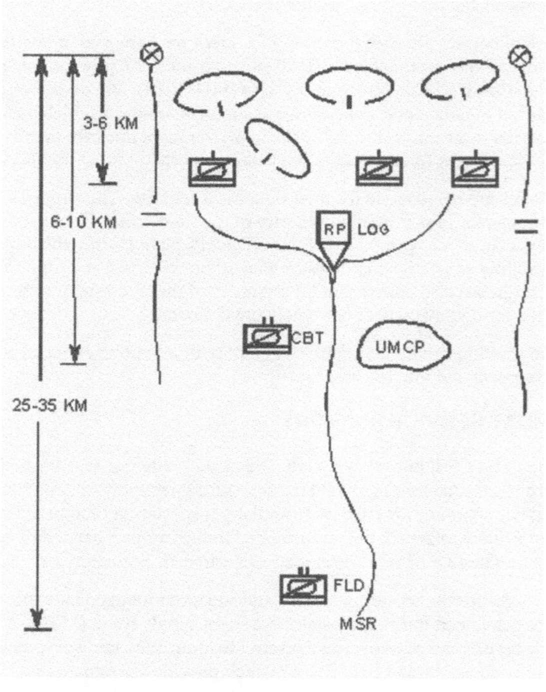

Figure 9-5. Squadron combat service support plan.

Class I (Rations).Meals ready to eat (MRE), C rations, are stocked on board each troop vehicle in a basic load prescribed by SOP (three-day supply). Class I is delivered daily by the supply sergeant as part of the

LOGPAC from the squadron field trains. Hot meals (B or tray-pack rations) are served when possible, but the normal ration cycle is C-C-C during combat operations. This requires resupply of at least 34 cases of MRE per day per troop. Water is a critical item, and must be resupplied daily. Soldiers may require up to 5 gallons of water per soldier per day, or more than 600 gallons for the entire troop. The troop can transport only 400 gallons of water on the water trailer. Soldiers should top off their water cans when possible. The supply sergeant should carry 5-gallon cans of water on his truck to supplement the needs of the troop. When he resupplies the troop, he exchanges the full cans for empty ones.

Class II (Supplies and Equipment).Class II items are requisitioned through the S4 and delivered by the supply sergeant as needed. The supply sergeant also maintains the following supplies, as a minimum, in his troop supply:

- One set of NBC overgarments and filter per soldier.
- One case of BA-30s.
- One mile of WD-1 wire.
- Additional batteries as necessary for night-vision devices and other equipment.
- Five sets of TA-50, to include personal clothing of various sizes, stored in duffel bags.
- Fifteen body bags.
- Map cases, grease pencils, permanent markers, etc.

Class III (POL).Class III is delivered by support platoon assets both as part of the troop LOGPAC and on an as-needed basis. Platoon sergeants report their Class III status to the command post daily, when status is requested, or when any Class III item is below a certain percentage, usually 59 percent.

Each troop vehicle carries a small basic load of packaged products as prescribed by SOP, and the fuel carriers normally have only small amounts of packaged products on board. Requests for large quantities of packaged products must be requested through the S4.

Class IV (Construction Materials). Class IV supplies are requested through the CTCP (combat trains command post) and delivered by the supply sergeant with the LOGPAC. Class IV supplies needed at troop level may include concertina wire, sandbags, and lumber.

Class V (Ammunition).The troop deploys with a basic load of Class V as prescribed by the unit for each type of vehicle. Ammunition status is reported by platoon sergeants to the command post daily, upon completion of enemy contact, or as needed. Normal resupply of Class V is delivered by the support platoon vehicles with the LOGPAC. Emergency resupply is requested through the first sergeant to the S4, and is delivered from the combat trains if it is available. Emergency resupply of Class V will normally consist of ammunition for major weapon systems, such as tank main guns, TOW missiles, and 25-mm guns in the heavy troop; and TOW missiles and caliber .50, 40-mm grenades, 7.62-mm, and individual weapons ammunition in the light troop.

Class VI (Personal Demand Items).Class VI items (such as soap, toothpaste, and cigarettes) are requested through the S4 by the first sergeant. They are usually picked up by the supply sergeant in the field trains and delivered as part of the LOGPAC.

Class VII (Major End Items).Class VII items, such as vehicles and night-vision devices, are automatically requested based on equipment shortages and battle losses. Items are delivered to the S4, who notifies the troop of the availability of the equipment. The equipment is delivered with a LOGPAC or brought forward immediately to an LRP where the first sergeant meets it and delivers it to the troop.

Class VIII (Medical Supplies).Class VIII is provided by the squadron medical platoon. The troop senior medic requests supplies through the squadron aid station. During intense combat, supplies may be pushed forward. They are then delivered to the combat trains where the troop medic picks them up from the squadron aid station, or are brought forward on ambulances.

Class IX (Repair Parts).The troop PLL clerk, who is in the field trains or the UMCP, works under the control of the squadron maintenance officer and the squadron maintenance technician. The PLL clerk requisitions, receives, and maintains records for all Class IX items. All troop PLLs are consolidated under the squadron maintenance officer during combat operations, and one or more PLL trucks remain in the UMCP at all times to provide a supply of Class IX for forward area support. When the troop maintenance sergeant needs a repair part, the request is sent through the squadron maintenance officer. The squadron maintenance officer determines if the part is available through the PLL, and fills the request if possible. If the part is not in the PLL, he then requests it through the squadron maintenance team in the field trains. They have the troop PLL clerk requisition the part for immediate use or replenish the part used from the PLL. The troop PLL clerk submits the requisition through squadron maintenance to the supporting maintenance unit. The part is delivered through squadron maintenance to the troop when the requisition is filled. The PLL clerk updates the records to reflect the requisition.

Maps.Maps are requested through the troop command post to the S4. As maps are available, the supply sergeant picks them up in the field trains and delivers them to the troop as part of a LOGPAC.

RESUPPLY OPERATIONS

Resupply of combat resources is accomplished using standardized procedures to rearm, refuel, and refit the troop as fast as possible to sustain its combat potential. There are two types of resupply operations—routine and supplementary. Methods of resupply are tailgate issue and service station.

Routine Resupply

Routine resupply operations include daily resupply of Classes I, III, V, and IX, and mail and other items needed by the troop. Routine resupply takes place when the troop is not in heavy contact, and may be conducted in an assembly area or behind troop positions when the troop is deployed in sector or zone. Class III is required more than once a day when the vehicles are in continuous operation.

Routine resupply is conducted using the LOGPAC from the field trains and the troop combat trains. The LOGPAC is organized in the field trains. Its composition is based on the troop's needs as reported to the S4, on requisitions, and on the availability of supplies. The troop supply sergeant reports to the S1 and S4 sections, squadron mess section, and squadron maintenance section to ensure all available supplies are picked up and loaded on trucks for troop resupply. The S4, support platoon leader, and troop supply sergeants assemble the LOGPACs for each troop by adding the troop supply trucks to the troop slice of Class III and V trucks from the support platoon. The support platoon leader leads the LOGPACs to the LRP, where the first sergeants meet them. Each first sergeant leads his LOGPAC to the troop resupply site. When it arrives, members of the troop combat trains guide the LOGPAC vehicles into position. Once the LOGPAC is established, the first sergeant reports to the command post that he is set. The troop commander or the XO coordinates resupply operations, and ensures all platoons and sections are resupplied. Resupply is conducted by one of two methods—tailgate issue or service station.

Tailgate Issue Method.The tailgate method is used in static positions such as assembly areas. Class III and V supply vehicles and other bulk-issue vehicles move from vehicle to vehicle to conduct resupply. The rest of the service support vehicles are centrally located in the troop area. Little or no movement is required by the combat vehicles. Personnel move to a centralized location to receive supplies, Class I, and mail. This method provides 360-degree security throughout the resupply operation; however, it is very time-consuming and requires an adequate road network for the wheeled supply vehicles to reach each vehicle. An example of this method is shown in Figure 9-6.

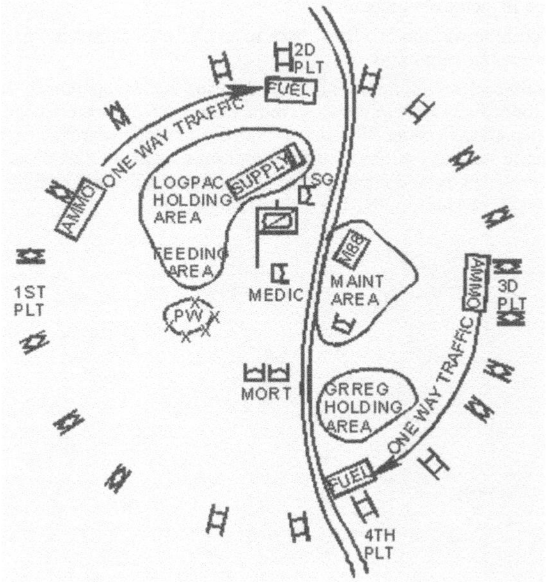

Figure 9-6. Tailgate issue method in a troop assembly area.

The following takes place during tailgate resupply:

- Combat vehicles remain in place. POL and ammunition trucks travel in a clockwise direction around the assembly area to each vehicle position, in turn, to conduct resupply.
- Crewmen rotate through the feeding area and pick up supplies, water, and mail.
- The first sergeant and platoon sergeants arrange for pickup of those killed in action (KIA) and their personal effects. The KIA are brought to a holding area near the medical aid station, but kept out of view.
- Ambulances pick up, treat, and evacuate seriously wounded soldiers. Other wounded soldiers are

carried to the ambulance or walk to it for emergency medical treatment.

- Prisoners are kept together and guarded. As soon as possible, they are moved to the squadron trains on a returning supply vehicle.
- The troop armorer, the radio repairman, and the organizational mechanics repair known problems and spot-check other vehicles.
- Vehicles needing maintenance are brought to the maintenance area.
- The first sergeant and platoon sergeants closely monitor the resupply operation.
- Empty LOGPAC vehicles are moved to a holding area, where they are loaded with KIA, prisoners of war (PW) and inoperative equipment.
- The supply sergeant moves the LOGPAC back to an LRP to link up with the support platoon leader and return to the field trains.

Service Station Method.Service station resupply is used during tactical operations. It is most effective when the troop is positioned in a zone or sector no more than 3 to 5 kilometers wide, such as is found in a defend in sector or defend from a troop BP mission. Platoons or sections are resupplied at the LOGPAC while the rest of the troop stays in position. The first sergeant sites the LOGPAC as shown in Figure 9-7. LOGPAC security is provided by soldiers from the combat trains who are not involved in the resupply, and by platoon vehicles that have completed or are awaiting resupply.

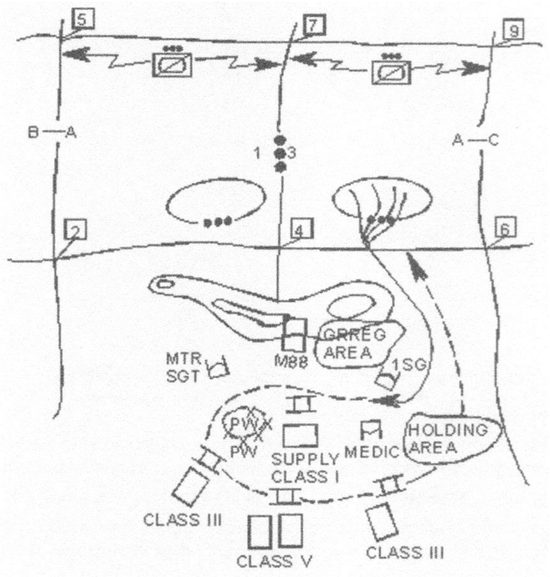

Figure 9-7. Service station method.

The following takes place during service station resupply:

- Vehicles of the first platoon or section enter the LOGPAC at a designated location and follow one-way traffic flow.
- The vehicle carrying the KIA in body bags moves to the holding area, out of view of the troop, and leaves the remains and personal effects.
- Soldiers requiring medical attention are delivered to medics and treated or prepared for evacuation.
- The platoon sergeant supervises the operation and coordinates face-to-face with the first sergeant for any special requirements.
- The platoon leader dismounts his vehicle and uses the first sergeant's vehicle and radio to contact the troop commander for orders and situation reports.
- Crews requiring unit-level maintenance remain in the maintenance holding area.
- Vehicles rotate through stations.
- Unit-level mechanics, the armorer, and the radio repairman repair known problems and spot-check other vehicles.
- Crews rotate to the supply truck to pick up mail, supplies, and Class I.
- The platoon leader and platoon sergeant conduct precombat inspection.
- When the platoon or section has completed resupply, it moves to its designated position.
- The rest of the platoons rotate individually through the LOGPAC for resupply.

At the end of the resupply operation, the troop LOGPAC returns to the LRP where it links up with the support platoon leader and returns to the field trains. In the field trains, the supply sergeant returns the fuel and ammunition carriers to the support platoon, returns ration-serving equipment to the mess team, delivers the KIA to the graves registration (GRREG) collection point, requests any additional supplies from the squadron supply sergeant, and returns to his position in the field trains.

Supplementary Resupply

In the heavy troop, daily routine resupply will not sustain the combat power of the M1/M3 fleet when operating without a break for 8 to 10 hours. The tanks will consume so much fuel that they will have to supplement routine resupply of Class III one or two times daily. The troop may also need to resupply Class V. Supplementary resupply can be conducted in a couple of ways depending on the mission and if the troop is in contact with the enemy. Regardless of the technique used, the commander must push the supplies forward to his platoons; the tempo of combat operations cannot be disrupted in order to resupply. Conduct emergency resupply when in heavy contact with the enemy, and prestock resupply when not. Once the supplies are brought forward, either the tailgate issue or service station methods are used.

Supplementary resupply must be planned when continuous operations are expected. Resupply must be requested through the S4 several hours in advance to ensure the service support assets are prepared to support the operation. The troop must quickly resupply once fuel trucks are available, and then return them to the field trains. There are few fuel carriers and drivers available in the squadron and these valuable assets cannot be wasted.

Platoon Prestock.When operating over wide frontages, such as during screen, zone reconnaissance, or movement to contact operations, break the supplies down into platoon packages and push the supplies forward to each platoon position. Position the vehicles behind the platoon and resupply individual vehicles or sections at a time using the service station method. If the situation allows, the resupply vehicles can move through the platoon position and resupply the platoon in place using the tailgate issue method.

Controlling platoon prestock is difficult because resupply vehicles are moving to several locations rather than remaining under the centralized control of the first sergeant. The platoon leader must also provide security for their own prestock (see Figure 9-8).

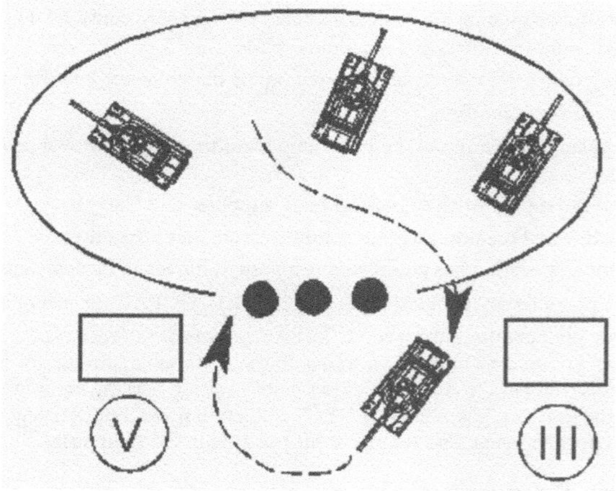

Figure 9-8. Platoon prestock.

Troop Prestock.Troop prestock is used to supplement routine resupply of Classes III and V. Troop prestock is used when the troop is operating in a narrow zone or sector. The resupplies are pushed forward to the troop position using either the service station or tailgate method. Each platoon rotates through the resupply site, using the service station method, while the rest of the troop remains in position; or each vehicle is resupplied in position using the tailgate method (see Figure 9-9).

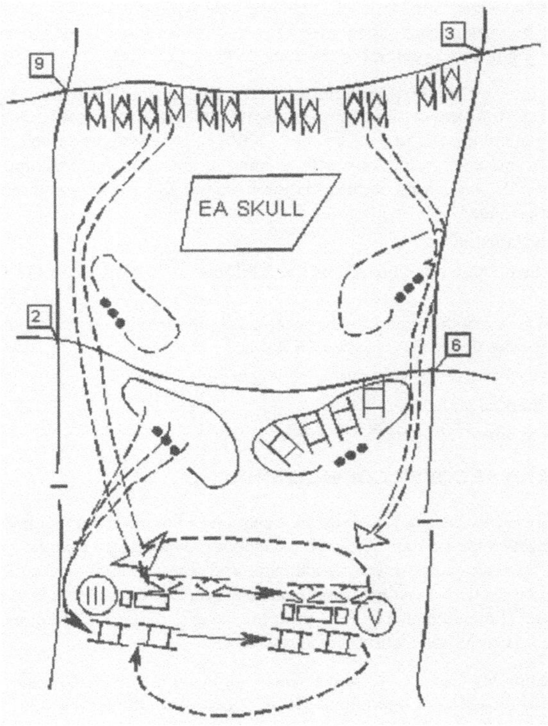

Figure 9-9. Troop prestock.

Variations of these methods can be used to meet the troop's situation. The troop must, however, rehearse these techniques to ensure it can quickly and efficiently conduct resupply operations and continue the mission.

Emergency Resupply

Emergency resupply normally involves only fuel and ammunition and is conducted while in contact with the enemy. The resupply begins at section and platoon level by redistributing ammunition between vehicles to cross-level loads. The platoon sergeant reports his need for emergency resupply to the first sergeant, who relays the request to the S4. The squadron combat trains maintain a small load of Class III and V for these situations. The S4 or support platoon leader coordinates a linkup between the squadron combat trains and the troop first sergeant. The first sergeant meets the resupply trucks and moves back to the troop area. The first sergeant and troop XO choose a resupply point that is just behind the troop position and masked by terrain from enemy direct fire and observation. If fuel is needed, the fuel truck is moved to the resupply point and vehicles or sections go there to refuel.

Resupply Site Selection.The LOGPAC site must be carefully chosen to provide responsive support for the troop, to support the movement of wheeled resupply vehicles, and to limit exposure to enemy fires. The XO selects the general area to be used by the LOGPAC. He knows about the current tactical situation, and can determine what the troop will be doing in the next couple of hours. The first sergeant selects the exact LOGPAC site based on information from the XO and map and ground reconnaissance. He selects a site that provides—

- Cover and concealment.
- Proximity to the platoon positions, from 3 to 8 kilometers behind the FLOT and center of the sector.
- A road or trail network that supports the wheeled resupply vehicles and the heaviest troop vehicles and allows one-way traffic flow to the LOGPAC.
- Enough room to disperse the vehicles.
- Reduction of thermal signature.
- Level enough to allow refueling.

MAINTENANCE AND RECOVERY OPERATIONS

The first sergeant supervises the troop maintenance section in the troop combat trains and the maintenance sergeant runs the section. The maintenance sergeant operates on the troop command net and stands by to provide responsive unit-level maintenance and recovery support for combat vehicles and radio equipment. When needed, squadron maintenance augments troop assets. The troop armorer is responsible for unit-level maintenance and repair of the troop's small arms weapons. Position him in the combat trains where he can provide battlefield support.

Maintenance Operations.Maintenance responsibilities begin at crew level. Operator-level PMCS must be performed before, during, and after operations. During operations, the crew begins the maintenance process when it identifies the fault. Once a problem develops, the vehicle commander determines support requirements, to include self-recovery, assistance from another vehicle, or assistance from troop or squadron maintenance, and takes the following actions:

- Reports the situation to the platoon sergeant, providing condition(s), location, and circumstances.
- Attempts to self-recover, if mired.
- Uses another vehicle to move to a secure location if exposed to enemy observation and fire. Uses smoke to screen the area between the mired vehicle and the enemy.

When repairs are beyond the capability of the crew, the platoon sergeant uses the troop command net to notify the first sergeant of the situation and to request assistance. The crew must maintain radio contact (if the radio is operational) on the platoon net and maintain local security. The maintenance sergeant monitors the radio traffic. He switches his radio to the platoon frequency to coordinate maintenance support and dispatches the appropriate assets to the "down" vehicle. The maintenance team checks whether the problem can be corrected in place within the time criteria (30 minutes in defensive situations and 2 hours in offensive situations). They move the vehicle to a more secure location, if necessary, and fix the vehicle. If repairs will take too long, the maintenance sergeant uses the squadron A/L net to notify the squadron maintenance officer who will coordinate for a handover of the equipment from troop to squadron at the LRP or along the MSR. The squadron maintenance officer needs the following information:

- Identification of the troop.
- Equipment involved.
- Location of equipment.
- Nature of damage.
- Pertinent information about the tactical situation, and any other necessary information.

Based on the information above, the squadron maintenance officer will designate a time and place for equipment handover.

The troop does not have the personnel, equipment, or time to create a holding area in the troop combat trains. If repairs will take more than the allowable time, evacuate the equipment to squadron.

Recovery Operations.When a vehicle cannot be repaired within the allowable time or is damaged beyond repair, the maintenance section must recover it and pull it to the MSR where it is turned over to squadron. Squadron maintenance will pull it to the UMCP or the field trains.

If a vehicle is catastrophic, or not able to be recovered based on the enemy situation, the commander may authorize the destruction of the vehicle. Before using thermite grenades to destroy a vehicle, remove as many valuable items as time permits, such as—

- Classified equipment and documents.
- Communications equipment.
- Weapons.
- Ammunition.

FIELD SERVICES

GRREG is performed by the field services platoon from corps, located in the regimental support area. The initial collection, identification, safeguarding of personal effects, and evacuation of the dead is the troop's responsibility.

When remains are discovered, be careful to preserve all items that may be used for identification. If metal identification tags (dog tags) are on the remains, do not remove them. Secure all personal effects in a bag or poncho and tie it to the remains. Place each casualty in a body bag, poncho, or shelter half and evacuate with the first available means of transportation, such as LOGPAC vehicles or disabled vehicles, to the squadron field trains. The supply sergeant receives all remains, and is responsible for turning them over to the GRREG collection point.

If the tactical and logistical situation makes evacuation impossible, emergency on-site burial is performed. On-site burial requires the permission of the squadron commander. If an on-site burial is performed, do the following:

- Complete two copies of DD Forms 551 and 1077 for each body. These forms are available through the squadron PAC. The first sergeant and supply sergeant should carry them.

- If there are several remains, engineer support may be needed to dig trenches.
- Dig a trench 6-1/2 feet wide and 3-1/2 feet deep. The length is determined by the number of remains.
- Remove one identification tag from each body, and string the tags on a wire in the order in which the remains are buried.
- Place the remains in the grave shoulder to shoulder.
- Bury all personal effects with the remains.
- Mark the ends of each row with a stake. Tag each stake to identify the marker as a grave. Indicate the length of the grave.
- Prepare an overlay that shows the location of the grave site.
- Cover the grave with earth removed from the trench.
- Forward the completed forms, overlay, and identification tags to squadron.

If the remains are contaminated, the grave site must be clearly marked and separated from noncontaminated grave sites. This must also be indicated on the grave-site overlay.

BATH AND LAUNDRY SERVICES

Bath and laundry services are provided by supply and service units from the corps support command. When available, these services are coordinated through the S4.

Section III. Personnel Service Support

PERSONNEL AND ADMINISTRATIVE

All troop administrative and personnel actions are handled by the squadron PAC in the field trains. The troop chain of command is responsible for ensuring that soldiers receive passes, leaves, promotions, awards, mail, legal assistance, financial services, and other personnel and welfare services on a fair and prompt basis. The first sergeant interfaces with PAC daily through the supply sergeant, at the administrative/logistics operations center with the S1/S4, or at the LRP.

Casualties and missing personnel are reported on DA Forms 1155 and 1156. These forms are initiated by the individual who witnessed the incident and are forwarded to the first sergeant. This action is the basis for notification of the next of kin and awarding benefits such as Serviceman's Group Life Insurance (SGLI), so accuracy is a must.

The commander should continuously manage the troop personnel situation to ensure trained personnel are manning key positions, crew shortages are filled with available personnel, and replacement personnel are trained and incorporated into crews. During combat, platoon leaders and platoon sergeants must balance their crews after casualties are evacuated. They must fill key positions on vehicles with the most qualified soldier. If necessary the commander may direct cross-leveling of personnel within the troop.

The troop submits daily personnel reports to the squadron as prescribed by SOP. The personnel reports are important because they allow the squadron commander to make tactical decisions for the employment of the troop. The reports also allow the S1, with the squadron commander's guidance, to properly distribute personnel replacements. Replacement personnel are received by the supply sergeant in the field trains. The supply sergeant in-processes the soldier into the troop, ensures he has the proper equipment, then delivers him to the first sergeant during the troop LOGPAC.

combat health support

First Aid.First aid is the responsibility of all soldiers; they use first aid, self-aid, buddy-aid, and combat lifesavers. All must be trained to take action after a soldier is wounded to keep him breathing, stop the bleeding, prevent shock, and dress the wound until medical personnel are available to treat the soldier. Training must include treatment of NBC casualties and crew evacuation drills to get soldiers out of a vehicle without creating further injury. Selected soldiers from each platoon will be trained in advanced medical skills in the combat lifesaver program.

Treatment.The first sergeant should position troop medical aid and evacuation teams on the battlefield where they can be most responsive. They will usually operate under the control and direction of the first sergeant in the troop combat trains. If the troop zone is pushing 10 kilometers wide, the troop may operate split combat trains with the first sergeant controlling medical evacuation on one side of the zone and the motor sergeant controlling the other half. The medics must know the locations of and routes to each platoon, the troop combat trains, the squadron combat trains, and each collection point.

Combat Stress Control.The psychological effects of combat on soldiers influence their ability to execute their missions. Individual self-control and self-discipline in the face of danger are maintained through unit discipline and firm leadership. The commander and his subordinate leaders are key to the mental toughness of soldiers. Commanders should be visible to soldiers and share their hardships with them. They should talk to their soldiers individually and as a troop to keep them informed. The commander's tone of voice on the radio must indicate that he has the situation under control. He should transmit the successful accomplishments of the troop and tell the soldiers they are doing a good job. Keep details about any casualties in the troop off the radio and ensure that soldiers who are killed in action are covered up and kept out of view of the troop. The chaplain can be a great asset in maintaining troop morale. Do not neglect the mental fitness of soldiers, because uncontrolled fear is contagious and can quickly lead to chaos.

Evacuation.To receive medical assistance, a platoon leader or platoon sergeant calls the first sergeant on the troop command net. The medic team monitors the call and goes to the platoon frequency to make any other necessary coordination. The medic team moves to the reported location to treat the soldier(s). The medics transport as many wounded soldiers as possible in their armored ambulance. Do not transport the dead with the wounded; evacuate them separately. If the medics determine the soldiers are seriously wounded, they either transport them directly to the squadron combat trains, or coordinate to hand them over to squadron at a collection point. Soldiers with slight wounds should not be evacuated unless it is necessary. They should be treated and returned to duty. Secure the seriously wounded soldiers on their litters to prevent any further injury during the evacuation.

As a general rule, do not use combat vehicles to evacuate the wounded; this reduces combat strength. Use combat vehicles only if absolutely necessary, and no other transportation is available.

Weapons and military equipment (except NBC protective equipment) of personnel being evacuated are secured with the platoon sergeant, first sergeant, or supply sergeant. Ensure the soldier keeps his protective mask and overgarments. When the situation permits, all equipment that was secured by the first sergeant or platoon sergeant is turned over to the supply sergeant.

disease and nonbattle injuries

More soldiers are lost in combat to illness and disease than to combat-related wounds. Maintaining the health and fighting fitness of the troop is a leadership responsibility. Personal hygiene, field sanitation, and rest must be incorporated into all troop operations. If these are ignored, the combat effectiveness of the troop will decrease rapidly.

UNIT MINISTRY TEAM

The unit ministry team (UMT), consisting of an appropriate number of chaplains and chaplain assistants, is assigned to provide unit, area, and denominational coverage to all troops in the cavalry squadron. UMT provides the following eight subfunctions of worship opportunities:

- Administration of sacraments, rites, and ordinances.
- Pastoral care and counseling.
- Development and management of the UMT.
- Management of material resources.
- Advising the commander on matters of religion, morale, and morals as affected by religion.
- Ministry in support of soldiers suffering from battle fatigue.
- Development of programs that enhance the total well-being of the soldier.
- Development of activities to enhance unit cohesion.

PERSONAL HYGIENE

The personal hygiene of the troop's soldiers is a leader's responsibility. Rules of hygiene must be observed to ward off disease and improve troop morale. Ensure soldiers wash and change their socks and underwear daily, if possible. All soldiers should shower as time and resources are available. The senior aidman plays a big part in personal hygiene by educating soldiers and by checking for signs of trench foot or immersion foot and frostbite.

Disease can quickly spread through the troop and incapacitate its soldiers. Watch for symptoms and ensure the soldiers' immunizations are current.

REST

Rest is extremely important for both commanders and their soldiers. When the troop is at REDCON 3 or 4, troop leadership should ensure a sleep plan is planned and executed.

FIELD SANITATION

Field sanitation is vital to the prevention of the spread of disease. The medics must assist in this effort by checking troop water supplies to ensure they are potable, and by ensuring mess utensils and equipment are properly cleaned. In static situations, such as when in assembly areas, soldiers will use slit trenches or latrines; at other times soldiers will use cat holes. Slit trenches and cat holes must be covered up after use to prevent the spread of disease.

Section IV. Prisoners of War

CAPTURE AND HANDLING

Maps, military documents, letters, and diaries obtained on the battlefield, as well as PWs, are valuable sources of combat intelligence. Scout platoons usually obtain information while searching the battlefield in the conduct of reconnaissance operations or after contact with the enemy. They should rapidly report this information to the troop commander for evacuation instructions. The scouts may be ordered to take the documents to the troop command post or to rendezvous with the first sergeant at a specific location to turn over the documents. Proper handling and evacuation of captured documents and equipment are important. The materials must be immediately turned over to trained intelligence personnel. In most cases, captured documents lose their value over time.

EVACUATION TECHNIQUES

In most situations, returning supply vehicles, aircraft, or troop headquarters vehicles can be used to evacuate PWs from the troop combat trains to the squadron collection point. The troop is responsible for guarding PWs until they are turned over to the S1. Wounded PWs are treated through normal medical channels, but are kept separate from US soldiers.

When support vehicles are not readily available and the troop must continue its operation, secure the PWs in a holding area such as a basement or compound. Notify the first sergeant and the S1 of the PW location and continue the mission. The first sergeant will pick up the PWs and complete their evacuation to the collection point.

Appendix A

Fratricide Prevention

Fratricide is as old as warfare itself, a complex problem that defies simple solutions. Fratricide can be broadly defined as the employment of friendly weapons and munitions, with the intent to kill the enemy or destroy his equipment or facilities, that results in unforeseen and unintentional death or injury to friendly personnel. This appendix focuses on actions leaders can take with current resources to reduce the risk of fratricide.

CONTENTS

Section I. Magnitude of the Problem

The modern battlefield is more lethal than any in history. The tempo of operations is rapid, and the nonlinear nature of the battlefield creates command and control challenges for all unit leaders.

The accuracy and lethality of modern weapons make it possible to engage and destroy targets at these extended acquisition ranges. At the same time, however, the ability of US forces to acquire targets using thermal imagery and other sophisticated sighting systems exceeds our ability to accurately identify these targets as friend or foe. As a result, friendly elements can be engaged unintentionally and destroyed in a matter of seconds.

Added to this is the problem of battlefield obscuration, which becomes a critical consideration whenever thermal sights are the primary source of target identification. Rain, dust, fog, smoke, and snow degrade identification capability by reducing the intensity and clarity of thermal images.

On the battlefield, positive visual identification cannot be the sole engagement criteria at ranges beyond 1,000 meters. Situational awareness is the key; it must be maintained throughout an operation.

Section II. Risk Identification and Preventive Measures

Reduction of fratricide risk begins during the planning phase of an operation and continues throughout preparation and execution. Risk identification must be conducted at all levels during each phase, and the results clearly communicated up and down the chain of command so risk assessment can begin. This section covers considerations that influence risk identification; it also focuses on measures the platoon leader can implement both to make the identification process more effective and to help prevent friendly fire incidents from occurring. Section III of this appendix covers the risk assessment process; Section IV lists additional fratricide reduction measures and guidelines.

Planning Phase

A plan that is thoroughly developed and understood helps to minimize fratricide risk. The following considerations help indicate the potential for fratricide in a given operation:

- The clarity of the enemy situation.
- The clarity of the friendly situation.
- The clarity of the commander's intent.
- The complexity of the operation.
- The planning time available at each level.

Graphics are a basic tool that commanders at all levels use to clarify their intent, add precision to their concept, and communicate their plan to subordinates. As such, graphics can be a very useful tool in reducing the risk of fratricide. Commanders at all levels must understand the definitions and purpose of operational graphics and the techniques of their employment.

Note. See FM 101-5-1 for the definitions of each type of graphic control measure.

Preparation Phase

The following factors may cause fratricide risks to become evident during rehearsals:

- Number and type of rehearsals.
- Training and proficiency levels of units and individuals.
- The habitual relationships between units conducting the operation.
- The physical readiness (endurance) of the troops conducting the operation.

Backbriefs and rehearsals are primary tools in identifying and reducing fratricide risk during the preparation phase. The following are some considerations for their use:

- Backbriefs ensure subordinates understand the commander's intent. They often highlight areas of confusion, complexity, or planning errors.
- The type of rehearsal conducted determines what types of risks are identified.
- Rehearsals should extend to all levels of command and involve all key players.

Execution Phase

During execution, in-stride risk assessment and reaction are necessary to overcome unforeseen fratricide risk situations. The following are factors to consider when assessing fratricide risks:

- Intervisibility between adjacent units.
- Amount of battlefield obscuration.
- Ability or inability to positively identify targets.
- Similarities and differences in equipment, vehicles, and uniforms among friendly and enemy forces.
- Vehicle density on the battlefield.
- The tempo of the battle.

Maintaining situational awareness at all levels and at all times is another key to fratricide reduction as an operation progresses. Units must develop and employ effective techniques and SOPs to aid leaders and crewmen in this process. These techniques include—

- Monitoring on the next higher net.
- Radio cross-talk between units.
- Accurate position reporting and navigation.
- Training and use/exchange of liaison officers (LO).

Section III. Risk Assessment

Risk assessment must be conducted whenever fratricide risk factors are identified. It must take place at all levels during each phase of operations. As with risk identification, the results of the assessment must be passed on to all levels of the chain of command so that fratricide reduction measures can be developed and implemented. Refer to Section IV for specific reduction measures.

Figure A-1 is a worksheet for evaluating fratricide risk in the context of mission requirements. The worksheet lists six mission-accomplishment factors that affect the risk of fratricide, along with related considerations for each factor. Assess the potential risk in each area as low, medium, or high, and assign a point value to each (one point for low risk, two for medium risk, three for high risk). Add the point values for the overall fratricide assessment score. Use the resulting score only as a guide, however. The final assessment must be based both on observable risk factors like those on the worksheet and on your "feel" for the intangible factors affecting the operation. Note that descriptive terms are listed only in the low- and high-risk columns of the worksheet. The assessment of each factor will determine whether the risk matches one of these extremes or lies somewhere between them as a medium risk.

FACTORS	LOW (1)	MEDIUM (2)	HIGH (3)
1. UNDERSTAND PLAN			
● Commander's Intent	Clear		Foggy
● Complexity	Simple		Complex
● Enemy Situation	Known		Unknown
● Friendly Situation	Clear		Unclear
● ROE	Clear		Unclear

2. ENVIRONMENT ● Intervisibility ● Obscuration ● Battle tempo ● Positive target ID	Favorable Clear Slow 100%		Unfavorable Obscured Fast 0%
3. CONTROL MEASURES ● Command relationships ● Audio ● Visual ● Graphic ● SOPs ● LOs ● Location/Navigation	Organic Loud/Clear Well Seen Standard Standard Proficient Sure		Joint/Combined Jammed Obscured Not understood Not used Untrained Unsure
4. EQUIPMENT (Compared to US) ● Friendly ● Enemy	Similar Different		Different Similar
5. TRAINING ● Individual proficiency ● Unit proficiency ● Rehearsal ● Habitual relationship ● Endurance	MOS Qual Trained Multiple Yes Alert		Untrained Untrained None No Fatigued
6. PLANNING TIME (1/3 - 2/3 Rule) ● Higher HQ ● Own HQ ● Lower HQ	Adequate Adequate Adequate		Inadequate Inadequate Inadequate
OVERALL FRATRICIDE ASSESSMENT	LOW 26-46%*	MEDIUM 42-62%*	HIGH 58-78%*
● Commander may use numbers as the situation dictates. ● Numbers alone may not give accurate fratricide risk.			

Figure A-1. Fratricide risk assessment worksheet.

Section IV. Fratricide Reduction Measures

The following measures are provided as a guide to actions that can reduce fratricide risk. They are not directive in nature, nor are they intended to restrict initiative. Apply them as appropriate based on the specific situation and METT-T factors.

- Identify and assess potential fratricide risks in the estimate of the situation. Express these risks in the OPORD or FRAGO.
- Maintain situational awareness, focusing on such areas as current intelligence; unit locations and dispositions; denial areas (minefields/FASCAM); contaminated areas, such as ICM and NBC; SITREPs; and METT-T factors.
- Ensure positive target identification. Review vehicle/weapon ID cards. Know at what ranges and under what conditions positive identification of friendly vehicles/weapons is possible.
- Establish a command climate that stresses fratricide prevention. Enforce fratricide prevention measures, emphasize the use of doctrinally sound tactics, techniques, and procedures. Ensure constant supervision in the execution of orders and the performance of all tasks and missions to standard.
- Recognize the signs of battlefield stress. Maintain unit cohesion by taking quick, effective action to alleviate it.
- Conduct individual, leader, and collective (unit) training covering fratricide awareness, target identification and recognition, and fire discipline.
- Develop a simple, decisive plan.
- Give complete and concise mission orders.
- Use SOPs that are consistent with doctrine to simplify mission orders. Periodically review and change SOPs as needed.
- Strive for maximum planning time for you and your subordinates.
- Use common language/vocabulary and doctrinally correct standard terminology and control measures, such as fire support coordination line, zone of engagement, and restrictive fire lines.
- Ensure thorough coordination is conducted.
- Plan for and establish effective communications.
- Plan for collocation of command posts whenever it is appropriate to the mission, such as during a passage of lines.
- Designate and employ LOs as appropriate.
- Ensure rules of engagement are clear.
- Include fratricide risk as a key factor in terrain analysis (OCOKA).
- Conduct rehearsals whenever the situation allows time to do so.
- Be in the right place at the right time. Use position location/navigation devices (GPS and POSNAV); know your location and the locations of adjacent units (left, right, leading, and follow-on); and synchronize tactical movement.
- Include discussion of fratricide incidents in after-action reports.

Section V. Fratricide Risk Considerations (OPORD Format)

This section, which parallels the five-paragraph OPORD, contains key factors and considerations in fratricide reduction. This is not a change to the OPORD format; rather, it should be used during OPORD development to ensure fratricide reduction measures are included in the order. It is not a strict guide. The factors and considerations are listed where they would likely appear in the OPORD, but they may warrant evaluation during preparation of other paragraphs.

1. Situation.

a. Enemy forces.

(1) Are there similarities among enemy and friendly equipment and uniforms that could lead to fratricide?
(2) What languages do enemy forces speak? Could these contribute to fratricide risk?

(3) What are the enemy's deception capabilities and his past record of deception activities?

(4) Do you know the locations of enemy forces?

b. Friendly forces.

(1) Among the allied forces, are there differences (or similarities with enemy forces) in language, uniform, and equipment that could increase fratricide risk during combined operations?
(2) Could differences in equipment and uniforms among US armed forces increase fratricide risk during joint operations?

(3) What differences in equipment and uniforms can be stressed to help prevent fratricide?

(4) What is the friendly deception plan?

(5) What are the locations of your unit and adjacent units (left, right, leading, follow-on)?
(6) What are the locations of neutrals and noncombatants?

c. Own forces.

(1) What is the status of training activities? What are the levels of individual, crew, and unit proficiency?

(2) Will fatigue be a factor for friendly forces during the operation? Has an effective sleep plan been developed?

(3) Are friendly forces acclimatized to the area of operations?

(4) What is the age (new, old, or mix) and condition of equipment in friendly units? What is the status of new equipment training?

(5) What are the expected MOPP requirements for the operation?

d. Attachments and detachments.

(1) Do attached elements understand pertinent information regarding enemy and friendly forces?
(2) Are detached elements supplied this pertinent information by their gaining units?

e. Weather.

(1) What are the expected visibility conditions (light data and precipitation) for the operation?
(2) What effect will heat and cold have on soldiers, weapons, and equipment?

f. Terrain.

(1) Do you know the topography and vegetation (such as urban, mountainous, hilly, rolling, flat, desert, swamp/marsh, prairie/steppe, jungle, dense forest, open woods) of the expected area of operations?
(2) Have you evaluated the terrain using the factors of OCOKA?

2. Mission. Is the mission, as well as all associated tasks and purposes, clearly understood?

3. Execution.

a. Task organization.

(1) Has the unit worked under this task organization before?
(2) Are SOPs compatible with the task organization (especially with attached units)?

(3) Are special markings or signals (for example, cats' eyes, chemlites, or panels) needed for positive identification of uniforms and equipment?

(4) What special weapons and/or equipment will be used? Do they look or sound like enemy weapons and/or equipment?

b. Concept of the operation.

(1) Maneuver. Are main and supporting efforts identified to ensure awareness of fratricide risks and prevention measures?

(2) Fires (direct and indirect).

(a) Are priorities of fires identified?

(b) Have target lists been developed?
(c) Has the fire execution matrix/overlay been developed?

(d) Have locations of denial areas (minefields, FASCAM) and contaminated areas (ICM, NBC) been identified?

(e) Are the locations of all supporting fire targets identified in the OPORD/OPLAN overlays?

(f) Are aviation and CAS targets clearly identified? Have signals been established to positively identify these targets for the aircraft? Have airspace coordination areas been developed? Have enemy air defense systems been suppressed?

(g) Has the direct-fire plan been developed and synchronized with the fire support plan?

(h) Have final protective fires been designated?

(i) Have you identified and verified sector limits?

(j) Have executors for each target been assigned and do they understand when and where to shoot? Do the shooters have "eyes on" the target?

(k) Are the observers surveyed in or are they using a map spot? Target location errors can cause big problems.

(l) Do all leaders and executors understand where the fire support coordination measures are and when they go into effect? Rehearsal is the key.

(m) Can the fire support officer hear what targets are being called on the maneuver nets?

(n) Have all targets been rehearsed with the executors and the field artillery battalion?
(o) Does the reinforcing or general support reinforcing field artillery have all the proper graphics and understand where they fit in? Did they attend the rehearsal?

(p) Have restrictions on specific munitions been established and does everyone know where they are planned and emplaced?

(3) Engineer tasks.

(a) Are friendly minefields, including FASCAM and ICM dud-contaminated areas, known?

(b) Are obstacles identified, along with the approximate time needed for reduction/breaching of each?

(4) Tasks to each subordinate unit. Are friendly forces identified, as appropriate, for each subordinate maneuver element?

(5) Tasks to CS/CSS units. Have locations of friendly forces been reported to CS/CSS units?

(6) Coordinating instructions.

(a) Will rehearsals be conducted? Are they necessary? Are direct and indirect fires included?

(b) Is a backbrief necessary?

(c) Are appropriate control measures clearly explained and illustrated in the OPORD and overlays? Have they been disseminated to everyone who has a need to know? What is the plan for using these control measures to synchronize the battle and prevent fratricide?

(d) Have target/vehicle identification drills been practiced?

(e) Do subordinate units know the immediate action, drill, or signal for "CEASE FIRE" or "I AM FRIENDLY" if they come under unknown or friendly fire? Is there a backup action?

(f) Is guidance in handling dud munitions, such as ICM and CBUs, included?

4. Service Support.

a. Are train locations and identification markings known by everyone?

b. Do medical and maintenance personnel know the routes between train units?

5. Command and Signal.

a. Command.

(1) What are the locations of the commander and key staff members?
(2) What is the chain of command and the succession of command?

b. Signal.

(1) Do instructions include backup code words and visual signals for all special and emergency events?

(2) Do instructions cover how to identify friendly forces to aircraft?

(3) Are SOI distributed to all units with a need to know, such as higher, lower, adjacent, leading, and follow-on elements?

Appendix B

Standing Operating Procedures

The intent of this appendix is to provide an outline as an aid for developing a squadron tactical SOP.

I. GENERAL

A. Purpose: This tactical SOP prescribes standard procedures for use during all combat situations and provides a comprehensive reference for conducting operations in a field environment.

B. Conformity: All assigned, attached and OPCON personnel will read and comply with the provisions of this tactical SOP.

II. BATTLE COMMAND

A. Command.

1. Organization.

a. Succession of command.

b. Cues for assuming command.

c. Operation of the command post.

(1) Shifts.

(2) Displacement/set-up/tear-down.

2. Troop-leading procedures.

a. Estimate input (checklist).

b. Precombat inspection (checklist).

c. Backbriefs.

d. Rehearsals.

e. Combat orders.

(1) Formats.

(2) Preparation.

(3) Reproduction.

(4) Dissemination.

3. Coordination with adjacent units (checklist).

4. Liaison with main body elements (checklist).

B. Control.

1. Combat graphics and symbols.

2. Control measures designation (numbering system).

3. Terrain index reference system (TIRS).

4. Operational terms.

5. Vehicle identification marking system.

6. Unit recognition signals.

7. Communication.

a. Net diagrams.

b. Fixed call signs.

c. Brevity codes/cue words.

d. Antijamming actions.

e. Alternate means.

8. Reports.

a. Battle.

b. Intelligence.

c. Logistical.

III. MANEUVER

A. Readiness condition (REDCON).

B. Quartering party.

C. Road marches.

D. Assembly areas.

E. Squadron organization for combat.

F. Formations.

G. Battle plays.

1. Actions on contact.

2. Fix and bypass.

3. Close assault.

4. In-stride breach.

5. Bridges/defile.

6. Formation changes.

7. Passage of lines.

8. Hasty defense.

9. Consolidate on the objective.

10. Ambush.

11. Other plays.

H. Reconnaissance operations.

1. Zone reconnaissance.

a. Graphics.

b. Critical tasks.

2. Area reconnaissance.

a. Graphics.

b. Critical tasks.

3. Route reconnaissance.

a. Graphics.

b. Critical tasks.

I. Security operations.

1. Screen.

a. Graphics.

b. Critical tasks.

2. Area security.

a. Route.

b. Convoy.

J. Offensive operations.

1. Movement to contact.

2. Hasty attack.

3. Offensive graphics.

K. Defensive operations (defensive graphics).

L. Limited visibility operations (checklist).

M. Break in action (checklist).

1. Redistribution of ammunition, personnel, and equipment.

2. Evacuation of casualties and PWs.

3. Redistribution of ammunition under fire.

N. Relief in place.

1. Relief in place graphics.

2. Critical tasks.

O. Air-ground team.

1. Command relationship.

2. Formations/techniques of movement.

IV. MOBILITY AND SURVIVABILITY

A. Mobility.

1. Standard tasks.

2. Standard priorities.

B. Countermobility.

1. Standard tasks.

2. Standard priorities.

3. Engineer target turnover (checklist).

C. Survivability.

1. Standard tasks.

2. Standard priorities.

3. Fighting position construction.

D. NBC defense.

1. NBC team organization and equipment.

2. Unit NBC equipment.

3. Defense against nuclear attack.

4. Defense against chemical/biological attack.

5. Decontamination.

6. NBC reconnaissance.

V. FIRE SUPPORT

A. Fire support request sequence.

B. Fire support planning and execution matrix format.

C. Graphics.

VI. AIR DEFENSE

A. Air defense warning and cue words.

B. Weapons control status and cue words.

VII. INTELLIGENCE/ELECTRONIC WARFARE

A. Intelligence.

1. Standard tasks.

2. Standard priorities.

B. Electronic warfare.

1. Standard tasks.

2. Standard priorities.

3. Countermeasures.

VIII. LOGISTICS

A. Resupply procedures.

1. LOGPAC procedures.

2. Battle loss actions.

B. Combat Health Support.

1. Emergency medical treatment.

2. Medical evacuation.

3. Field hygiene and sanitation.

4. Combat stress control.

5. Routine sick call.

C. Maintenance support.

1. Battle damage assessment and repair.

2. Exchange criteria.

3. Cannibalization criteria.

4. Destruction criteria.

5. Maintenance repair time guidelines.

D. Personnel.

1. Replacements.

2. Accountability.

3. Personnel actions.

E. Combat trains.

1. Layout.

2. Operations.

IX. SAFETY

Appendix C

Operations Order

Operations order format

A sample OPORD is outlined below.

1. Situation.

a. Enemy.

(1) Weather (note effects - include light data).

(a) Light data.
(b) Weather forecast for operation.
(c) Effects of weather and light data on operations.

- Trafficability.
- Visibility.
- Effect on lasers/thermals.
- Effect on air operations.

(2) Terrain.

(a) Obstacles, hills, valleys, road types and conditions, streams, rivers, bridges, towns.
(b) Avenues of approach.

- Size unit supported.
- Start and end point.
- Objective.

(c) Key terrain.

(d) Observation.
(e) Cover and concealment.
(f) Engagement areas.
(g) Effect of terrain on the operation.

(3) Enemy forces.

(a) Identification of enemy forces.

(b) Activity of enemy forces.
(c) Location of enemy units.
(d) Disposition of enemy forces.
(e) Strength of enemy forces.
(f) Composition of enemy forces, to include type of equipment.
(g) Other enemy information critical to the upcoming operation, to include the following:

- Chemical and nuclear capabilities.
- Air defense artillery (ADA).
- Aviation, including helicopters.
- Electronic warfare.

(h) Enemy courses of action.
(i) Probable enemy course of action.

b. Friendly.

(1) Mission of higher headquarters (squadron) and commander's intent.
(2) Mission of adjacent units (left, right, front, rear).
(3) Mission of other organic units in higher headquarters.
(4) Mission of reserves in higher headquarters.
(5) Mission of supporting units who are in direct support/reinforcing (DS/R) to higher headquarters (field artillery, engineer, ADA).
(6) Which element from higher headquarters has priority of fires.
(7) Reinforcing units with a reinforcing/general support (R/GS) role to supporting units.
(8) Close air support (CAS) and number of sorties allocated to higher headquarters.

c. Attachments and detachments to the troop.

2. Mission. The *who, what, when, where,* and *why* for troop/company. State the essential task or tasks to be accomplished by the entire unit, to include on-order missions. Clearly define the troop's objective.

3. Execution.

a. Concept of the Operation. This paragraph explains the commander's intent by stating the *purpose*, *method*, and *endstate* of the operation. The *purpose* tells the "why" of the operation. The *method* tells how the commander visualizes achieving success with respect to the troop as a whole and the utilization of any combat multipliers, in general terms. The *endstate* tells what the final disposition of forces will be and how the endstate will facilitate future operations. For example--

We will conduct the reconnaissance to determine the location, disposition, and composition of enemy main defensive belt in order to facilitate movement through zone, and penetration of the main defensive belt by 52d ID (Mech), allowing 10th Corps to continue offensive operations to the north. We will conduct a forward passage of lines, and perform zone reconnaissance in a troop vee with the engineer platoon conducting a detailed route reconnaissance of Route Cherry behind the scout platoons. Initially movement will be rapid and aggressive until we make contact with the enemy main defensive belt, then transition to a more deliberate, detailed reconnaissance. Indirect fires will be used to provide immediate suppression and assist in the development of the situation by scout platoons. Success is determining the disposition, composition, and location of the enemy main defensive belt then determining the axis and point for the 52d ID (Mech) penetration of the enemy's defense. Our endstate is the troop screening along the enemy main defensive belt ready to pass elements of the 52d ID (Mech) forward for their attack.

(1) Scheme of Maneuver. How major units (maneuver and combat support) will be employed; including movement techniques, bypass criteria (if any), and engagement/disengagement criteria. See the following example.

C Troop moves at 130430 from TAA Viper along Route Dog in troop column. Order of march on Route Dog is Red, Blue, Mortars, FIST, Engineers, Black 6, White, Green, TOC, and combat trains.

C Troop crosses LD at 0500 with Red on left (south), Blue on right (north), Mortars, Engineers, White, Green, TOC, and combat trains moving center of zone (troop vee).

Scout platoons reconnoiter in zone abreast, establishing contact at designated troop internal and flank contact points. Upon enemy contact, develop the situation at platoon level, initially with indirect fire, and report. Destroy squad- and section-sized light-armored and soft-skinned vehicles, within capability. Anything beyond capability, be prepared to assist hasty attack by White and/or Green. Bypass only enemy dismount elements squad-sized or smaller. Bypass all built-up areas larger than 1 kilometer square not on Route Cherry. Upon contact with enemy main defensive belt, conduct dismounted patrols to ascertain enemy disposition, composition, orientation, and location of obstacles.

Mortars cross LD and establish MFP 10 following Blue. Maneuver from MFP 11 to MFP 12 through MFP 25 in sequence. Maintain ability to range at least one-third maximum range forward of lead scouts. Key movement on scouts reporting crossing of phase lines throughout zone.

Engineers cross LD following the FIST and conduct a detailed route reconnaissance of Route Cherry from SP to RP. Maneuver on Route Cherry approximately 500 meters rearward of scouts. Upon identification of enemy's main defensive belt, be prepared to assist scouts with obstacle reconnaissance. White and Green cross LD in platoon columns, White leading. Deploy into platoon wedges forward of LD, terrain permitting. Maneuver in center of troop zone, Green following White by 500 meters. White and Green maneuver approximately one phase line behind Red and Blue through hide positions T-44 through T-52 in sequence. Key movement on scouts reporting crossing of phase lines throughout zone. Scouts will assist in positioning tanks for conduct of hasty attacks by fire or by fire and maneuver upon approval of the troop commander.
TOC and combat trains cross LD in column, following Green. TOC maneuvers center troop zone, establishing positions H-44 through H-48 in sequence. Remains approximately two phase lines rearward of lead scouts. Trains move center troop zone in column, establishing positions T-27 through T-32 in sequence. Remain approximately two phase lines rearward of lead scouts. Key movement on scouts reporting crossing of phase lines throughout zone.

(2) Fires.

(a) Purpose for field artillery and mortar fires. (How will indirect fires support the scheme of maneuver?)

(b) Preparation starting time, duration and description of any fires landing in the area of operations.
(c) Allocation of final protective fires (FPF).
(d) Which element will have priority of fires.
(e) Number of priority targets allocated and who will control them.
(f) Special fires, restrictions, allocation/use of smoke, illumination, or CAS.

(g) Description of scheduled fires (offensive).
(h) Reference to fire support or target annexes.
(3) Obstacles, mines, fortifications.

(a) Priority of engineer effort.

(b) Priority of engineer mission.
(c) Obstacle overlay.
(d) Obstacle list.
(e) Logistical constraints.
(f) On-order missions.
b. Specific tasks to subordinate units. List specific missions in "battle sequence" for each subordinate unit, including attached units. Include movement techniques, flank coordination requirements, other details, and any "be-prepared" missions.

c. Coordinating instructions.
(1) Movement instructions.

(2) Time schedule of events.

- Rehearsal.
- Backbrief times.
- Precombat inspection time.
- First movement time.

(3) Passage of lines.

- Contact points.
- Passage points.
- Lanes and identification.

(4) Mission-oriented protection posture (MOPP) level.

(5) Operational exposure guide (OEG).
(6) Actions on contact.
(7) Actions at danger areas.
(8) Priority intelligence requirements (PIR) and intelligence requirements (IR).
(9) Effective time of attachment or detachment.
(10) Rally points.
(11) Rules of engagement (ROE).
(12) Priority targets for direct-fire weapons.
(13) Uniform and equipment, to include weapon and ammunition.
(14) Any changes regarding battlecarry and battlesight ranges.
(15) Air defense posture and weapons control status.
(16) Any instructions not provided in concept of the operation or specific instructions.

4. Service Support.
a. General.

(1) Organization of trains (squadron and troop).

- Combat.
- Field.

(2) Location of trains (initial and subsequent).

(3) Movement of trains (movement instructions).

b. Materiel and services.
(1) Supply.

- Class I: Time and type.
- Class III: Time, location, and method.
- Class V: Time, amount, and type.
- LOGPAC instructions (include squadron logistics release point [LRP]).

(2) Transportation. Location of main supply route.

(3) Services.

- Mortuary affairs.
- Clothing exchange and bath.
- Maintenance.

c. Medical evacuation and treatment.
(1) Location of squadron aid station.

(2) Displacement of squadron aid station.
(3) Location of regimental clearing station.
(4) Aero-medical evacuation information.
(5) Location of ambulance exchange points.
(6) Handling of contaminated wounded personnel.

d. Personnel.
(1) Prisoner of war (PW) handling and disposition instructions.

(2) PW guard instructions.
(3) Location of PW collection point.
(4) Instructions for interaction with local civil populace (ROE).
(5) Number of expected replacements.
(6) Cross-leveling procedures.
e. Miscellaneous.

5. Command and Signal.

a. Command.

(1) Location of troop commander/troop TOC/squadron command post.
(2) Succession of command.

b. Signal.

(1) SOI index and edition in effect.

- Key frequencies.
- Key call signs.
- Current item number identifier.

(2) KY-57 fill and change over.

(3) Current sign and countersign.
(4) Code words.
(5) Action if jamming or "hot mike" occurs.
(6) Pyrotechnics use.
(7) Periods and conditions of listening silence.

Appendix D

Operations Other Than War

Operations other than war are military activities during peacetime and conflict that do not necessarily involve armed clashes between two organized forces. A cavalry troop will conduct reconnaissance, security, offensive, defensive, and other tactical operations in support of operations other than war. Some examples of operations other than war that a cavalry troop may be involved in are as follows:

PEACETIME CONFLICT
Counterdrug Strikes and raids

Disaster relief Peace enforcement

Civil support Support to insurgency

Peace-building Anti-terrorism

Nation assistance Peacekeeping

Noncombatant evacuation

operations (NEO)

The primary operational constraints during operations other than war are defined by the rules of engagement (ROE). These rules range from the use of force for self-defense only to total impartiality when applying force. The rules of engagement for peacetime will be more restrictive than those for conflict. The troop commander should consider the following in relation to the rules of engagement:

- The degree of force used must only be sufficient to achieve the task at hand and prevent, as far as possible, loss of human life or serious injury.
- Leaders at all levels must ensure that the ROE do not limit the soldiers' ability to defend themselves.
- Leaders must never tape over magazines to keep soldiers from chambering rounds.
- Leaders must ensure all soldiers understand what conditions must be met before chambering rounds.
- The ROE must be realistic, simple, and easy to understand. Develop a single card that clearly outlines the ROE and issue it to all soldiers for reference, keeping in mind that the card in itself is not the answer. Soldiers must know and understand the ROE.
- During peacekeeping, forces have no mandate to prevent violations of peace agreements by the active use of force. (Observe and report only.) To maintain the peace, units may be positioned

between belligerents. Troop commanders must realize that soldiers are being placed at risk. The protection of the troop must be emphasized.

- Peace enforcement missions allow the active use of force. The peacetime ROE resemble the ROE for hostilities (wartime).
- The formulation of ROE should consider the cultural differences of multinational forces.
- Soldiers should be trained in the ROE, using tactical vignettes or simulated events. Train soldiers to avoid unnecessary collateral damage to property.

Actions on contact and the troop's ability to develop the situation will be affected by the rules of engagement for the theater in which the troop is operating. The troop commander should rehearse battle drills based on those rules.

In preparing for operations other than war, a troop may require specialized training. They could benefit from training in the following areas:

- Nature of peacekeeping.
- Regional orientation/culture of belligerents.
- Negotiating skills.
- Mine/booby trap/unexploded ordnance training.
- Checkpoint operations.
- Investigating and reporting.
- Information collection.
- Patrolling.
- Media inter-relationships.
- Establishing a buffer zone.
- Supervising a truce or cease-fire.
- Contributing to maintenance of law and order.
- Demilitarizing cities or geographical areas.
- Border surveillance/security.

Glossary

AAR	after-action review
A&O	assault and obstacle
ACE	armored combat earthmover
ACR	armored cavalry regiment
ACT	air cavalry troop
ADA	air defense artillery
A/L	administrative/logistics
ALO	air liaison officer
ALOC	administrative/logistics operations center
AO	area of operations
APERS	antipersonnel (ammunition)
AT	antitank
APC	armored personnel carrier
arty	artillery
AVLB	armored vehicle launched bridge
BHL	battle handover line
BP	battle position
BUA	built-up area
C2	command and control
cal	caliber
CAS	close air support
cbt	combat
CBU	cluster bomb unit
CEOI	communications-electronics operation instructions
CEV	combat engineer vehicle
CDR	commander
CFV	cavalry fighting vehicle
cGy/hr	centigray(s) per hour
CHEMWARN	chemical warning (message sent in the event of a friendly strike)
COA	course of action
COMSEC	communications security
COT	counter obstacle team
CP	command post
CS	combat support
CSS	combat service support

CTCP	combat trains command post
DA	Department of the Army
DS	direct support
DS/R	direct support/reinforcing
EA	engagement area
engr	engineer
EPW	enemy prisoner of war
evac	evacuate; evacuation
EW	electronic warfare
1SG	first sergeant
FA	field artillery
FASCAM	family of scatterable mines
FAC	forward air controller
FEBA	forward edge of the battle area
FIST	fire support team
FLOT	forward line of own troops
FM	frequency modulation (radio); field manual
FPF	final protective fires
FRAGO	fragmentary order
FS	fire support
FSCL	fire support coordination line
FSE	fire support element
FSO	fire support officer
FSN	fire support net
GLD	ground laser designator
GPS	global positioning system
GRREG	graves registration
GSR	ground surveillance radar
HC	heavy concentration (smoke)
HE	high explosive
HEMTT	heavy expanded mobile tactical truck
HHT	headquarters and headquarters troop
HMMWV	high mobility multipurpose wheeled vehicle
HQ	headquarters
hvy	heavy
ICM	improved conventional munitions

ID	identification
IPB	intelligence preparation of the battlefield
IR	intelligence requirements
KIA	killed in action
km	kilometers
LACR	light armored cavalry regiment
LC	line of contact
LD	line of departure
ldr	leader
LID	light infantry division
LO	liaison officer
LOA	limit of advance
LOGPAC	logistics package
LRP	logistics release point
lt	light
maint	maintenance
MBA	main battle area
MCOO	modified combined obstacle overlay
mech	mechanized
METT-T	mission, enemy, terrain, troops and time available
MFP	mortar firing point
MICLIC	mine clearing line charge
MOPP	mission-oriented protection posture
mort	mortar
MRB	motorized rifle battalion
MRC	motorized rifle company
MRE	meals ready to eat
MRP	motorized rifle platoon
MSR	main supply route
MTOE	modified table(s) of organization and equipment

NAI	named area(s) of interest
NBC	nuclear, biological, chemical
NCO	noncommissioned officer
NCOIC	noncommissioned officer in charge
NEO	noncombatant evacuation operations
NET	new equipment training
NPL	no penetration line
NUCWARN	nuclear warning (message sent in the event of a friendly strike)
obj	objective
OCOKA	observation and fields of fire, cover and concealment, obstacles, key terrain, avenues of approach (considerations in evaluating terrain as part of METT-T)
OEG	operational exposure guidance
OI	operations and intelligence
OIC	officer in charge
OP	observation post
OPCON	operational control
OPLAN	operation plan
OPORD	operations order
OPSEC	operations security
PAC	personnel and administration center
PD	point of departure
PEWS	platoon early warning system
PIR	priority intelligence requirements
PL	phase line
PLL	prescribed load list
plt	platoon
PMCS	preventive maintenance checks and services
POL	petroleum, oils, and lubricants
POSNAV	position navigation
PSG	platoon sergeant
PW	prisoner of war

recon	reconnaissance; reconnoiter
REDCON	readiness condition
RFA	restricted fire area
RFL	restricted fire line
R/GS	reinforcing/general support
ROE	rules of engagement
ROM	refuel on the move
RP	release point
rte	route
S1	adjutant (US Army)
S2	intelligence officer (US Army)
S3	operations and training officer (US Army)
S4	supply officer (US Army)
SCO	squadron commander
sec ldr	section leader
SEE	small emplacement excavator
SGLI	Serviceman's Group Life Insurance
sgt	sergeant
SIDPERS	Standardization Installation/Division Personnel System
SIGSEC	signal security
SITREP	situation report
SOI	signal operation instructions
SOP	standing operating procedure
SP	start point
sqd ldr	squad leader
TAA	tactical assembly area
TAC CP	tactical command post
TACFIRE	tactical fire direction system
TC	tank commander
TIRS	terrain index reference system
TOC	tactical operations center
TOE	table(s) of organization and equipment
TOW	tube-launched, optically-tracked, wire-guided (missile)
TRADOC	Training and Doctrine Command
TRP/trp	target reference point; troop
TTP	tactics, techniques, and procedures
UMCP	unit maintenance collection point

UMT	unit ministry team
US	United States
veh	vehicle
WIA	wounded in action
wrkr	wrecker
XO	executive officer

References

SOURCES USED

These are the sources quoted or paraphrased in this publication.

Army Publications

FM 17-15	Tank Platoon. 7 October 1987.
FM 17-95	Cavalry Operations. 19 September 1991.
FM 17-95-10	Armored Cavalry Regiment and Squadron. 22 September 1993.
FM 17-98	Scout Platoon. 9 September 1994.
FM 71-1	Tank and Mechanized Infantry Company. 22 November 1988.

Command Publications

Command publications cannot be obtained through Armywide resupply channels. Determine availability by contacting the address below.

Commander
USAARMC
ATTN: ATZK-IMO-RS
Fort Knox, KY 40121-5000

FKSM 17-97-3	Cavalry Troop Common SOP. 15 July 1994.

READINGS RECOMMENDED

These readings contain relevant supplemental information.

Army Publications

ARTEP 17-57-10-MTP	Mission Training Plan for the Scout Platoon. 27 December 1988.
ARTEP 17-237-10-MTP	Mission Training Plan for the Tank Platoon. 3 October 1988.
ARTEP 17-487-30-MTP	Mission Training Plan for the Regimental Armored Cavalry Troop. 3 September 1991.

DA Form 1155	Witness Statement on Individual. June 1966.
DA Form 1156	Casualty Feeder Report. June 1966.
DA Form 1971-R	Radiological Data Sheet Monitoring and Point Technique. September 1986.
DA Form 2028	Recommended Changes to Publications and Blank Forms. February 1974.

DD Form 551	Record of Interment. August 1984.
DD Form 1077	Collecting Point Register of Deceased Personnel. July 1984.

FM 1-114	Tactics, Techniques, and Procedures for the Regimental Aviation Squadron. 20 February 1991.
FM 1-116	Tactics, Techniques, and Procedures for the Air Cavalry/Reconnaissance Troop. 20 February 1991.
FM 3-3	Chemical and Biological Contamination Avoidance. 16 November 1992.
FM 3-4	NBC Protection. 20 May 1992.
FM 3-5	NBC Decontamination. 23 July 1992.
FM 3-50	Smoke Operations. 4 December 1990.
FM 3-100	NBC Defense, Chemical Warfare, Smoke, and Flame Operations.

	23 May 1991.
FM 3-101	Chemical Staffs and Units. 22 April 1987.
FM 5-36	Route Reconnaissance and Classification. 10 May 1985.
FM 5-100	Engineer Combat Operations. 22 November 1988.
FM 5-250	Explosives and Demolitions. 15 June 1992.
FM 6-20	Fire Support in the AirLand Battle. 17 May 1988.
FM 7-91	Tactical Employment of Antiarmor Platoons, Companies, and Battalions. 30 September 1987.
FM 7-98	Operations in a Low-Intensity Conflict. 19 October 1992.
FM 8-55	Planning for Health Service Support. 15 February 1985.
FM 10-63-1	Graves Registration Handbook. 17 July 1986.
FM 17-12-1-1	Tank Gunnery (Abrams), Volume I. 19 March 1993.
FM 17-12-1-2	Tank Gunnery (Abrams), Volume II. 19 March 1993.
FM 17-12-8	Light Cavalry Gunnery. 28 October 1994.
FM 17-18	Light Armor Operations. 8 March 1994.
FM 20-32	Mine/Countermine Operations. 30 September 1992.
FM 21-10	Field Hygiene and Sanitation. 22 November 1988.
FM 21-10-1	Unit Field Sanitation Team. 11 October 1989.
FM 21-11	First Aid for Soldiers. 27 October 1988.
FM 22-100	Military Leadership. 31 July 1990.

FM 23-1	Bradley Fighting Vehicle Gunnery. 1 March 1991.
FM 34-3	Intelligence Analysis. 15 March 1990.
FM 44-1	US Army Air Defense Artillery Employment. 9 May 1983.
FM 44-3	Air Defense Artillery Employment: Chaparral/Vulcan/Stinger. 15 June 1984.
FM 44-8	Small Unit Self-Defense Against Air Attack. 30 December 1981.
FM 44-18-1	Stinger Team Operations. 31 December 1984.
FM 71-2	The Tank and Mechanized Infantry Battalion Task Force. 27 September 1988.
FM 100-2-1	Soviet Army Operations and Tactics. 16 July 1984.
FM 100-2-3	The Soviet Army: Troops, Organization, and Equipment. 6 June 1991.
FM 100-5	Operations. 14 June 1993.
FM 101-5-1	Operational Terms and Symbols. 21 October 1985.
TC 6-40A	Field Artillery Automated Cannon Gunnery. 21 April 1989.

Command Publications

Contact the address shown on page References-1 for these command publications.

FKSM 17-15-3	Tank Platoon SOP. February 1989.

FKSM 17-98-3	Scout Platoon SOP. November 1994.
FKSM 17-98-4	Scout Platoon Leader's Notebook.

Standardization Agreements (STANAGs)

To requisition these documents, complete DD Form 1425 and send it to—

Naval Publications and Forms Center
5801 Tabor Avenue
Philadelphia, PA 19120

STANAG 2002	Warning Signs for the Marking of Contaminated or Dangerous Land Areas, Complete Equipment, Supplies and Stores. 2 December 1989.
STANAG 2014	Operation Orders, Warning Orders and Administrative/Logistics Orders. 14 August 1991.
STANAG 2029	Method of Describing Ground Locations, Areas and Boundaries. 10 June 1991.
STANAG 2036	Land Mine Laying, Marking, Recording and Reporting Procedures. 29 April 1988.
STANAG 2044	Procedures for Dealing with Prisoners of War (PW). 12 July 1988.
STANAG 2047	Emergency Alarms of Hazard or Attack (NBC and Air Attack Only). 10 October 1990.
STANAG 2082	Relief of Combat Troops. 24 July 1991.
STANAG 2083	Commanders' Guide on Nuclear Radiation Exposure of Groups.

10 September 1990.

STANAG 2084 Handling and Reporting of Captured Enemy Equipment and Documents. 26 June 1986.

STANAG 2103 Reporting Nuclear Detonations, Biological and Chemical Attacks, and Predicting and Warning of Associated Hazards and Hazard Areas. 10 June 1991.

STANAG 2104 Friendly Nuclear Strike Warning. 12 December 1989.

STANAG 2112 NBC Reconnaissance. 5 August 1986.

STANAG 2113 Denial of a Unit's Military Equipment and Supplies to an Enemy. 11 July 1991.

STANAG 2129 Identification of Land Forces on the Battlefield. 16 May 1989.

STANAG 2889 Marking of Hazardous Areas and Routes Through Them. 8 February 1990.

*FIELD MANUAL
NO. 4-25.11
NAVY TACTICAL
REFERENCE
PUBLICATION
NO. 4-02.1
AIR FORCE MANUAL
NO. 44-163(I)

HEADQUARTERS
DEPARTMENT OF THE ARMY,
THE NAVY, AND THE AIR FORCE
Washington, DC, 23 December 2002

FIRST AID

TABLE OF CONTENTS

*This publication supersedes FM 21-11, 27 October 1988

PREFACE

This manual meets the first aid training needs of individual service members. Because medical personnel will not always be readily available, the nonmedical service members must rely heavily on their own skills and knowledge of life-sustaining methods to survive on the integrated battlefield. This publication outlines both self-aid and aid to other service members (buddy aid). More importantly, it emphasizes prompt and effective action in sustaining life and preventing or minimizing further suffering and disability. First aid is the emergency care given to the sick, injured, or wounded before being treated by medical personnel. The term *first aid* can be defined as "urgent and immediate lifesaving and other measures, which can be performed for casualties by nonmedical personnel when medical personnel are not immediately available." Nonmedical service members have received basic first aid training and should remain skilled in the correct procedures for giving first aid. This manual is directed to *all* service members. The procedures discussed apply to all types of casualties and the measures described are for use by both male and female service members.

This publication is in consonance with the following North Atlantic Treaty Organization (NATO) International Standardization Agreements (STANAGs) and American, British. Canadian, and Australian Quadripartite Standardization Agreements (QSTAGs).

TITLE	STANAG	QSTAG
Medical Training in First Aid, Basic Hygiene and Emergency Care	2122	535
First Aid Kits and Emergency Medical Care Kits	2126	
Medical First Aid and Hygiene Training in NBC Operations	2358	
First Aid Material for Chemical Injuries	2871	

These agreements are available on request, using Department of Defense (DD) Form 1425 from the Standardization Documents Order Desk, 700 Robins Avenue, Building 4, Section D, Philadelphia, Pennsylvania 19111-5094.

Unless this publication states otherwise, masculine nouns and pronouns do not refer exclusively to men.

Use of trade or brand names in this publication is for illustrative purposes only and does not imply endorsement by the Department of Defense (DOD).

The proponent for this publication is the US Army Medical Department Center and School. Submit comments and recommendations for the improvement of this publication directly to the **Commander, US Army Medical Department Center and School, ATTN: MCCS-FCD-L, 1400 East Grayson Street, Fort Sam Houston, Texas 78234-5052**.

CHAPTER 1

FUNDAMENTAL CRITERIA FOR FIRST AID

"The fate of the wounded rests in the hands of the ones who apply the first dressing."

Nicholas Senn (1898) (49th President of the American Medical Association)

1-1. General

When a nonmedical service member comes upon an unconscious or injured service member, he must accurately evaluate the casualty to determine the first aid measures needed to prevent further injury or death. He should seek medical assistance as soon as possible, but he should not interrupt the performance of first aid measures. To interrupt the first aid measures may cause more harm than good to the casualty. Remember that in a chemical environment, the service member should not evaluate the casualty until the casualty has been masked. After performing first aid, the service member must proceed with the evaluation and continue to monitor the casualty for development of conditions which may require the performance of necessary basic lifesaving measures, such as clearing the airway, rescue breathing, preventing shock, and controlling bleeding. He should continue to monitor the casualty until relieved by medical personnel.

Service members may have to depend upon their first aid knowledge and skills to save themselves (self-aid) or other service members (buddy aid/ combat lifesaver). They may be able to save a life, prevent permanent disability, or reduce long periods of hospitalization by knowing **WHAT** to do, **WHAT NOT** to do, and **WHEN** to seek medical assistance.

NOTE

The prevalence of various body armor systems currently fielded to US service members, and those in development for future fielding, may present a temporary obstacle to effective evaluation of an injured service member. You may have to *carefully remove* the body armor from the injured service member to complete the evaluation or administer first aid. Begin by removing the outermost hard or soft body armor components (open, unfasten or cut the closures, fasteners, or straps), then remove any successive layers in the same manner. Be sure to follow other notes, cautions and warnings regarding procedures in contaminated situations and when a broken back or neck is suspected. Continue to evaluate.

1-2. Terminology

To enhance the understanding of the material contained in this publication, the following terms are used—

- *Combat lifesaver.* This is a US Army program governed by Army Regulation (AR) 350-41. The combat lifesaver is a member of a nonmedical unit selected by the unit commander for additional training beyond basic first aid procedures (referred to as *enhanced first aid*). A minimum of one individual per squad, crew, team, or equivalent-sized unit should be trained. The primary duty of this individual does not change. The additional duty of combat lifesaver is to provide enhanced first aid for injuries based on his training before the trauma specialist (military occupational specialty [MOS] 91W) arrives. The combat lifesaver's training is normally provided by medical personnel assigned, attached, or in direct support (DS) of the unit. The senior medical person designated by the commander manages the training program.

- *Trauma Specialist (US Army) or Hospital Corpsman (HM).* A medical specialist trained in emergency medical treatment (EMT) procedures and assigned or attached in support of a combat or combat support unit or marine forces.

- *Casualty evacuation.* Casualty evacuation (CASEVAC) is a term used by nonmedical units to refer to the movement of casualties aboard nonmedical vehicles or aircraft. See also the term *transported* below. Refer to FM 8-10-6 for additional information.

> **CAUTION**
>
> Casualties transported in this manner do not receive en route medical care.

- *Enhanced first aid (US Army).* Enhanced first aid is administered by the combat lifesaver. It includes measures, which require an additional level of training above self-aid and buddy aid, such as the initiation of intravenous (IV) fluids.

- *Medical evacuation.* Medical evacuation is the timely, efficient movement of the wounded, injured, or ill service members from the battlefield and other locations to medical treatment facilities (MTFs). Medical personnel provide en route medical care during the evacuation. Once the casualty has entered the medical stream (trauma specialist, hospital corpsman, evacuation crew, or MTF), the role of first aid in the care of the casualty ceases and the casualty becomes the responsibility of the health service support (HSS) chain. Once he has entered the HSS chain he is referred to as a *patient*.

- *First aid measures*. Urgent and immediate lifesaving and other measures, which can be performed for casualties (or performed by the casualty himself) by nonmedical personnel when medical personnel are not immediately available.

- *Medical treatment*. Medical treatment is the care and management of wounded, injured, or ill service members by medically trained (MOS-trained) HM, and area of concentration (AOC) personnel. It may include EMT, advanced trauma management (ATM), and resuscitative and surgical intervention.

- *Medical treatment facility*. Any facility established for the purpose of providing medical treatment. This includes battalion aid stations, Level II facilities, dispensaries, clinics, and hospitals.

- *Self-aid/buddy aid*. Each individual service member is trained to be proficient in a variety of specific first aid procedures. This training enables the service member or a buddy to apply immediate first aid measures to alleviate a life-threatening situation.

- *Transported*. A casualty is moved to an MTF in a nonmedical vehicle without en route care provided by a medically-trained service member (such as a Trauma Specialist or HM). First aid measures should be continually performed while the casualty is being transported. If the casualty is acquired by a dedicated medical vehicle with a medically-trained crew, the role of first aid ceases and the casualty becomes the responsibility of the HSS chain, and is then referred to as a *patient*. This method of transporting a casualty is also referred to as *CASEVAC*.

1-3. Understanding Vital Body Functions for First Aid

In order for the service member to learn to perform first aid procedures, he must have a basic understanding of what the vital body functions are and what the result will be if they are damaged or not functioning.

a. Breathing Process. All humans must have oxygen to live. Through the breathing process, the lungs draw oxygen from the air and put it into the blood. The heart pumps the blood through the body to be used by the cells that require a constant supply of oxygen. Some cells are more dependent on a constant supply of oxygen than others. For example, cells of the brain may die within 4 to 6 minutes without oxygen. Once these cells die, they are lost forever since they do not regenerate. This could result in permanent brain damage, paralysis, or death.

b. Respiration. Respiration occurs when a person inhales (oxygen is taken into the body) and then exhales (carbon dioxide [CO_2] is expelled from the body). Respiration involves the—

- *Airway.* The airway consists of the nose, mouth, throat, voice box, and windpipe. It is the canal through which air passes to and from the lungs.

- *Lungs.* The lungs are two elastic organs made up of thousands of tiny air spaces and covered by an airtight membrane. The *bronchial tree* is a part of the lungs.

- *Rib cage.* The rib cage is formed by the muscle-connected ribs, which join the spine in back, and the breastbone in front. The top part of the rib cage is closed by the structure of the neck, and the bottom part is separated from the abdominal cavity by a large dome-shaped muscle called the *diaphragm* (Figure 1-1). The diaphragm and rib muscles, which are under the control of the respiratory center in the brain, automatically *contract* and *relax*. *Contraction* increases and *relaxation* decreases the size of the rib cage. When the rib cage increases and then decreases, the air pressure in the lungs is first less and then more than the atmospheric pressure, thus causing the air to rush into and out of the lungs to equalize the pressure. This cycle of inhaling and exhaling is repeated about 12 to 18 times per minute.

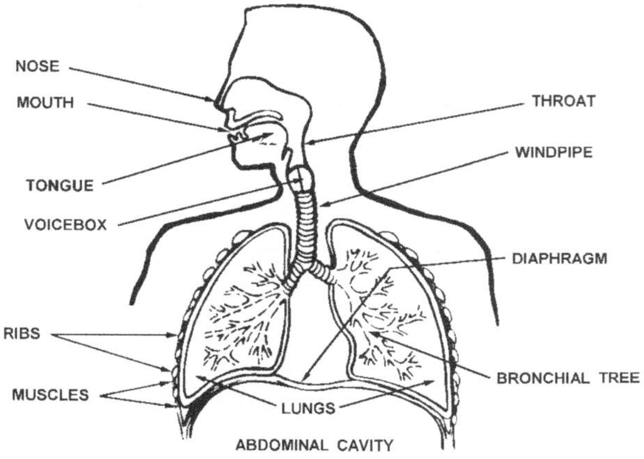

Figure 1-1. Airway, lungs, and rib cage.

c. *Blood Circulation.* The heart and the blood vessels (arteries, veins, and capillaries) circulate blood through the body tissues. The heart is divided into two separate halves, each acting as a pump. The left side pumps oxygenated blood (bright red) through the arteries into the capillaries; nutrients and oxygen pass from the blood through the walls of the capillaries into the cells. At the same time waste products and CO_2 enter the capillaries. From the capillaries the oxygen poor blood is carried through the veins to the right side of the heart and then into the lungs where it expels the CO_2 and picks up oxygen. Blood in the veins is dark red because of its low oxygen content. Blood does not flow through the veins in spurts as it does through the arteries. The entire system of the heart, blood vessels, and lymphatics is called the *circulatory system.*

(1) *Heartbeat.* The heart functions as a pump to circulate the blood continuously through the blood vessels to all parts of the body. It contracts, forcing the blood from its chambers; then it relaxes, permitting its chambers to refill with blood. The rhythmical cycle of contraction and relaxation is called the *heartbeat.* The normal heartbeat is from 60 to 80 beats per minute.

(2) *Pulse.* The heartbeat causes a rhythmical expansion and contraction of the arteries as it forces blood through them. This cycle of expansion and contraction can be felt (monitored) at various points in the body and is called the *pulse.* The common points for checking the pulse are at the—

- Side of the neck (*carotid*).
- Groin (*femoral*).
- Wrist (*radial*).
- Ankle (*posterior tibial*).

(*a*) *Carotid pulse.* To check the carotid pulse, feel for a pulse on the side of the casualty's neck closest to you. This is done by placing the tips of your first two fingers beside his Adam's apple (Figure 1-2).

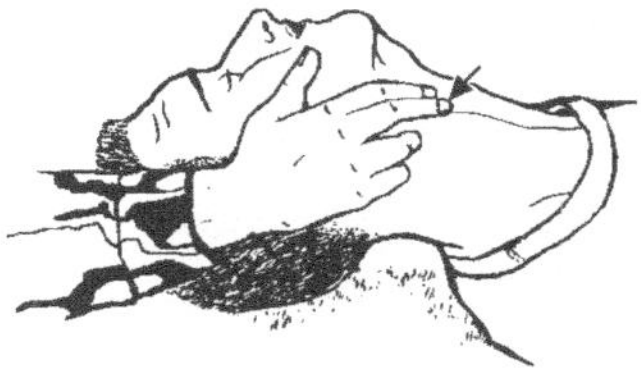

Figure 1-2. Carotid pulse.

(*b*) *Femoral pulse*. To check the femoral pulse, press the tips of your first two fingers into the middle of the groin (Figure 1-3).

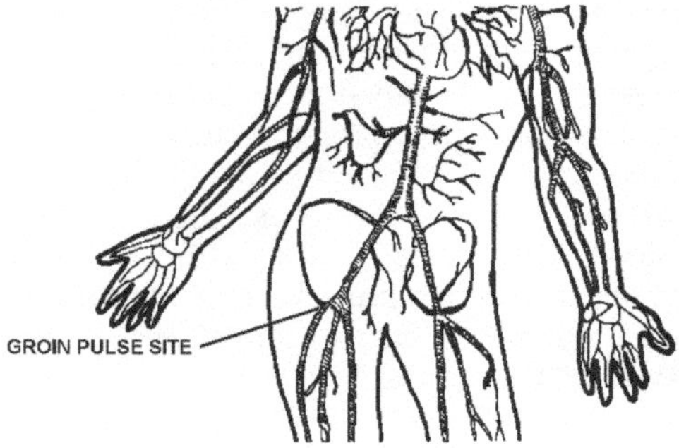

Figure 1-3. Femoral pulse.

(*c*) *Radial pulse*. To check the radial pulse, place your first two fingers on the thumb side of the casualty's wrist (Figure 1-4).

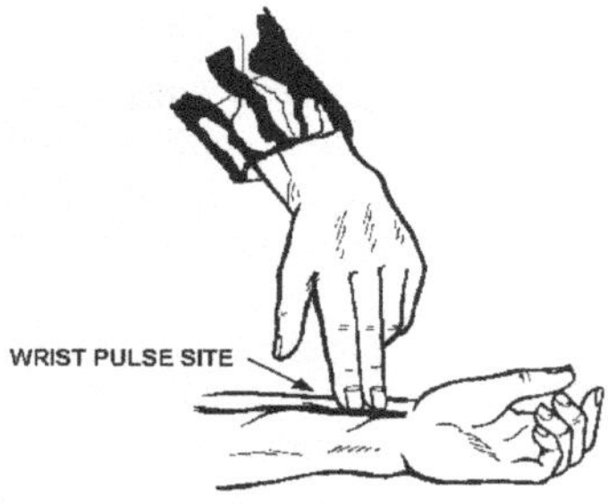

Figure 1-4. Radial pulse.

(*d*) *Posterior tibial pulse*. To check the posterior tibial pulse, place your first two fingers on the inside of the ankle (Figure 1-5).

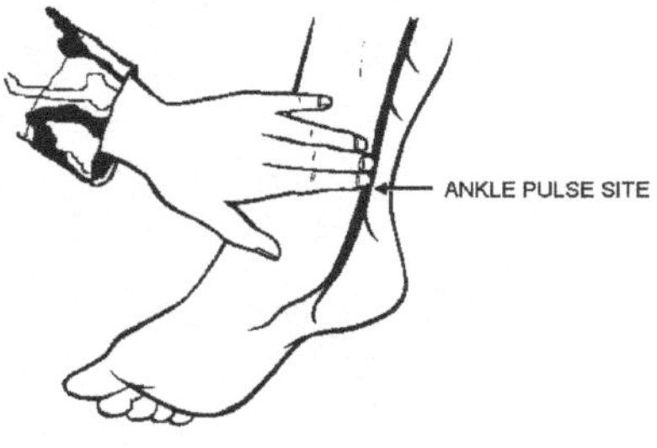

Figure 1-5. Posterior tibial pulse.

NOTE

DO NOT use your thumb to check a casualty's pulse because you may confuse the beat of your pulse with that of the casualty.

1-4. Adverse Conditions

a. Lack of Oxygen. Human life cannot exist without a continuous intake of oxygen. Lack of oxygen rapidly leads to death. First aid involves knowing how to open the airway and restore breathing.

b. Bleeding. Human life cannot continue without an adequate volume of blood circulating through the body to carry oxygen to the tissues. An important first aid measure is to stop the bleeding to prevent the loss of blood.

c. Shock. Shock means there is an inadequate blood flow to the vital tissues and organs. Shock that remains uncorrected may result in death even though the injury or condition causing the shock would not otherwise be fatal. Shock can result from many causes, such as loss of blood, loss of fluid from deep burns, pain, and reaction to the sight of a wound or blood. First aid includes preventing shock, since the casualty's chances of survival are much greater if he does not develop shock. Refer to paragraphs 2-21 through 2-24 for a further discussion of shock.

d. Infection. Recovery from a severe injury or a wound depends largely upon how well the injury or wound was initially protected. Infections result from the multiplication and growth (spread) of harmful microscopic organisms (sometimes referred to as germs). These harmful microscopic organisms are in the air, water, and soil, and on the skin and clothing. Some of these organisms will immediately invade (contaminate) a break in the skin or an open wound. The objective is to keep wounds clean and free of these organisms. A good working knowledge of basic first aid measures also includes knowing how to dress a wound to avoid infection or additional contamination.

1-5. Basics of First Aid

Most injured or ill service members are able to return to their units to fight or support primarily because they are given appropriate and timely first aid followed by the best medical care possible. Therefore, all service members must remember the basics.

- Check for **BREATHING**: Lack of oxygen intake (through a compromised airway or inadequate breathing) can lead to brain damage or death in very few minutes.

- Check for **BLEEDING**: Life cannot continue without an adequate volume of blood to carry oxygen to tissues.

- Check for **SHOCK**: Unless shock is prevented, first aid performed, and medical treatment provided, death may result even though the injury would not otherwise be fatal.

1-6. Evaluating a Casualty

a. The time may come when you must instantly apply your knowledge of first aid measures. This could occur during combat operations, in training situations, or while in a nonduty status. Any service member observing an unconscious and/or ill, injured, or wounded person must carefully and skillfully evaluate him to determine the first aid measures required to prevent further injury or death. He should seek help from medical personnel as soon as possible, but must not interrupt his evaluation of the casualty or fail to administer first aid measures. A second service member may be sent to find medical help. One of the cardinal principles for assisting a casualty is that you (the initial rescuer) must continue the evaluation and first aid measures, as the tactical situation permits, until another individual relieves you. If, during any part of the evaluation, the casualty exhibits the conditions (such as shock) for which the service member is checking, the service member must stop the evaluation and immediately administer first aid. In a chemical environment, the service member should not evaluate the casualty until both the individual and the casualty have been masked. If it is suspected that a nerve agent was used, administer the casualty's own nerve agent antidote autoinjector. After providing first aid, the service member must proceed with the evaluation and continue to monitor the casualty for further complications until relieved by medical personnel.

WARNING

Do not use your own nerve agent antidote autoinjector on the casualty.

NOTE

Remember, when evaluating and/or administering first aid to a casualty, you should seek medical aid as soon as possible. **DO NOT** stop first aid measures, but if the situation allows, send another service member to find medical aid.

b. To evaluate a casualty, perform the following steps:

(1) *Check the casualty for responsiveness.* This is done by gently shaking or tapping him while calmly asking, "Are you OK?" Watch for a response. If the casualty does not respond, go to step (2). If the casualty responds, continue with the evaluation.

(*a*) If the casualty is conscious, ask him where he feels different than usual or where it hurts. Ask him to identify the location of pain if he can, or to identify the area in which there is no feeling.

(*b*) If the casualty is conscious but is choking and cannot talk, stop the evaluation and begin first aid measures. Refer to paragraphs 2-10 and 2-11 for specific information on opening the airway.

WARNING

If a broken back or neck is suspected, do not move the casualty unless his life is in immediate danger (such as close to a burning vehicle). Movement may cause permanent paralysis or death.

(2) *Check for breathing.* (Refer to paragraph 2-6 for this procedure.)

(*a*) If the casualty is breathing, proceed to step (3).

(*b*) If the casualty is not breathing, stop the evaluation and begin first aid measures to attempt to ventilate the casualty. Attempt to open the airway, if an airway obstruction is apparent, clear the airway obstruction, then ventilate (see paragraphs 2-10 and 2-11).

(*c*) After successfully ventilating the casualty, proceed to step (3).

(3) *Check for pulse.* (Refer to paragraph 1-3*c*(2) for specific methods.) If a pulse is present and the casualty is breathing, proceed to step (4).

(*a*) If a pulse is present, but the casualty is still not breathing, start rescue breathing.

(*b*) If a pulse is not present, seek medical personnel for help.

(4) *Check for bleeding*. Look for spurts of blood or blood-soaked clothes. Also check for *both* entry and exit wounds. If the casualty is bleeding from an open wound, stop the evaluation and begin first aid procedures as follows for a—

(*a*) Wound of the arm or leg (refer to paragraphs 2-16 through 2-18 for information on putting on a field or pressure dressing).

(*b*) Partial or complete amputation, apply dressing (refer to paragraph 2-16 to 2-18) and then apply tourniquet if bleeding is not stopped (refer to paragraph 2-20 for information on putting on a tourniquet).

(*c*) Open head wound (refer to paragraph 3-10 for information on applying a dressing to an open head wound).

(*d*) Open chest wound (refer to paragraph 3-5 for information on applying a dressing to an open chest wound).

(*e*) Open abdominal wound (refer to paragraph 3-7 for information on applying a dressing to an open abdominal wound).

WARNING

In a chemically contaminated area, do not expose the wounds. Apply field dressing and then pressure dressing over wound area as needed.

(5) *Check for shock*. (Refer to paragraph 2-24 for first aid measures for shock.) If the signs and symptoms of shock are present, stop the evaluation, and begin first aid measures immediately. The following are the nine signs and symptoms of shock.

(*a*) Sweaty but cool skin (clammy skin).

(*b*) Paleness of skin. (In dark-skinned service members look for a grayish cast to the skin.)

(*c*) Restlessness or nervousness.

(*d*) Thirst.

(*e*) Loss of blood (bleeding).

(*f*) Confusion (does not seem aware of surroundings).

(*g*) Faster than normal breathing rate.

(*h*) Blotchy or bluish skin, especially around the mouth.

(*i*) Nausea or vomiting.

WARNING

Leg fractures must be splinted before elevating the legs as a first aid measure for shock.

(6) *Check for fractures.*

(*a*) Check for the following signs and symptoms of a back or neck injury and perform first aid procedures as necessary.

- Pain or tenderness of the back or neck area.
- Cuts or bruises on the back or neck area.
- Inability of a casualty to move or decreased sensation to extremities (paralysis or numbness).
  - Ask about ability to move (paralysis).
  - Touch the casualty's arms and legs and ask whether he can feel your hand (numbness).
- Unusual body or limb position.

(*b*) Immobilize any casualty suspected of having a back or neck injury by doing the following:

- Tell the casualty not to move.
- If a back injury is suspected, place padding (rolled or folded to conform to the shape of the arch) under the natural arch of the casualty's back. (For example, a blanket/poncho may be used as padding.)

WARNING

Do not move casualty to place padding.

- If a neck injury is suspected, immediately immobilize (manually) the head and neck. Place a roll of cloth under the casualty's neck, and put weighted boots (filled with dirt or sand) or rocks on both sides of his head.

(*c*) Check the casualty's arms and legs for open or closed fractures.

- Check for *open* fractures by looking for—
  - Bleeding.
  - Bones sticking through the skin.
  - Check for pulse.
- Check for *closed* fractures by looking for—
  - Swelling.
  - Discoloration.
  - Deformity.
  - Unusual body position.
  - Check for pulse.

(*d*) Stop the evaluation and begin first aid measures if a fracture to an arm or leg is suspected. Refer to Chapter 4 for information on splinting a suspected fracture.

(*e*) Check for signs/symptoms of fractures of other body areas (for example, shoulder or hip) and provide first aid as necessary.

(7) *Check for burns*. Look carefully for reddened, blistered, or charred skin; also check for singed clothing. If burns are found, stop the evaluation and begin first aid procedures. Refer to paragraph 3-9 for information on giving first aid for burns.

NOTE

Burns to the upper torso and face may cause respiratory complications. When evaluating the casualty, look for singed nose hair, soot around the nostrils, and listen for abnormal breath sounds or difficulty breathing.

(8) *Check for possible head injury.*

(*a*) Look for the following signs and symptoms:

- Unequal pupils.
- Fluid from the ear(s), nose, mouth, or injury site.
- Slurred speech.
- Confusion.
- Sleepiness.
- Loss of memory or consciousness.
- Staggering in walking.
- Headache.
- Dizziness.
- Nausea or vomiting.
- Paralysis.
- Convulsions or twitches.
- Bruising around the eyes and behind the ears.

(*b*) If a head injury is suspected, continue to watch for signs which would require performance of rescue breathing, first aid measures for shock, or control of bleeding; seek medical aid. Refer to paragraph 3-10 for information on first aid measures for head injuries.

CHAPTER 2

BASIC MEASURES FOR FIRST AID

2-1. General

Several conditions that require immediate attention are an inadequate airway, lack of breathing, and excessive loss of blood (circulation). A casualty without a clear airway or who is not breathing may die from lack of oxygen. Excessive loss of blood may lead to shock, and shock can lead to death; therefore, you must act immediately to control the loss of blood. All wounds are considered to be contaminated, since infection-producing organisms (germs) are always present on the skin and clothing, and in the soil, water, and air. Any missile or instrument (such as a bullet, shrapnel, knife, or bayonet) causing a wound pushes or carries the germs into that wound. Infection results as these organisms multiply. That a wound is contaminated does not lessen the importance of protecting it from further contamination. You must dress and bandage a wound as soon as possible to prevent further contamination.

NOTE

> It is also important that you attend to any airway, breathing, or bleeding problems **IMMEDIATELY** because these problems, if left unattended, may become life threatening.

Section I. OPEN THE AIRWAY AND RESTORE BREATHING

2-2. Breathing Process

All humans must have oxygen to live. Through the breathing process, the lungs draw oxygen from the air and put it into the blood. The heart pumps the blood through the body to be used by the cells that require a constant supply of oxygen. Some cells are more dependent on a constant supply of oxygen than others. For example, cells of the brain may die within 4 to 6 minutes without oxygen. Once these cells die, they are lost forever since they do not regenerate. This could result in permanent brain damage, paralysis, or death.

2-3. Assessment of and Positioning the Casualty

a. **CHECK** for responsiveness (Figure 2-1A)—establish whether the casualty is conscious by gently shaking him and asking, "Are you OK?"

b. **CALL** for help (Figure 2-1B).

c. **POSITION** the unconscious casualty so that he is lying on his back and on a firm surface (Figure 2-1C).

WARNING
If the casualty is lying on his chest (prone position), cautiously roll the casualty as a unit so that his body does not twist (which may further complicate a back, neck, or spinal injury).

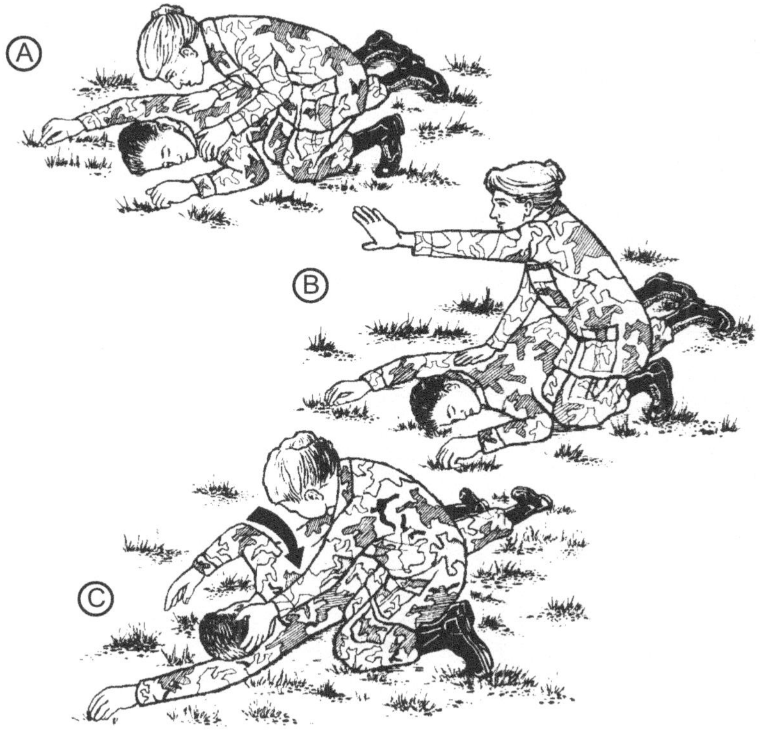

Figure 2-1. Assessment (Illustrated A—C).

(1) Straighten the casualty's legs. Take the casualty's arm that is nearest to you and move it so that it is straight and above his head. Repeat the procedure for the other arm.

(2) Kneel beside the casualty with your knees near his shoulders (leave space to roll his body) (Figure 2-1B). Place one hand behind his head and neck for support. With your other hand, grasp the casualty under his far arm (Figure 2-1C).

(3) Roll the casualty towards you using a steady, even pull. His head and neck should stay in line with his back.

(4) Return the casualty's arms to his side. Straighten his legs. Reposition yourself so that you are now kneeling at the level of the casualty's shoulders. However, if a neck injury is suspected and the jaw-thrust technique will be used, kneel at the casualty's head, looking towards his feet.

2-4. Opening the Airway of an Unconscious or Not Breathing Casualty

The tongue is the single most common cause of an airway obstruction (Figure 2-2). In most cases, simply using the head-tilt/chin-lift technique can clear the airway. This action pulls the tongue away from the air passage in the throat (Figure 2-3).

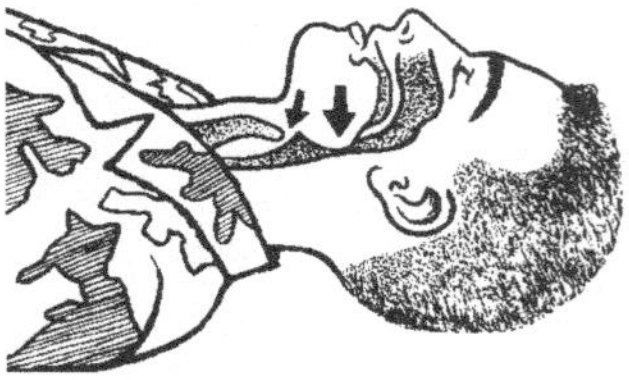

Figure 2-2. Airway blocked by tongue.

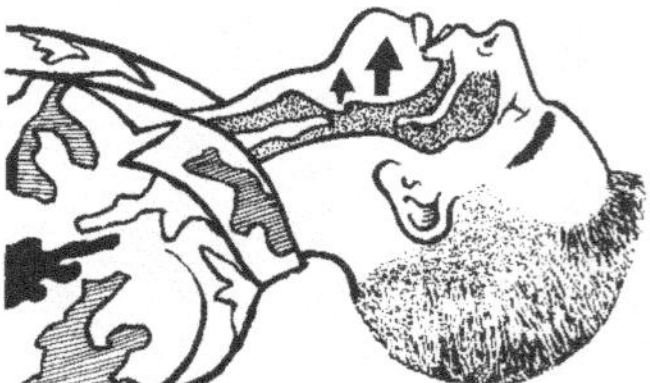

Figure 2-3. Airway opened by extending neck.

a. Call for help and then position the casualty. Move (roll) the casualty onto his back (Figure 2-1C). (Refer to paragraph 2-3*c* for information on positioning the casualty.)

NOTE

Perform finger sweep. If foreign material or vomitus is visible in the mouth, it should be removed, but do not spend an excessive amount of time doing so.

b. Open the airway using the jaw-thrust or head-tilt/chin-lift technique.

CAUTION

The head-tilt/chin-lift technique is an important procedure in opening the airway; however, use extreme care because excess force in performing this maneuver may cause further spinal injury. In a casualty with a suspected neck injury or severe head trauma, the safest approach to opening the airway is the jaw-thrust technique because in most cases it can be accomplished without extending the neck.

(1) *Perform the jaw-thrust technique.* The jaw-thrust may be accomplished by the rescuer grasping the angles of the casualty's lower jaw and lifting with both hands, one on each side, displacing the jaw forward and up (Figure 2-4). The rescuer's elbows should rest on the surface on which the casualty is lying. If the lips close, the lower lip can be retracted with the thumb. If mouth-to-mouth breathing is necessary, close the nostrils by placing your cheek tightly against them. The head should be carefully supported without tilting it backwards or turning it from side to side. If this is unsuccessful, the head should be tilted back very slightly. The jaw-thrust is the safest first approach to opening the airway of a casualty who has a suspected neck injury because in most cases it can be accomplished without extending the neck.

Figure 2-4. Jaw-thrust technique of opening airway.

(2) *Perform the head-tilt/chin-lift technique*. Place one hand on the casualty's forehead and apply firm, backward pressure with the palm to tilt the head back. Place the fingertips of the other hand under the bony part of the lower jaw and lift, bringing the chin forward. The thumb should not be used to lift the chin (Figure 2-5).

NOTE

The fingers should not press deeply into the soft tissue under the chin because the airway may be obstructed.

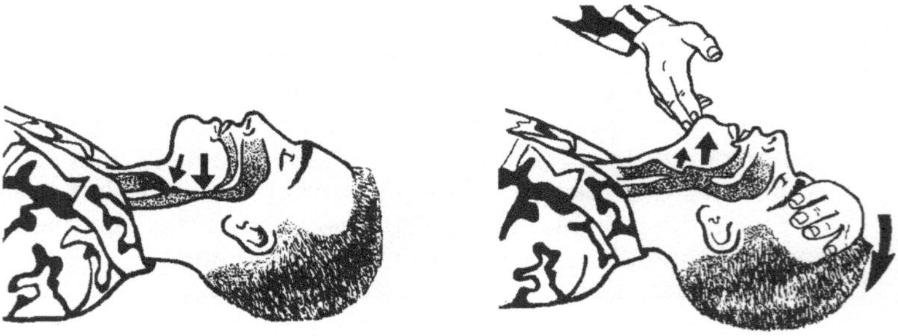

Figure 2-5. Head-tilt/chin-lift technique of opening airway.

(3) *Check for breathing (while maintaining an airway).* After establishing an open airway, it is important to maintain that airway in an open position. Often the act of just opening and maintaining the airway will allow the casualty to breathe properly. Once the rescuer uses one of the techniques to open the airway (jaw-thrust or head-tilt/chin-lift), he should maintain that head position to keep the airway open. Failure to maintain the open airway will prevent the casualty from receiving an adequate supply of oxygen. Therefore, while maintaining an open airway the rescuer should check for breathing by observing the casualty's chest and performing the following actions within 3 to 5 seconds:

(*a*) **LOOK** for the chest to rise and fall.

(*b*) **LISTEN** for air escaping during exhalation by placing your ear near the casualty's mouth.

(*c*) **FEEL** for the flow of air on your cheek (see Figure 2-6).

(*d*) **PERFORM** rescue breathing if the casualty does not resume breathing spontaneously.

NOTE

If the casualty resumes breathing, monitor and maintain the open airway. He should be transported to an MTF, as soon as practical.

2-5. Rescue Breathing (Artificial Respiration)

a. If the casualty does not promptly resume adequate spontaneous breathing after the airway is open, rescue breathing (artificial respiration) must be started. Be calm! Think and act quickly! The sooner you begin rescue breathing, the more likely you are to restore the casualty's breathing. If you are in doubt whether the casualty is breathing, give artificial respiration, since it can do no harm to a person who is breathing. If the casualty is breathing, you can feel and see his chest move. If the casualty is breathing, you can feel and hear air being expelled by putting your hand or ear close to his mouth and nose.

b. There are several methods of administering rescue breathing. The mouth-to-mouth method is preferred; however, it cannot be used in all situations. If the casualty has a severe jaw fracture or mouth wound or his jaws are tightly closed by spasms, use the mouth-to-nose method.

2-6. Preliminary Steps—All Rescue Breathing Methods

a. Establish unresponsiveness. Call for help. Turn or position the casualty.

b. Open the airway.

c. Check for breathing by placing your ear over the casualty's mouth and nose, and looking toward his chest.

(1) **LOOK** for rise and fall of the casualty's chest (Figure 2-6).

(2) **LISTEN** for sounds of breathing.

(3) **FEEL** for breath on the side of your face. If the chest does not rise and fall and no air is exhaled, then the casualty is not breathing.

(4) **PERFORM** rescue breathing if the casualty is not breathing.

NOTE

Although the rescuer may notice that the casualty is making respiratory efforts, the airway may still be obstructed and opening the airway may be all that is needed. If the casualty resumes breathing, the rescuer should continue to maintain an open airway.

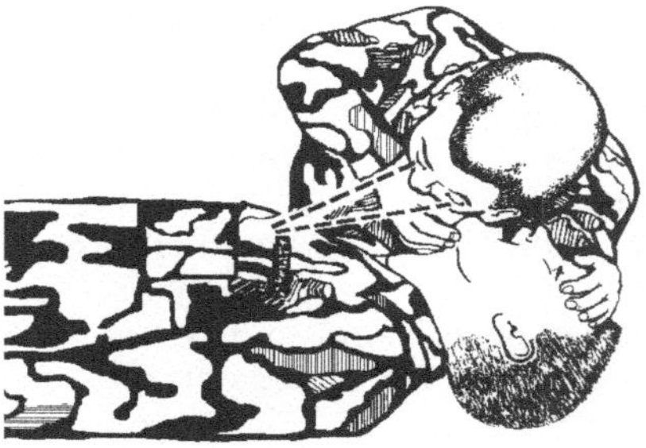

Figure 2-6. Check for breathing.

2-7. Mouth-to-Mouth Method

In this method of rescue breathing, you inflate the casualty's lungs with air from your lungs. This can be accomplished by blowing air into the person's mouth. The mouth-to-mouth rescue breathing method is performed as follows:

a. If the casualty is not breathing, place your hand on his forehead, and pinch his nostrils together with the thumb and index finger of this hand. Let this same hand exert pressure on his forehead to maintain the backward head tilt and maintain an open airway. With your other hand, keep your fingertips on the bony part of the lower jaw near the chin and lift (Figure 2-7).

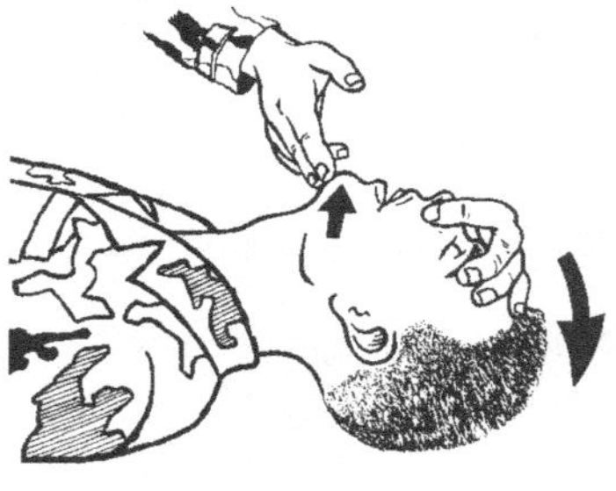

Figure 2-7. Head tilt/chin lift.

NOTE

If you suspect the casualty has a neck injury and you are using the jaw-thrust technique, close the nostrils by placing your cheek tightly against them.

b. Take a deep breath and place your mouth (in an airtight seal) around the casualty's mouth (Figure 2-8). (If the injured person is small, cover both his nose and mouth with your mouth, sealing your lips against the skin of his face.)

Figure 2-8. Rescue breathing.

c. Blow two full breaths into the casualty's mouth (1 to 1 1/2 seconds per breath), taking a breath of fresh air each time before you blow. Watch out of the corner of your eye for the casualty's chest to rise. If the chest rises, sufficient air is getting into the casualty's lungs. Therefore, proceed as described in step (1). If the chest does not rise, do the following (*a*, *b*, and *c* below) and then attempt to ventilate again.

(1) Take corrective action immediately by reestablishing the airway. Make sure that air is not leaking from around your mouth or out of the casualty's pinched nose.

(2) Reattempt to ventilate.

(3) If the chest still does not rise, take the necessary action to open an obstructed airway (paragraph 2-10).

NOTE

If the initial attempt to ventilate the casualty is unsuccessful, reposition the casualty's head and repeat rescue breathing. Improper chin and head positioning is the most common cause of difficulty with ventilation. If the casualty cannot be ventilated after repositioning the head, proceed with foreign-body airway obstruction maneuvers (see paragraph 2-10).

(4) After giving two slow breaths, which cause the chest to rise, attempt to locate a pulse on the casualty. Feel for a pulse on the side of the casualty's neck closest to you by placing the first two fingers (index and middle fingers) of your hand on the groove beside the casualty's Adam's apple (carotid pulse) (Figure 2-9). (Your thumb should not be used for pulse taking because you may confuse your pulse beat with that of the casualty.) Maintain the airway by keeping your other hand on the casualty's forehead. Allow 5 to 10 seconds to determine if there is a pulse.

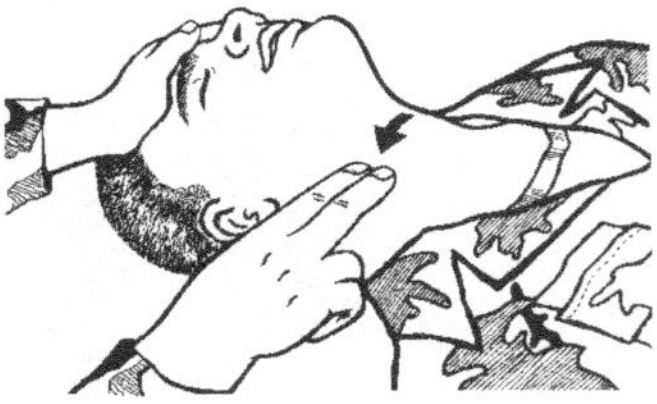

Figure 2-9. Placement of fingers to detect pulse.

(*a*) If signs of circulation are present and a pulse is found and the casualty is breathing—**STOP**; allow the casualty to breathe on his own. If possible, keep him warm and comfortable.

(*b*) If a pulse is found and the casualty is not breathing, continue rescue breathing.

(*c*) If a pulse is not found, seek medically trained personnel for help as soon as possible.

2-8. Mouth-to-Nose Method

Use this method if you cannot perform mouth-to-mouth rescue breathing because the casualty has a severe jaw fracture or mouth wound or his jaws are tightly closed by spasms. The mouth-to-nose method is performed in the same way as the mouth-to-mouth method except that you blow into the nose while you hold the lips closed with one hand at the chin. You then remove your mouth to allow the casualty to exhale passively. It may be necessary to separate the casualty's lips to allow the air to escape during exhalation.

2-9. Heartbeat

If a casualty's heart stops beating, you must immediately seek medical help. **SECONDS COUNT!** Stoppage of the heart is soon followed by cessation of respiration unless it has occurred first. Be calm! Think and act! When a casualty's heart has stopped, there is no pulse at all; the person is unconscious and limp, and the pupils of his eyes are open wide. When evaluating a casualty or when performing the preliminary steps of rescue breathing, feel for a pulse. If you DO NOT detect a pulse, seek medical help.

2-10. Airway Obstructions

In order for oxygen from the air to flow to and from the lungs, the upper airway must be unobstructed.

a. Upper airway obstructions often occur because—

(1) The casualty's tongue falls back into his throat while he is unconscious. The tongue *falls back* and *obstructs* the airway, it is not swallowed by the casualty.

NOTE

Ensure the correct positioning and maintenance of the open airway for an injured or unconscious casualty.

(2) Foreign bodies become lodged in the throat. These obstructions usually occur while eating. Choking on food (usually meat) is associated with—

- Attempting to swallow large pieces of poorly chewed food.
- Drinking alcohol.
- Slipping dentures.

(3) The contents of the stomach are regurgitated and may block the airway.

(4) Blood clots may form as a result of head and facial injuries.

b. Upper airway obstruction may cause either partial or complete airway blockage.

(1) *Partial airway obstruction.* The casualty may still have an air exchange. A good air exchange means that the casualty can cough

forcefully, though he may be wheezing between coughs. You, the rescuer, should not interfere, and should encourage the casualty to cough up the object obstructing his airway on his own. A poor air exchange may be indicated by weak coughing with a high pitched noise between coughs. Further, the casualty may show signs of shock (paragraph 1-6*b*[5]) indicating a need for oxygen. You should assist the casualty and treat him as though he had a complete obstruction.

(2) *Complete airway obstruction.* A complete obstruction (no air exchange) is indicated if the casualty cannot speak, breathe, or cough at all. He may be clutching his neck and moving erratically. In an unconscious casualty, a complete obstruction is also indicated if after opening his airway you cannot ventilate him.

2-11. Opening the Obstructed Airway—Conscious Casualty

Clearing a conscious casualty's airway obstruction can be performed with the casualty either standing or sitting and by following a relatively simple procedure.

WARNING

Once an obstructed airway occurs, the brain will develop an oxygen deficiency resulting in unconsciousness. Death will follow rapidly if breathing is not promptly restored.

a. Ask the casualty if he can speak or if he is choking. Check for the universal choking sign (Figure 2-10).

Figure 2-10. Universal sign of choking.

b. If the casualty can speak, encourage him to attempt to cough; the casualty still has a good air exchange. If he is able to speak or cough effectively, DO NOT interfere with his attempts to expel the obstruction.

c. Listen for high pitched sounds when the casualty breathes or coughs (poor air exchange). If there is poor air exchange or no breathing, CALL FOR HELP and immediately deliver manual thrusts (either an abdominal or chest thrust).

NOTE

The manual thrust with the hands centered between the waist and the rib cage is called an abdominal thrust (or Heimlich maneuver). The chest thrust (the hands are centered in the middle of the breastbone) is used only for an individual in the advanced stages of pregnancy, in the markedly obese casualty, or if there is a significant abdominal wound.

(1) Apply abdominal thrusts. This can be accomplished by using the following procedures:

(*a*) Stand behind the casualty and wrap your arms around his waist.

(*b*) Make a fist with one hand and grasp it with the other. The thumb side of your fist should be against the casualty's abdomen, in the midline and slightly above the casualty's navel, but well below the tip of the breastbone (Figure 2-11).

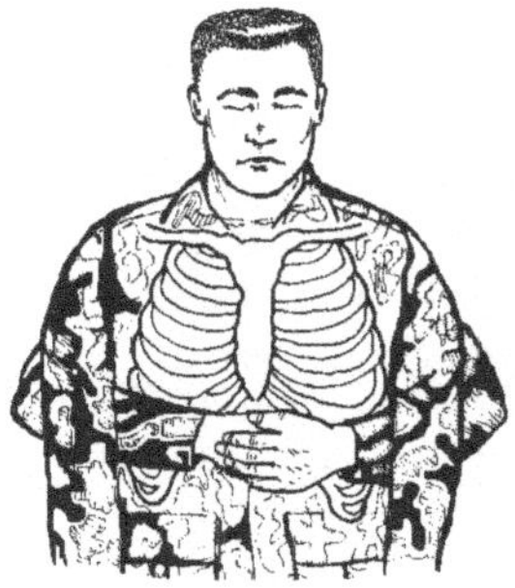

Figure 2-11. Anatomical view of abdominal thrust procedure.

(*c*) Press the fists into the abdomen with a quick backward and upward thrust (Figure 2-12).

Figure 2-12. Profile view of abdominal thrust.

(*d*) Each thrust should be a separate and distinct movement.

NOTE

Continue performing abdominal thrusts until the obstruction is expelled or the casualty becomes unresponsive.

(*e*) If the casualty becomes unresponsive, call for help as you proceed with steps to open the airway, and perform rescue breathing. (Refer to paragraph 2-7 for information on how to perform mouth-to-mouth resuscitation.)

(2) Apply chest thrusts. An alternate technique to the abdominal thrust is the chest thrust. This technique is useful when the casualty has an abdominal wound, when the casualty is pregnant, or when the casualty is so large that you cannot wrap your arms around the abdomen. To apply chest thrusts with casualty sitting or standing:

(*a*) Stand behind the casualty and wrap your arms around his chest with your arms under his armpits.

(*b*) Make a fist with one hand and place the thumb side of the fist in the middle of the breastbone (take care to avoid the tip of the breastbone and the margins of the ribs).

(*c*) Grasp the fist with the other hand and exert thrusts (Figure 2-13).

Figure 2-13. Profile view of chest thrust.

(*d*) Each thrust should be delivered slowly, distinctly, and with the intent of relieving the obstruction.

(*e*) Perform chest thrusts until the obstruction is expelled or the casualty becomes unresponsive.

(*f*) If the casualty becomes unresponsive, call for help as you proceed with steps to open the airway and perform rescue breathing.

2-12. Opening the Obstructed Airway—Casualty Lying Down or Unresponsive

The following procedures are used to expel an airway obstruction in a casualty who is lying down, who becomes unconscious, or who is found unconscious (the cause unknown):

- If a conscious casualty who is choking becomes unresponsive, call for help, open the airway, perform a finger sweep, and attempt rescue breathing (paragraphs 2-4 through 2-8). If you still cannot administer rescue breathing due to an airway blockage, then remove the airway obstruction using the procedures as in *b* below.

- If a casualty is unresponsive when you find him (the cause unknown), assess or evaluate the situation, call for help, position the casualty on his back, open the airway, establish breathlessness, and attempt to perform rescue breathing (paragraphs 2-4 through 2-8).

a. Open the airway and attempt rescue breathing (refer to paragraph 2-7 for information on how to perform mouth-to-mouth resuscitation).

b. If still unable to ventilate the casualty, perform 6 to 10 manual (abdominal or chest) thrusts.

(1) To perform the abdominal thrusts:

(*a*) Kneel astride the casualty's thighs (Figure 2-14).

Figure 2-14. Abdominal thrust on unresponsive casualty.

(*b*) Place the heel of one hand against the casualty's abdomen (in the midline slightly above the navel but well below the tip of the breastbone). Place your other hand on top of the first one. Point your fingers toward the casualty's head.

(*c*) Press into the casualty's abdomen with a quick, forward and upward thrust. You can use your body weight to perform the maneuver. Deliver each thrust quickly and distinctly.

(*d*) Repeat the sequence of abdominal thrusts, finger sweep, and rescue breathing (attempt to ventilate) as long as necessary to remove the object from the obstructed airway.

(*e*) If the casualty's chest rises, proceed to feeling for pulse.

(2) To perform chest thrusts:

(*a*) Place the unresponsive casualty on his back, face up, and open his mouth. Kneel close to the side of the casualty's body.

1. Locate the lower edge of the casualty's ribs with your fingers. Run the fingers up along the rib cage to the notch (Figure 2-15A).

2. Place the middle finger on the notch and the index finger next to the middle finger on the lower edge of the breastbone. Place the heel of the other hand on the lower half of the breastbone next to the two fingers (Figure 2-15B).

3. Remove the fingers from the notch and place that hand on top of the positioned hand on the breastbone, extending or interlocking the fingers (Figure 2-15C).

4. Straighten and lock your elbows with your shoulders directly above your hands without bending the elbows, rocking, or allowing the shoulders to sag. Apply enough pressure to depress the breastbone 1 1/2 to 2 inches, then release the pressure completely (Figure 2-15D). Do this 6 to 10 times. Each thrust should be delivered quickly and distinctly. See Figure 2-16 for another view of the breastbone being depressed.

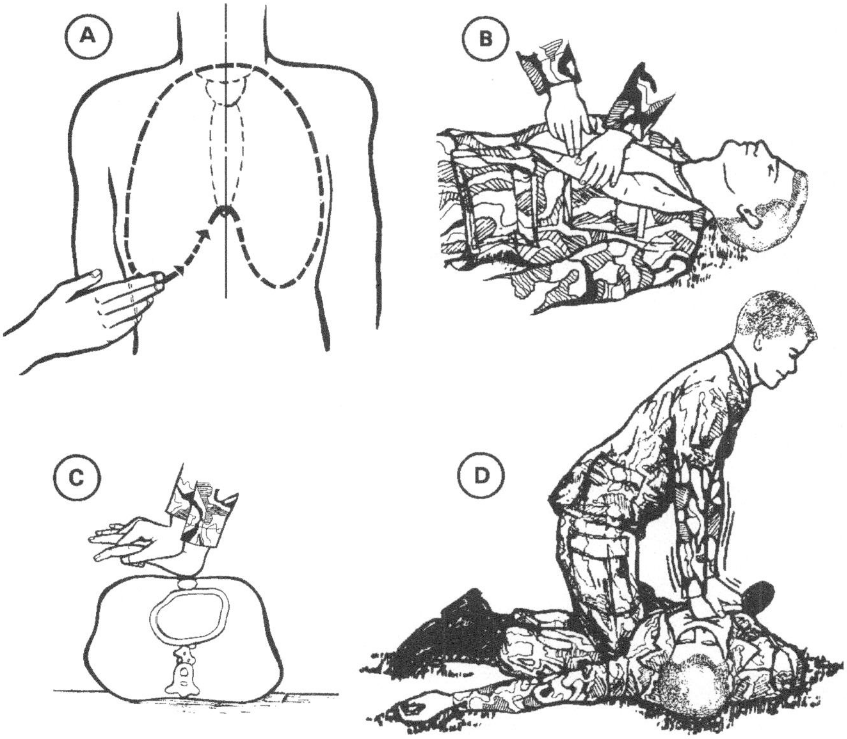

Figure 2-15. Hand placement for chest thrust (Illustrated A-D).

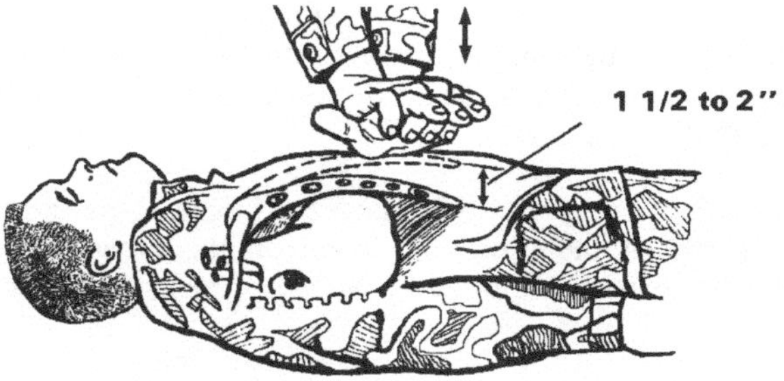

Figure 2-16. Breastbone depressed 1 1/2 to 2 inches.

(*b*) Repeat the sequence of chest thrust, finger sweep, and rescue breathing as long as necessary to clear the object from the obstructed airway. See paragraph (3) below.

(*c*) If the casualty's chest rises, proceed to feeling for his pulse.

(3) If you still cannot administer rescue breathing due to an airway obstruction, then remove the airway obstruction using the procedures in steps (*a*) and (*b*) below.

(*a*) Place the casualty on his back, face up, turn the unresponsive casualty as a unit, and call out for help.

(*b*) Perform finger sweep, keep casualty face up, use tongue-jaw lift to open mouth.

1. Open the casualty's mouth by grasping both his tongue and lower jaw between your thumb and fingers and lifting (tongue-jaw lift) (Figure 2-17). If you are unable to open his mouth, cross your fingers and thumb (crossed-finger method) and push his teeth apart (Figure 2-18) by pressing your thumb against his upper teeth and pressing your finger against his lower teeth.

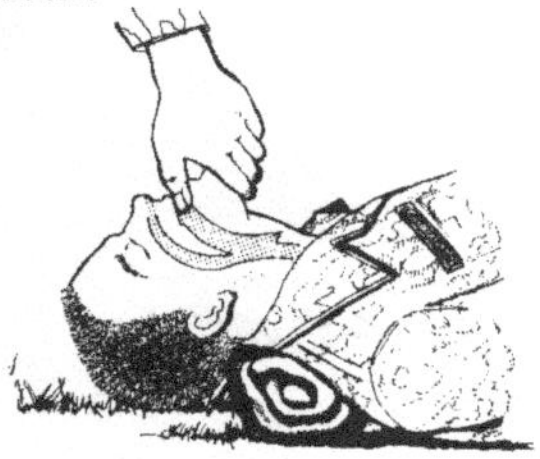

Figure 2-17. Opening casualty's mouth (tongue-jaw lift).

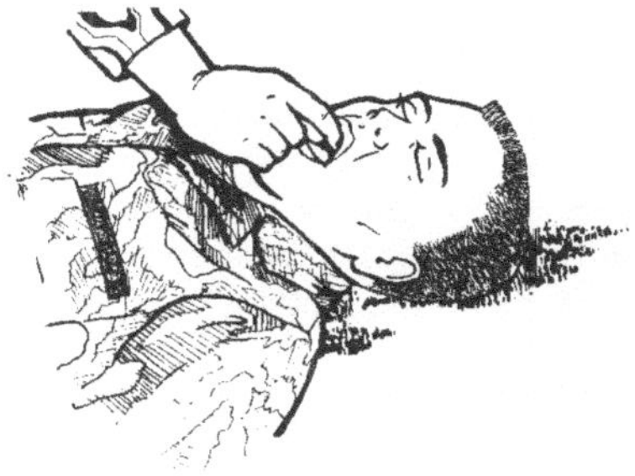

Figure 2-18. Opening casualty's mouth (crossed-finger method).

2. Insert the index finger of the other hand down along the inside of his cheek to the base of the tongue. Use a hooking motion from the side of the mouth toward the center to dislodge the foreign body (Figure 2-19).

Figure 2-19. Using finger to dislodge a foreign body.

WARNING

Take care not to force the object deeper into the airway by pushing it with the finger.

Section II. STOP THE BLEEDING AND PROTECT THE WOUND

2-13. General

The longer a service member bleeds from a major wound, the less likely he will be able to survive his injuries. It is, therefore, important that the first aid provider promptly stop the external bleeding.

2-14. Clothing

In evaluating the casualty for location, type, and size of the wound or injury, cut or tear his clothing and carefully expose the entire area of the wound. This procedure is necessary to properly visualize injury and avoid further contamination. Clothing stuck to the wound should be left in place to avoid further injury. DO NOT touch the wound; keep it as clean as possible.

> **WARNING**
>
> **DO NOT REMOVE protective clothing in a chemical environment. Apply dressings over the protective clothing.**

2-15. Entrance and Exit Wounds

Before applying the dressing, carefully examine the casualty to determine if there is more than one wound. A missile may have entered at one point and exited at another point. The *EXIT* wound is usually *LARGER* than the entrance wound.

> **WARNING**
>
> **The casualty should be continually monitored for development of conditions which may require the performance of necessary basic lifesaving measures, such as clearing the airway and mouth-to-mouth resuscitation. All open (or penetrating) wounds should be checked for a point of entry and exit and first aid measures applied accordingly.**

> **WARNING**
>
> **If the missile lodges in the body (fails to exit), DO NOT attempt to remove it or probe the wound. Apply a dressing. If there is an object extending from (impaled in) the wound, DO NOT remove the object. Apply a dressing around the object and use additional improvised bulky materials/dressings (use the cleanest material available) to build up the area around the object to stabilize the object and prevent further injury. Apply a supporting bandage over the bulky materials to hold them in place.**

2-16. Field Dressing

a. Use the casualty's field dressing; remove it from the wrapper and grasp the tails of the dressing with both hands (Figure 2-20).

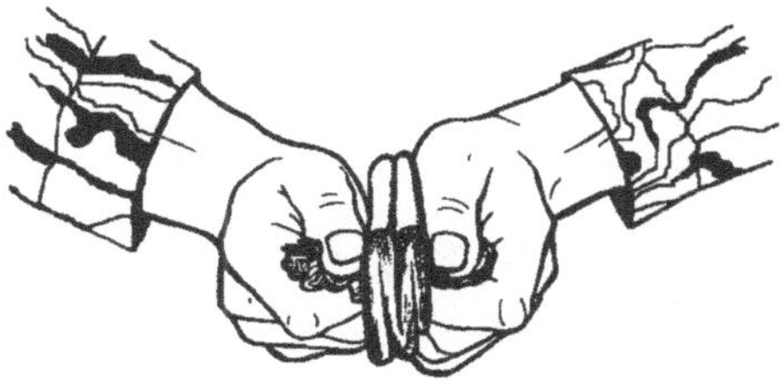

Figure 2-20. Grasping tails of dressing with both hands.

> **WARNING**
>
> **DO NOT touch the white (sterile) side of the dressing, and DO NOT allow it to come in contact with any surface other than the wound.**

b. Hold the dressing directly over the wound with the white side down. Pull the dressing open (Figure 2-21) and place it directly over the wound (Figure 2-22).

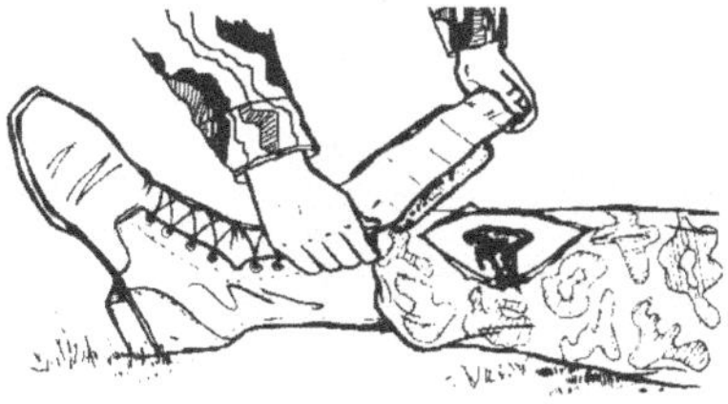

Figure 2-21. Pulling dressing open.

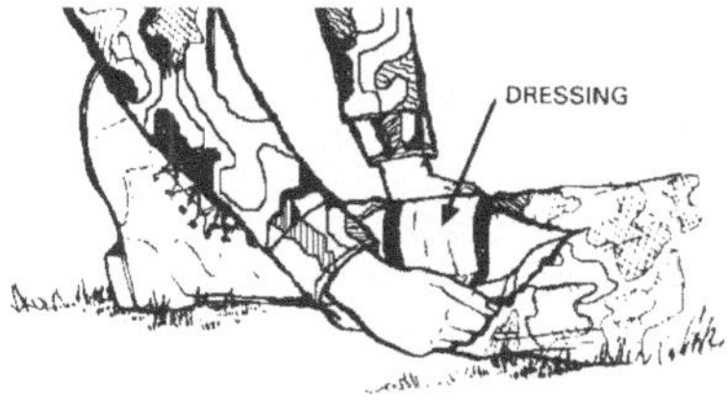

Figure 2-22. Placing dressing directly on wound.

c. Hold the dressing in place with one hand. Use the other hand to wrap one of the tails around the injured part, covering about one-half of the dressing (Figure 2-23). Leave enough of the tail for a knot. If the casualty is able, he may assist by holding the dressing in place.

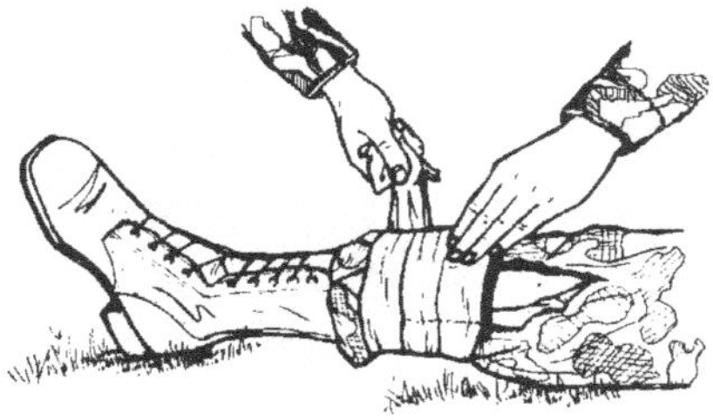

Figure 2-23. Wrapping tail of dressing around injured part.

d. Wrap the other tail in the opposite direction until the remainder of the dressing is covered. The tails should seal the sides of the dressing to keep foreign material from getting under it.

e. Tie the tails into a nonslip knot over the outer edge of the dressing (Figure 2-24). **DO NOT TIE THE KNOT OVER THE WOUND**. In order to allow blood to flow to the rest of an injured limb, tie the dressing firmly enough to prevent it from slipping but without causing a tourniquet-like effect; that is, the skin beyond the injury should not becomes cool, blue, or numb.

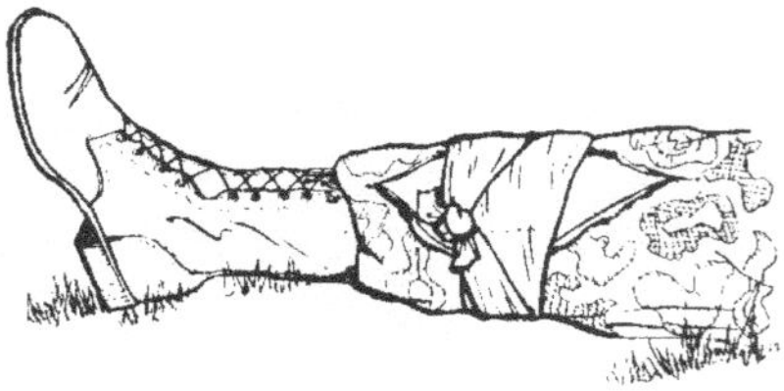

Figure 2-24. Tails tied into nonslip knot.

2-17. Manual Pressure

a. If bleeding continues after applying the sterile field dressing, direct manual pressure may be used to help control bleeding. Apply such pressure by placing a hand on the dressing and exerting firm pressure for 5 to 10 minutes (Figure 2-25). The casualty may be asked to do this himself if he is conscious and can follow instructions.

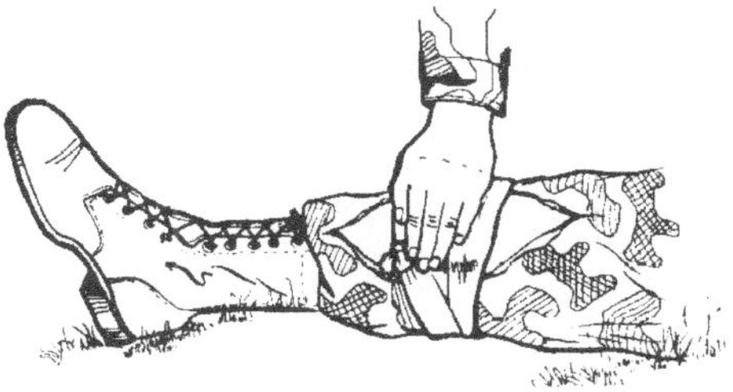

Figure 2-25. Direct manual pressure applied.

b. Elevate an injured limb slightly above the level of the heart to reduce the bleeding (Figure 2-26).

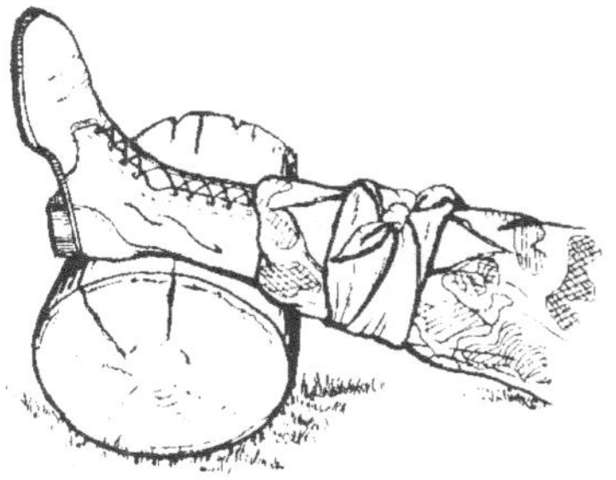

Figure 2-26. Injured limb elevated.

WARNING

DO NOT elevate a suspected fractured limb unless it has been properly splinted.

c. If the bleeding stops, check shock; administer first aid for shock as necessary. If the bleeding continues, apply a pressure dressing.

2-18. Pressure Dressing

Pressure dressings aid in blood clotting and compress the open blood vessel. If bleeding continues after the application of a field dressing, manual pressure, and elevation, then a pressure dressing must be applied as follows:

a. Place a wad of padding on top of the field dressing, directly over the wound (Figure 2-27). Keep the injured extremity elevated.

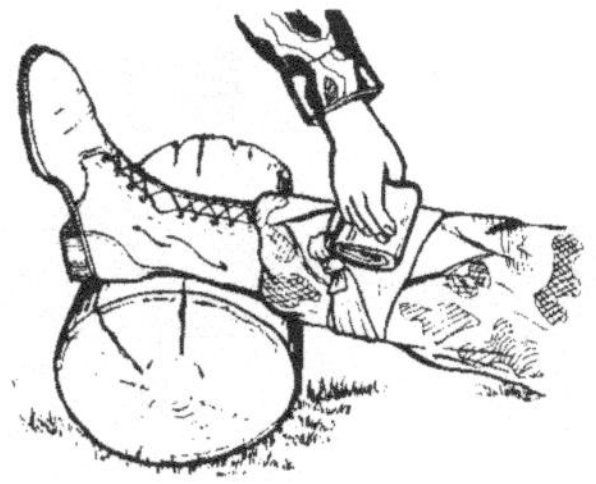

Figure 2-27. Wad of padding on top of field dressing.

NOTE

Improvised bandages may be made from strips of cloth. These strips may be made from T-shirts, socks, or other garments.

b. Place an improvised dressing (or cravat, if available) over the wad of padding (Figure 2-28). Wrap the ends tightly around the injured limb, covering the previously placed field dressing (Figure 2-29).

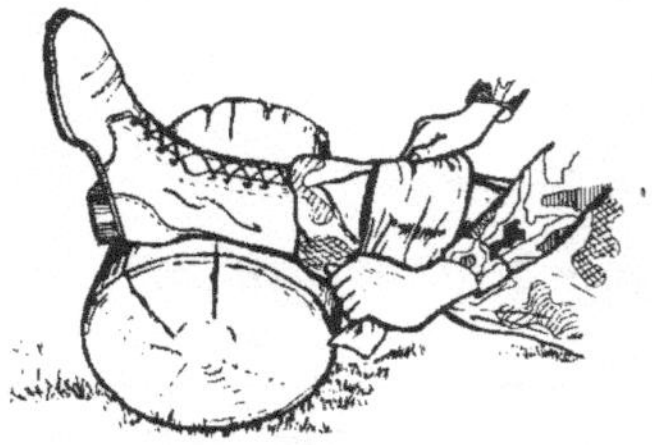

Figure 2-28. Improvised dressing over wad of padding

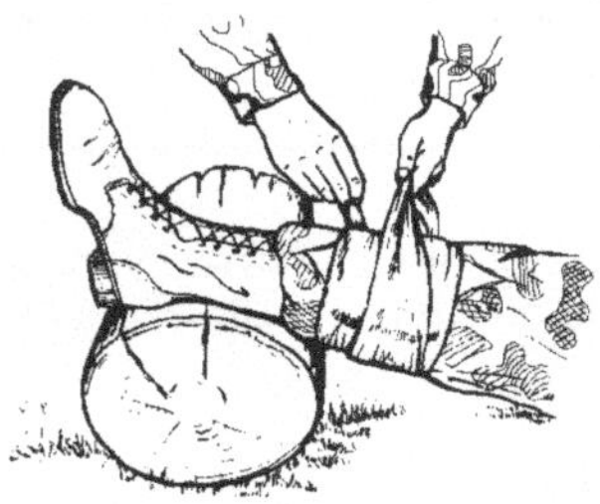

Figure 2-29. Ends of improvised dressing wrapped tightly around limb.

c. Tie the ends together in a nonslip knot, directly over the wound site (Figure 2-30). DO NOT tie so tightly that it has a tourniquet-like effect. If bleeding continues and all other measures have failed, or if the limb is severed, then apply a tourniquet. Use the tourniquet as a **LAST RESORT**. When the bleeding stops, check for shock; administer first aid for shock as necessary.

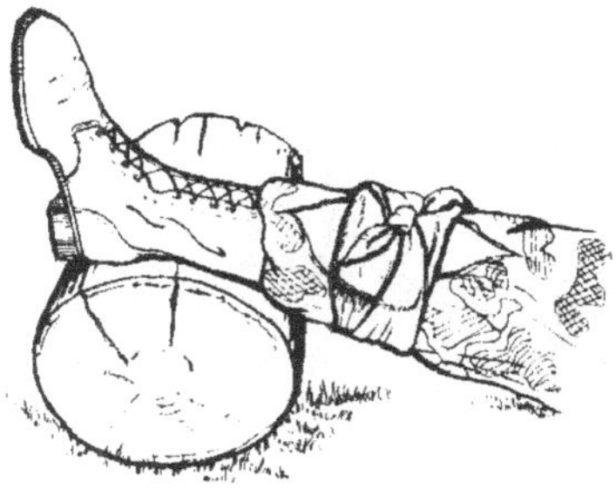

Figure 2-30. Ends of improvised dressing tied together in nonslip knot.

NOTE

Distal end of wounded extremities (fingers and toes) should be checked periodically for adequate circulation. The dressing must be loosened if the extremity becomes cool, blue, or numb.

NOTE

If bleeding continues and all other measures have failed (dressings and covering wound, applying direct manual pressure, elevating the limb above the heart level, and applying a pressure dressing while maintaining limb elevation) *then apply digital pressure* (see paragraph 2-19).

2-19. Digital Pressure

Digital pressure (often called "pressure points") is an alternative method to control bleeding. This method uses pressure from the fingers, thumbs, or hands to press at the site or point where a main artery supplying the wounded area lies near the skin surface or over bone (Figure 2-31). This pressure may help shut off or slow down the flow of blood from the heart to the wound and is used in combination with direct pressure and elevation. It may help in instances where bleeding is not easily controlled, where a pressure dressing has not yet been applied, or where pressure dressings are not readily available.

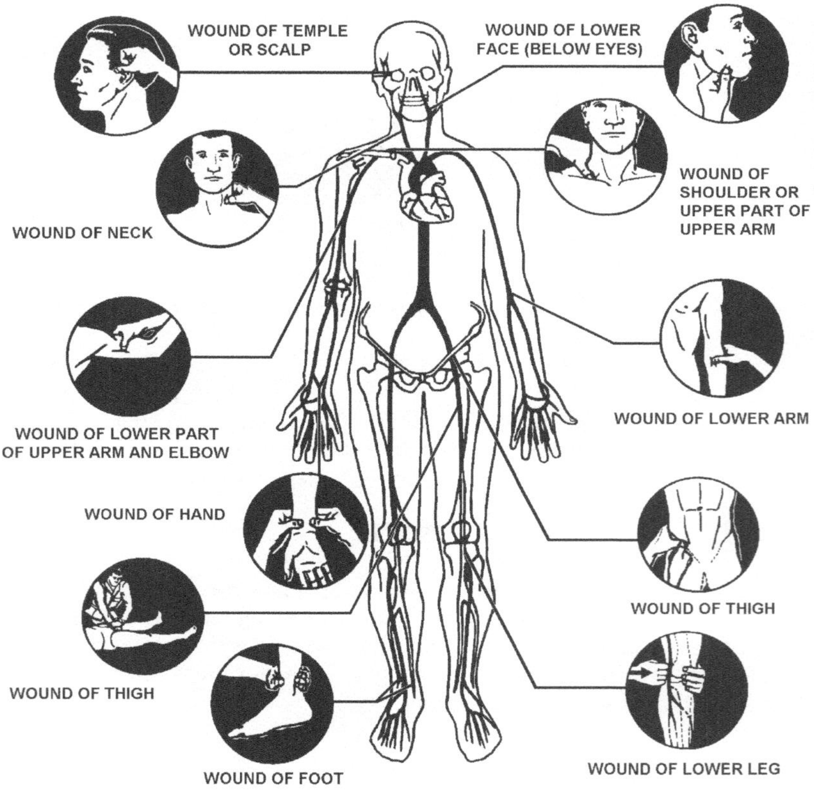

Figure 2-31. Digital pressure (pressure with fingers, thumbs or hands).

2-20. Tourniquet

> **DANGER**
>
> **A tourniquet is only used on an arm or leg where there is a danger of the casualty losing his life (bleeding to death).**

A tourniquet is a constricting band placed around an arm or leg to control bleeding. A service member whose arm or leg has been completely amputated may not be bleeding when first discovered, but a tourniquet should be applied anyway. This absence of bleeding is due to the body's normal defenses (contraction or clotting of blood vessels) as a result of the amputation, but after a period of time bleeding will start as the blood vessels relax or the clot may be knocked loose by moving the casualty. Bleeding from a major artery of the thigh, lower leg, or arm and bleeding from multiple arteries (which occurs in a traumatic amputation) may prove to be beyond control by manual pressure. If the pressure dressing (see paragraph 2-18, above) under firm hand pressure becomes soaked with blood and the wound continues to bleed, apply a tourniquet.

WARNING

Casualty should be continually monitored for development of conditions which may require the performance of necessary basic lifesaving measures, such as: clearing the airway, performing mouth-to-mouth resuscitation, preventing shock, and/or bleeding control. All open (or penetrating) wounds should be checked for a point of entry or exit and treated accordingly.

The tourniquet should not be used unless a pressure dressing has failed to stop the bleeding or an arm or leg has been cut off. On occasion, tourniquets have injured blood vessels and nerves. If left in place too long, a tourniquet can cause loss of an arm or leg. Once applied, it must stay in place, and the casualty must be taken to the nearest MTF as soon as possible. *DO NOT loosen or release a tourniquet after it has been applied as release could precipitate bleeding and potentially lead to shock.*

a. Improvising a Tourniquet. In the absence of a specially designed tourniquet, a tourniquet may be made from a strong, pliable material, such as gauze or muslin bandages, clothing, or cravats. An improvised tourniquet is used with a rigid stick-like object. To minimize skin damage, ensure that the improvised tourniquet is at least 2 inches wide.

WARNING

The tourniquet must be easily identified or easily seen.

WARNING

DO NOT use wire or shoestring for a tourniquet band.

b. Placing the Improvised Tourniquet.

(1) Place the tourniquet around the limb, between the wound and the body trunk (or between the wound and the heart). Never place it directly over a wound, a fracture, or joint. Tourniquets, for maximum effectiveness, should be placed on the upper arm or above the knee on the thigh (Figure 2-32).

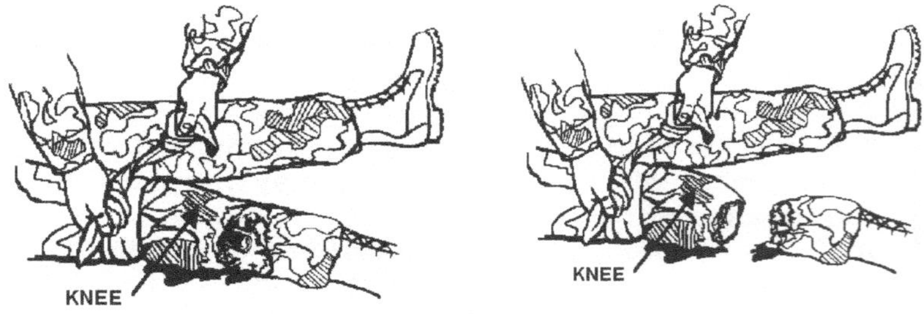

Figure 2-32. Tourniquet above knee.

(2) The tourniquet should be well-padded. If possible, place the tourniquet over the smoothed sleeve or trouser leg to prevent the skin from being pinched or twisted. If the tourniquet is long enough, wrap it around the limb several times, keeping the material as flat as possible. Damaging the skin may deprive the surgeon of skin required to cover an amputation. Protection of the skin also reduces pain.

c. Applying the Tourniquet.

(1) Tie a half-knot. (A half-knot is the same as the first part of tying a shoe lace.)

(2) Place a stick (or similar rigid object) on top of the half-knot (Figure 2-33).

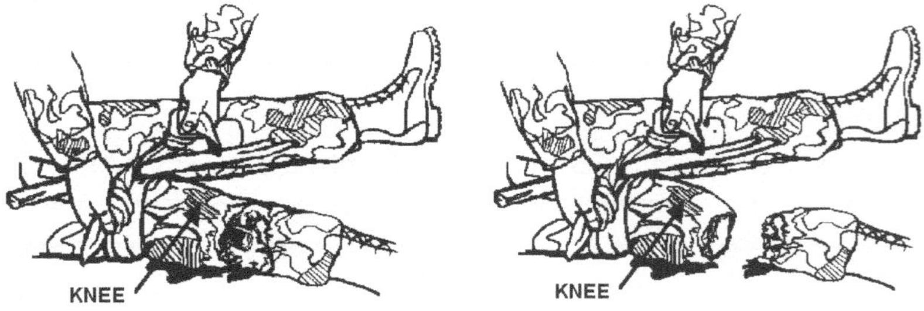

Figure 2-33. Rigid object on top of half-knot.

(3) Tie a full knot over the stick (Figure 2-34).

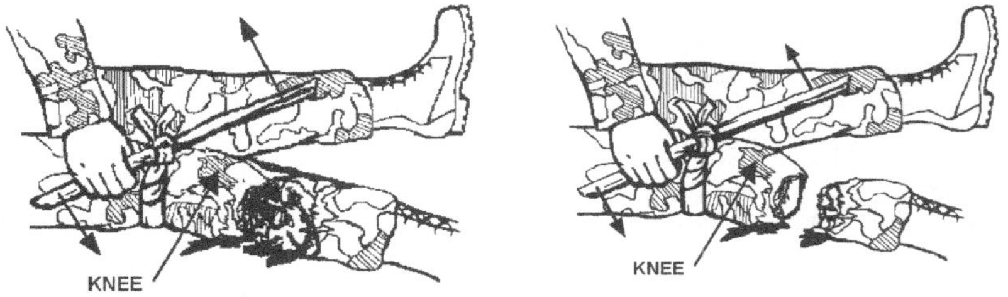

Figure 2-34. Full knot over rigid object.

(4) Twist the stick (Figure 2-35) until the tourniquet is tight around the limb and/or the bright red bleeding has stopped. In the case of amputation, dark oozing blood may continue for a short time. This is the blood trapped in the area between the wound and tourniquet.

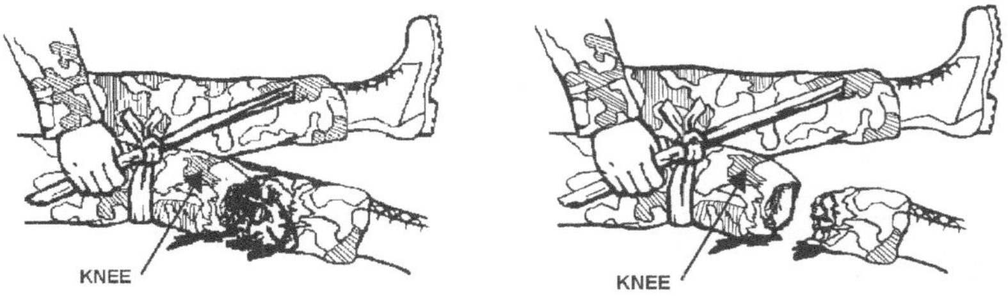

Figure 2-35. Stick twisted.

(5) Fasten the tourniquet to the limb by looping the free ends of the tourniquet over the ends of the stick. Then bring the ends around the limb to prevent the stick from loosening. Tie them together on the side of the limb (Figure 2-36).

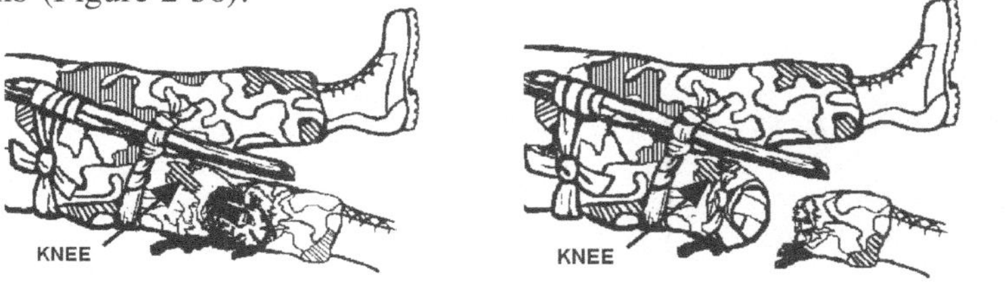

Figure 2-36. Tie free ends on side of limb.

NOTE

Other methods of securing the stick may be used as long as the stick does not unwind and no further injury results.

NOTE

If possible, save and transport any severed (amputated) limbs or body parts with (but out of sight of) the casualty.

(6) DO NOT cover the tourniquet—you should leave it in full view. If the limb is missing (total amputation), apply a dressing to the stump. All wounds should have a dressing to protect the wound from contamination.

(7) Mark the casualty's forehead with a "T" and the time to indicate a tourniquet has been applied. If necessary, use the casualty's blood to make this mark.

(8) Check and treat for shock.

(9) Seek medical aid.

CAUTION

Only appropriately skilled medical personnel may adjust or otherwise remove/release the tourniquet in the appropriate setting.

Section III. CHECK FOR SHOCK AND ADMINISTER FIRST AID MEASURES

2-21. General

The term *shock* has a variety of meanings. In medicine, it refers to a collapse of the body's cardiovascular system which includes an inadequate supply of blood to the body's tissues. Shock stuns and weakens the body. When the normal blood flow in the body is upset, death can result. Early recognition and proper first aid may save the casualty's life.

2-22. Causes and Effects

a. There are three basic mechanisms associated with shock. These are—

- The heart is damaged and fails to work as a pump.

- Blood loss (heavy bleeding) causes the volume of fluid within the vascular system to be insufficient.

- The blood vessels dilate (open wider) so that the blood within the system (even though it is a normal volume [the casualty is not bleeding or dehydrated]) is insufficient to provide adequate circulation within the body.

b. Shock may be the result of a number of conditions. These include—

- Dehydration.

- Allergic reaction to foods, drugs, insect stings, and snakebites.

- Significant loss of blood.

- Reaction to the sight of a wound, blood, or other traumatic scene.

- Traumatic injuries, such as—

  - Burns.

  - Gunshot or shrapnel wounds.

  - Crush injuries.

  - Blows to the body (which can cause broken bones or damage to internal organs).

  - Head injuries.

  - Penetrating wounds (such as from a knife, bayonet, or missile).

2-23. Signs and Symptoms of Shock

Examine the casualty to see if he has any of the following signs and symptoms:

- Sweaty but cool skin (clammy skin).

- Weak and rapid pulse.
- Paleness of skin (in dark-skinned individuals they may have a grayish look to their skin).
- Restlessness, nervousness.
- Thirst.
- Loss of blood (bleeding).
- Confusion (or loss of awareness).
- Faster-than-normal breathing rate.
- Blotchy or bluish skin (especially around the mouth and lips).
- Nausea and/or vomiting.

2-24. First Aid Measures for Shock

In the field, the first aid procedures administered for shock are identical to procedures that would be performed *to prevent shock*. When treating a casualty, assume that shock is present or will occur shortly. By waiting until actual signs and symptoms of shock are noticeable, the rescuer may jeopardize the casualty's life.

a. Position the Casualty. (DO NOT move the casualty or his limbs if suspected fractures have not been splinted. See Chapter 4 for details.)

(1) Move the casualty to cover, if cover is available and the situation permits.

(2) Lay the casualty on his back.

NOTE

A casualty in shock from a chest wound or one who is experiencing breathing difficulty, may breathe easier in a sitting position. If this is the case, allow him to sit upright, but monitor carefully in case his condition worsens.

(3) Elevate the casualty's feet higher than the level of his heart. Use a stable object (field pack or rolled up clothing) so that his feet will not slip off (Figure 2-37).

WARNING

DO NOT elevate legs if the casualty has an unsplinted broken leg, head injury, or abdominal injury.

Figure 2-37. Clothing loosened and feet elevated.

WARNING

Check casualty for leg fracture(s) and splint, if necessary, before elevating his feet. For a casualty with an abdominal wound, place his knees in an upright (flexed) position.

(4) Loosen clothing at the neck, waist, or wherever it may be binding.

CAUTION

DO NOT loosen or remove protective clothing in a chemical environment.

(5) Prevent chilling or overheating. The key is to maintain body temperature. In cold weather, place a blanket or other like item over him to keep him warm and under him to prevent chilling (Figure 2-38). However, if a tourniquet has been applied, leave it exposed (if possible). In hot weather, place the casualty in the shade and protect him from becoming chilled; however, avoid the excessive use of blankets or other coverings.

Figure 2-38. Body temperature maintained.

(6) Calm the casualty. Throughout the entire procedure of providing first aid for a casualty, the rescuer should reassure the casualty and keep him calm. This can be done by being authoritative (taking charge) and by showing self-confidence. Assure the casualty that you are there to help him.

(7) Seek medical aid.

b. Food and/or Drink. When providing first aid for shock, DO NOT give the casualty any food or drink. If you must leave the casualty or if he is unconscious, turn his head to the side to prevent him from choking if he vomits (Figure 2-39).

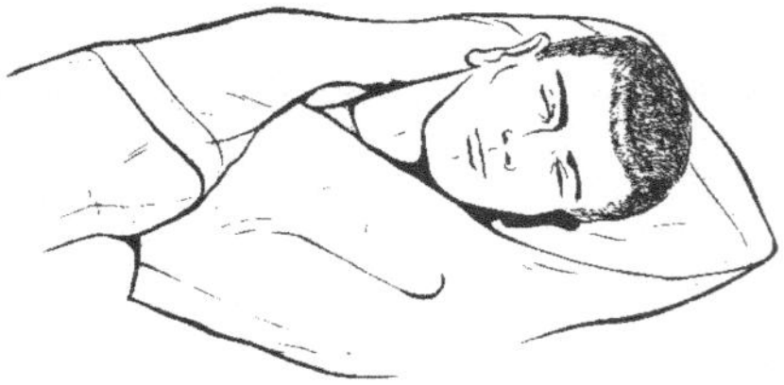

Figure 2-39. Casualty's head turned to side.

c. Evaluate Casualty. Continue to evaluate the casualty until medical personnel arrives or the casualty is transported to an MTF.

CHAPTER 3

FIRST AID FOR SPECIFIC INJURIES

3-1. General

Basic lifesaving steps are discussed in Chapters 1 and 2; they apply to first aid measures for all injuries. Some wounds and burns will require special precautions and procedures when applying these measures. This chapter discusses specific first aid procedures for wounds of the head, face, and neck; chest and stomach wounds; and burns. It also discusses the techniques for applying dressings and bandages to specific parts of the body.

3-2. Head, Neck, and Facial Injuries

a. Head Injuries.

(1) Head injuries range from minor abrasions or cuts on the scalp to severe brain injuries that may result in unconsciousness and sometimes death. Head injuries are classified as open or closed wounds. An open wound is one that is visible, has a break in the skin, and usually has evidence of bleeding. A closed wound may be visible (such as a depression in the skull) or the first aid provider may not be able to see any apparent injury (such as internal bleeding). Some head injuries result in unconsciousness; however, a service member may have a serious head wound and still be conscious. Casualties with head and neck injuries should be treated as though they also have a spinal injury. The casualty should not be moved until the head and neck is stabilized unless he is in immediate danger (such as close to a burning vehicle).

(2) Prompt first aid measures should be initiated for casualties with suspected head and neck injuries. The conscious casualty may be able to provide information on the extent of his injuries. However, as a result of the head injury, he may be confused and unable to provide accurate information. The signs and symptoms a first aid provider might observe are—

- Nausea and vomiting.
- Convulsions or twitches.
- Slurred speech.
- Confusion and loss of memory. (Does he know who he is? Does he know where he is? Does he know what day it is?)
- Recent unconsciousness.

- Dizziness.
- Drowsiness.
- Blurred vision, unequal pupils, or bruising (black eyes).
- Paralysis (partial or full).
- Complaint of headache.
- Bleeding or other fluid discharge from the scalp, nose, or ears.
- Deformity of the head (depression or swelling).
- Staggering while walking.

b. Neck Injuries. Neck injuries may result in heavy bleeding. Apply pressure above and below the injury, *but do not interfere with the breathing process*, and attempt to control the bleeding. Apply a dressing. Always evaluate the casualty for a possible neck fracture/spinal cord injury; if suspected, seek medical treatment immediately.

NOTE

Establish and maintain the airway in cases of facial or neck injuries. If a neck fracture or spinal cord injury is suspected, immobilize the injury and, if necessary, perform basic life support measures.

c. Facial Injuries. Soft tissue injuries of the face and scalp are common. Abrasions (scrapes) of the skin cause no serious problems. Contusions (injury without a break in the skin) usually cause swelling. A contusion of the scalp looks and feels like a lump. Laceration (cut) and avulsion (torn away tissue) injuries are also common. Avulsions are frequently caused when a sharp blow separates the scalp from the skull beneath it. Because the face and scalp are richly supplied with blood vessels (arteries and veins), wounds of these areas usually bleed heavily.

3-3. General First Aid Measures

a. General Considerations. The casualty with a head injury (or suspected head injury) should be continually monitored for the development of conditions that *may require* basic lifesaving measures. After initiating first aid measures, request medical assistance and evacuation. If dedicated medical evacuation assets are not available, transport the casualty to an MTF as soon as the situation permits. The first aid provider should not attempt to remove a protruding object from the head or give the casualty anything to eat or drink. Further, the first aid provider should be prepared to—

- Clear the airway.
- Control bleeding (external).
- Administer first aid measures for shock.
- Keep the casualty warm.
- Protect the wound.

b. Unconscious Casualty. An unconscious casualty does not have control of all of his body's functions and may choke on his tongue, blood, vomitus, or other substances. (Refer to Figure 2-39.)

(1) *Breathing*. The brain requires a constant supply of oxygen. A bluish (or in an individual with dark skin—grayish) color of skin around the lips and nail beds indicates that the casualty is not receiving enough oxygen. Immediate action must be taken to clear the airway, to position the casualty on his side, or to initiate rescue breathing.

(2) *Bleeding*. Bleeding from a head injury usually comes from blood vessels within the scalp. Bleeding can also develop inside the skull or within the brain. In most instances visible bleeding from the head can be controlled by application of the field first aid dressing.

CAUTION

DO NOT attempt to put unnecessary pressure on the wound or attempt to push any brain matter back into the head (skull). **DO NOT** apply a pressure dressing.

c. Concussion. If an individual receives a heavy blow to the head or face, he may suffer a brain concussion (an injury to the brain that involves a temporary loss of some or all of the brain's ability to function). For example, the casualty may not breathe properly for a short period of time, or he may become confused and stagger when he attempts to walk. Symptoms of a concussion may only last for a short period of time. However, if a casualty is suspected of having suffered a concussion, he should be transported to an MTF as soon as conditions permit.

d. Convulsions. Convulsions (seizures/involuntary jerking) may occur even after a mild head injury. When a casualty is convulsing, protect him from hurting himself. Take the following measures:

(1) Ease him to the ground if he is standing or sitting.

(2) Support his head and neck.

(3) Maintain his airway.

(4) Protect him from further injury (such as hitting close-by objects).

NOTE

DO NOT forcefully hold the arms and legs if they are jerking because this can lead to broken bones. **DO NOT** force anything between the casualty's teeth—especially if they are tightly clenched because this may obstruct the casualty's airway. Maintain the casualty's airway if necessary.

e. Brain Damage. In *severe* head injuries where brain tissue is protruding, *leave the wound alone*; carefully place a loose moistened dressing (moistened with sterile normal saline if available) and also a first aid dressing over the tissue to protect it from further contamination. ***DO NOT** remove or disturb any foreign matter that may be in the wound.* Position the casualty so that his head is higher than his body. Keep him warm and *seek medical assistance immediately*.

NOTE

If there is an object extending from the wound, **DO NOT** remove the object. Improvise bulky dressings from the cleanest material available and place this material around the protruding object for support, then apply the field dressing.

3-4. Chest Wounds

Blunt trauma, bullet or missile wounds, stab wounds, or falls may cause chest injuries. These injuries can be serious and may cause death quickly if first aid is not administered in a timely manner. A casualty with a chest injury may complain of pain in the chest or shoulder area; he may have difficulty breathing. His chest may not rise normally when he breathes. The injury may cause the casualty to cough up blood and to have a rapid or a weak heartbeat. A casualty with an open chest wound has a punctured chest wall. The sucking sound heard when he breathes is caused by air leaking into his chest cavity. This particular type of wound is dangerous and will collapse the injured lung (Figure 3-1). Breathing becomes difficult for the casualty because the wound is open. The service members life may depend upon how quickly you apply an occlusive dressing over the wound (refer to paragraph 3-5).

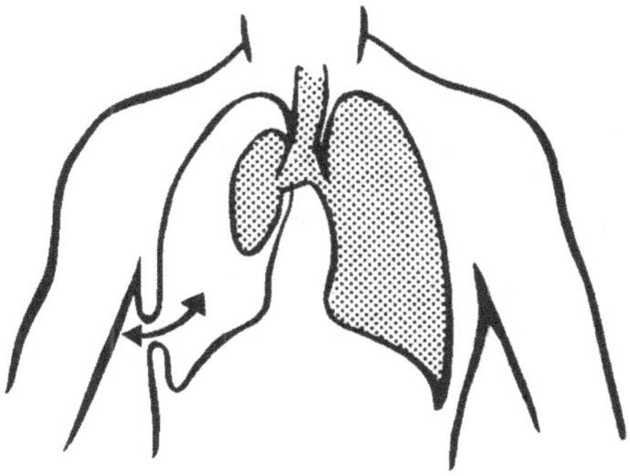

Figure 3-1. Collapsed lung.

3-5. First Aid for Chest Wounds

a. Evaluate the Casualty. Be prepared to perform first aid measures. These measures may include clearing the airway, rescue breathing, treatment for shock, and/or bleeding control.

b. Expose the Wound. If appropriate, cut or remove the casualty's clothing to expose the wound. Remember, **DO NOT** remove clothing that is stuck to the wound because additional injury may result. **DO NOT** attempt to clean the wound.

NOTE

Examine the casualty to see if there is an entry and exit wound. If there are two wounds (entry, exit), perform the same procedure for both wounds. Treat the more serious (heavier bleeding, larger) wound first. It may be necessary to improvise a dressing for the second wound by using strips of cloth, such as a torn T-shirt, or whatever material is available. Also, listen for sucking sounds to determine if the chest wall is punctured.

CAUTION

If there is an object impaled in the wound, **DO NOT** remove it. Apply a dressing around the object and use additional improvised bulky materials/dressings (use the cleanest materials available) to build up the area around the object. Apply a supporting bandage over the bulky materials to hold them in place.

CAUTION

DO NOT REMOVE protective clothing in a chemical environment. Apply dressings *over* the protective clothing.

c. Open the Casualty's Field Dressing Plastic Wrapper. In cases where there is a sucking chest wound, the plastic wrapper is used with the field dressing to create an occlusive dressing. If a plastic wrapper is not available, or if an additional wound needs to be treated; cellophane, foil, the casualty's poncho, or similar material may be used. The covering should be wide enough to extend 2 inches or more beyond the edges of the wound in all directions.

(1) Tear open one end of the casualty's plastic wrapper covering the field dressing. Be careful not to destroy the wrapper and **DO NOT** touch the inside of the wrapper.

(2) Remove the inner packet (field dressing).

(3) Complete tearing open the empty plastic wrapper using as much of the wrapper as possible to create a flat surface.

d. Place the Wrapper Over the Wound. Place the inside surface of the plastic wrapper directly over the wound *when the casualty exhales* and hold it in place (Figure 3-2). The casualty may hold the plastic wrapper in place if he is able.

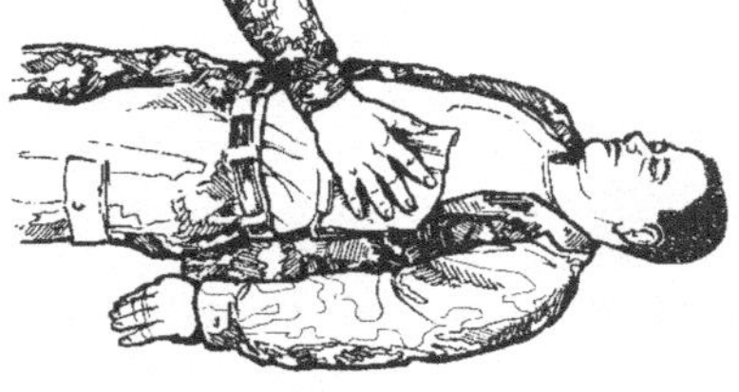

Figure 3-2. Open chest wound sealed with an occlusive dressing.

e. Apply the Dressing to the Wound.

(1) Use your free hand and shake open the field dressing (Figure 3-3).

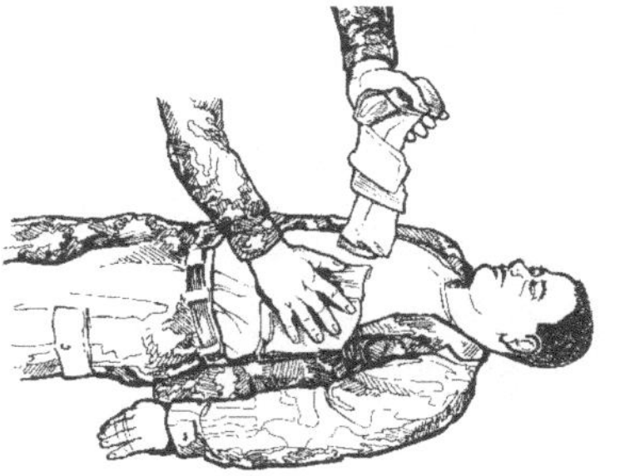

Figure 3-3. Shaking open the field dressing.

(2) Place the white side of the dressing on the plastic wrapper covering the wound (Figure 3-4).

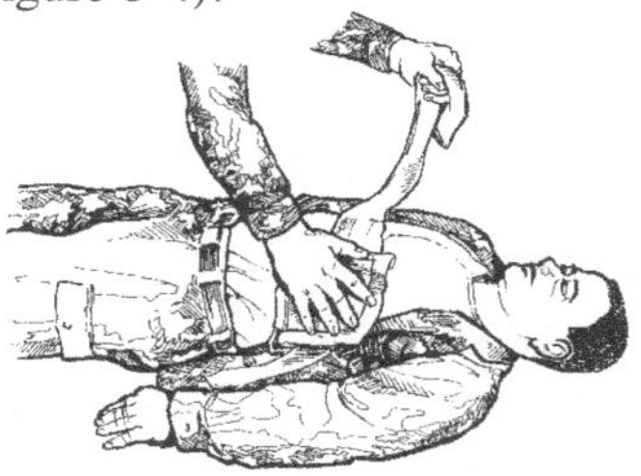

Figure 3-4. Field dressing placed on plastic wrapper.

NOTE

Use the casualty's field dressing, not your own.

(3) Have the casualty breathe normally.

(4) While maintaining pressure on the dressing, grasp one tail of the field dressing with the other hand and wrap it around the casualty's back. If tape is available, tape three sides of the plastic wrapper to the chest wall to provide occlusive type dressing. Leave one side untapped to provide emergency escape for air that may build up in the chest. If tape is not available, secure wrapper on three sides with field dressing leaving the fourth side as a flap.

(5) Wrap the other tail in the opposite direction, bringing both tails over the dressing (Figure 3-5).

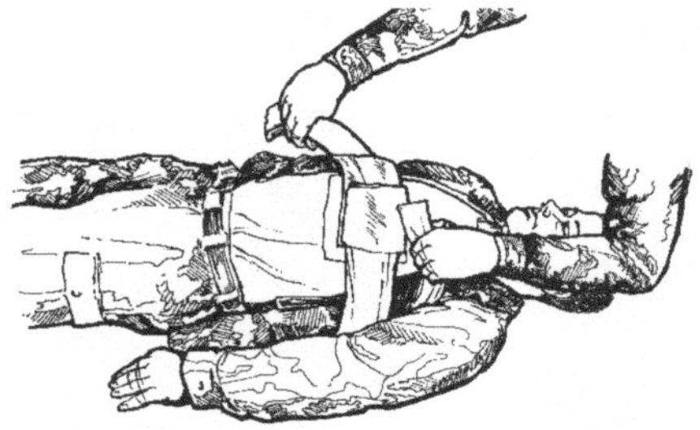

Figure 3-5. Tails of field dressing wrapped around casualty in opposite direction.

(6) Tie the tails into a square knot in the center of the dressing *after* the casualty exhales and *before* he inhales. This will aid in maintaining pressure on the bandage after it has been tied (Figure 3-6). Tie the dressing firmly enough to secure the dressing without interfering with the casualty's breathing.

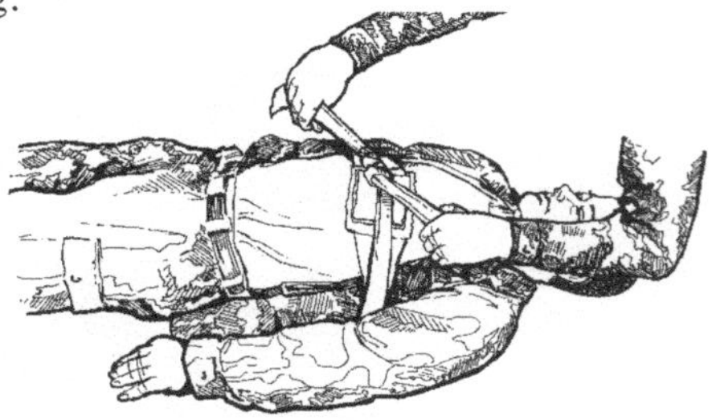

Figure 3-6. Tails of dressing tied into square knot over center of dressing.

NOTE

When practical, apply direct manual pressure over the dressing for 5 to 10 minutes to help control the bleeding.

f. *Position the Casualty*. Position the casualty on his injured side or in a sitting position, whichever makes breathing easier (Figure 3-7).

Figure 3-7. Casualty positioned (lying) on injured side.

g. Seek Medical Assistance. Contact medical personnel.

WARNING

If an occlusive dressing has been improperly placed, air may enter the chest cavity with no means of escape. This causes a life-threatening condition called *tension pneumothorax*. If the casualty's condition (for example, difficulty breathing, shortness of breath, restlessness, or blueness/grayness of the skin) worsens after placing the dressing, quickly lift or remove, and then replace the occlusive dressing.

3-6. Abdominal Wounds

The most serious abdominal wound is one in which an object penetrates the abdominal wall and pierces internal organs or large blood vessels. In these instances, bleeding may be severe and death can occur rapidly.

3-7. First Aid for Abdominal Wounds

a. Evaluate the Casualty. Be prepared to perform basic first aid measures. Always check for both entry and exit wounds. If there are two wounds (entry and exit), treat the wound that appears more serious first (for example, the heavier bleeding, protruding organs, larger wound, and so forth). It may be necessary to improvise dressings for the second wound by using strips of cloth, a T-shirt, or the cleanest material available.

b. Position the Casualty. Place and maintain the casualty on his back with his knees in an upright (flexed) position (Figure 3-8). The knees-up position helps relieve pain, assists in the treatment of shock, prevents further exposure of the bowel (intestines) or abdominal organs, and helps relieve abdominal pressure by allowing the abdominal muscles to relax.

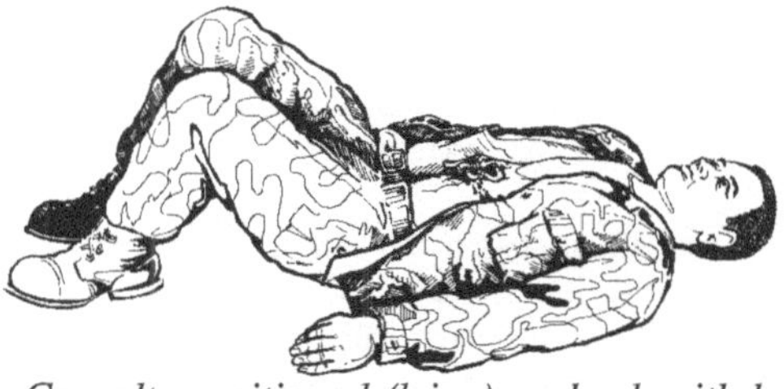

Figure 3-8. Casualty positioned (lying) on back with knees (flexed) up.

c. Expose the Wound.

(1) Remove the casualty's loose clothing to expose the wound. However, **DO NOT** attempt to remove clothing that is stuck to the wound; removing it may cause further injury.

> **CAUTION**
>
> **DO NOT REMOVE** protective clothing in a chemical environment. Apply dressings *over* the protective clothing.

(2) Gently pick up any organs that may be on the ground. Do this with a clean, dry dressing or with the cleanest available material. Place the organs on top of the casualty's abdomen (Figure 3-9).

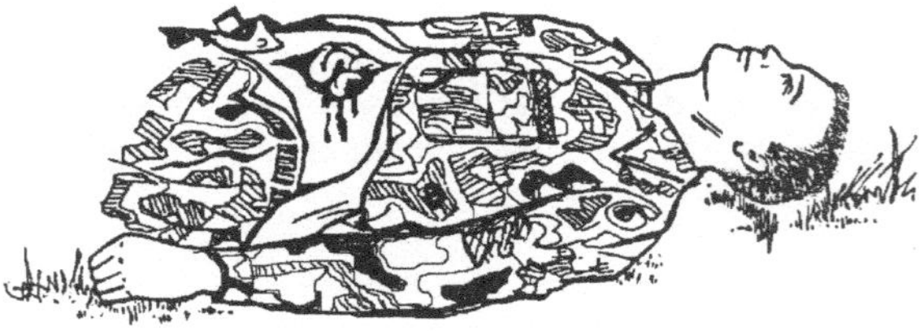

Figure 3-9. Protruding organs placed near wound.

NOTE

DO NOT probe, clean, or try to remove any foreign object from the abdomen. **DO NOT** touch with bare hands any exposed organs. **DO NOT** push organs back inside the body.

d. Apply the Field Dressing. Use the casualty's field dressing, not your own. If the field dressing is not large enough to cover the entire wound, the plastic wrapper from the dressing may be used to cover the wound first (placing the field dressing on top). Open the plastic wrapper carefully without touching the inner surface. If necessary, other improvised dressings may be made from clothing, blankets, or the cleanest materials available.

> **WARNING**
>
> **If there is an object extending from the wound, DO NOT remove it. Place as much of the wrapper over the wound as possible without dislodging or moving the object. DO NOT place the wrapper over the object.**

(1) Grasp the tails in both hands.

(2) Hold the dressing with the white side down directly over the wound. **DO NOT** touch the white (sterile) side of the dressing or allow anything except the wound to come in contact with it.

(3) Pull the dressing open and place it directly over the wound (Figure 3-10). If the casualty is able, he may hold the dressing in place.

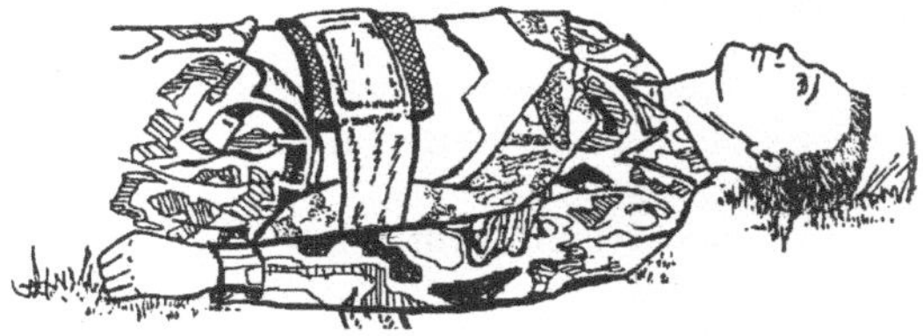

Figure 3-10. Dressing placed directly over the wound.

(4) Hold the dressing in place with one hand and use the other hand to wrap one of the tails around the body.

(5) Wrap the other tail in the opposite direction until the dressing is completely covered. Leave enough of the tail for a knot.

(6) Loosely tie the tails with a square knot at the casualty's side (Figure 3-11).

Figure 3-11. Dressing applied and tails tied with a square knot.

WARNING

When the dressing is applied, DO NOT put pressure on the wound or exposed internal parts, because pressure could cause further injury (vomiting, ruptured intestines, and so forth). Therefore, tie the dressing ties (tails) loosely at casualty's side, not directly over the dressing.

(7) Tie the dressing firmly enough to prevent slipping without applying pressure to the wound site (Figure 3-12).

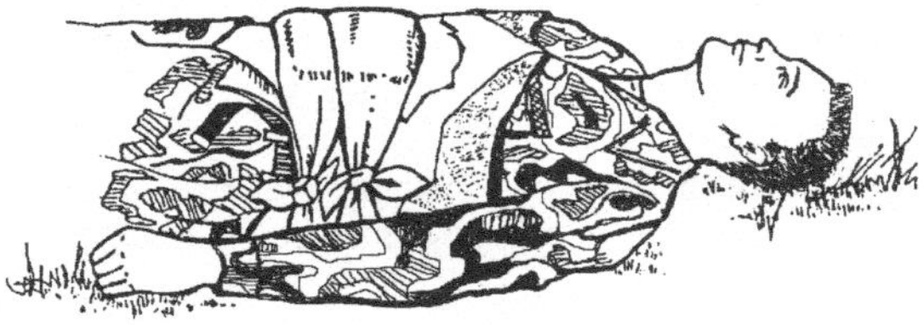

Figure 3-12. Field dressing covered with improvised material and loosely tied.

Field dressings can be covered with improvised reinforcement material (cravats, strips of torn T-shirt, or other cloth) for additional support and protection. Tie improvised bandage on the opposite side of the dressing ties firmly enough to prevent slipping but without applying additional pressure to the wound.

CAUTION

DO NOT give casualties with abdominal wounds food or water (moistening the lips is allowed).

e. Seek Medical Assistance. Notify medical personnel.

3-8. Burn Injuries

Burns often cause extreme pain, scarring, or even death. Before administering first aid, you must be able to recognize the type of burn. There are four types of burns:

- Thermal burns caused by fire, hot objects, hot liquids, and gases; or by nuclear blast or fireball.
- Electrical burns caused by electrical wires, current, or lightning.
- Chemical burns caused by contact with wet or dry chemicals or white phosphorus (WP)—from marking rounds and grenades.
- Laser burns (eye [ocular] injury).

3-9. First Aid for Burns

a. Eliminate the Source of the Burn. The source of the burn must be eliminated before any evaluation of the casualty can occur and first aid administered.

(1) Quickly remove the casualty from danger and cover the *thermal burn* with any large nonsynthetic material, such as a field jacket. If the casualty's clothing is still on fire, roll the casualty on the ground to smother (put out) the flames (Figure 3-13).

Figure 3-13. Casualty covered and rolled on ground.

CAUTION

Synthetic materials, such as nylon, may melt and cause further injury.

(2) Remove the *electrical burn* casualty from the electrical source by turning off the electrical current. **DO NOT** attempt to turn off the electricity if the source is not close by. Speed is critical, so **DO NOT** waste unnecessary time. If the electricity cannot be turned off, wrap any *nonconductive* material (*dry* rope, clothing, wood, and so forth) around the casualty's back and shoulders and drag the casualty away from the electrical source (Figure 3-14). **DO NOT** make body-to-body contact with the casualty or touch any wires because you could also become an electrical burn casualty.

Figure 3-14. Casualty removed from electrical source (using nonconductive material).

WARNING

High voltage electrical burns may cause temporary unconsciousness, difficulties in breathing, or difficulties with the heart (heartbeat).

(3) Remove the *chemical* from the *burned casualty*. Remove *liquid* chemicals by flushing with as much water as possible. Remove *dry* chemicals by brushing off loose particles (**DO NOT** use the bare surface of your hand because you could become a chemical burn casualty) and then flush with large amounts of water, if available. If large amounts of water are not available, then **NO** water should be applied because small amounts of water applied to a dry chemical burn may cause a chemical reaction. When WP strikes the skin, smother with a wet cloth or mud. Keep WP covered with a wet material to exclude air; this should help prevent the particles from burning.

(4) Remove the *laser burn* casualty from the source. When removing the casualty from the laser beam source, be careful not to enter the

beam or you may become a casualty. Never look directly at the beam source and if possible, wear appropriate eye protection.

NOTE

After the casualty is removed from the source of the burn, he should be evaluated for conditions requiring basic first aid measures.

b. Expose the Burn. Cut and gently lift away any clothing covering the burned area, without pulling clothing over the burns. Leave in place any clothing that is stuck to the burn. If the casualty's hands or wrists have been burned, remove jewelry if possible without causing further injury (rings, watches, and so forth) and place in his pockets. This prevents the necessity to cut off jewelry since swelling usually occurs as a result of a burn.

CAUTION

DO NOT lift or cut away clothing if in a chemical environment. Apply the dressing directly over the casualty's protective clothing. **DO NOT** attempt to decontaminate skin where blisters have formed.

c. Apply a Field Dressing to the Burn.

(1) Grasp the tails of the casualty's dressing in both hands.

(2) Hold the dressing directly over the wound with the white side down, pull the dressing open, and place it directly over the wound. **DO NOT** touch the white (sterile) side of the dressing or allow anything except the wound to come in contact with it. If the casualty is able, he may hold the dressing in place.

(3) Hold the dressing in place with one hand and use the other hand to wrap one of the tails around the limbs or the body.

(4) Wrap the other tail in the opposite direction until the dressing is completely covered.

(5) Tie the tails into a square knot over the outer edge of the dressing. The dressing should be applied lightly over the burn. Ensure that dressing is applied firmly enough to prevent it from slipping.

NOTE

Use the cleanest improvised dressing material available if a field dressing is not available or if it is not large enough for the entire wound.

d. Take the Following Precautions:

- **DO NOT** place the dressing over the face or genital area.
- **DO NOT** break the blisters.
- **DO NOT** apply grease or ointments to the burns.
- For electrical burns, check for both an entry and exit burn from the passage of electricity through the body. Exit burns may appear on any area of the body despite location of entry burn.
- For burns caused by wet or dry chemicals, flush the burns with large amounts of water and cover with a dry dressing.
- For burns caused by WP, flush the area with water, then cover with a wet material, dressing, or mud to exclude the air and keep the WP particles from burning.
- For laser burns, apply a field dressing.
- If the casualty is conscious and not nauseated, give him small amounts of water.

e. Seek Medical Assistance. Notify medical personnel.

3-10. Dressings and Bandages

a. Head Wounds.

(1) *Position the casualty.*

WARNING

DO NOT move the casualty if you suspect he has sustained a neck, spine, or head injury (which produces any signs or symptoms other than minor bleeding).

- If the casualty has a minor (superficial) scalp wound and is conscious:

  - Have the casualty sit up (unless other injuries prohibit or he is unable to).

  - If the casualty is lying down and is not accumulating fluids or drainage in his throat, elevate his head slightly.

  - If the casualty is bleeding from or into his mouth or throat, turn his head to the side or position him on his side so that the airway will be clear. Avoid putting pressure on the wound and place him on his uninjured side (Figure 3-15).

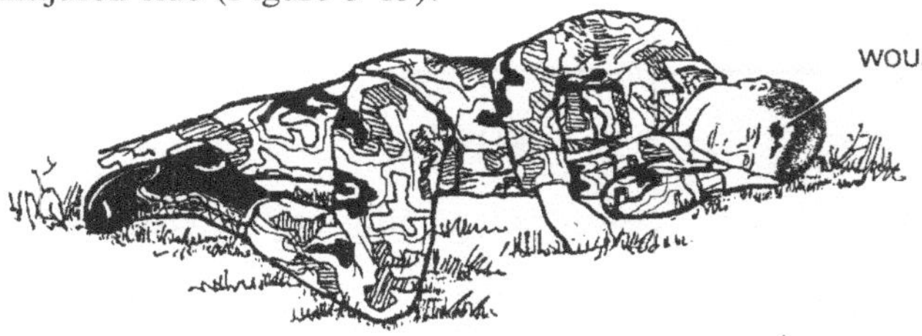

Figure 3-15. Casualty lying on side opposite injury.

- If the casualty is unconscious or has a severe head injury, then suspect and treat him as having a potential neck or spinal injury, *immobilize and **DO NOT** move the casualty.*

NOTE

If the casualty is choking or vomiting or is bleeding from or into his mouth (thus compromising his airway), position him on his uninjured side to allow for drainage and to help keep his airway clear.

WARNING

If it is necessary to turn a casualty with a suspected neck/spine injury; roll the casualty gently onto his side, keeping the head, neck, and body aligned while providing support for the head and neck. DO NOT roll the casualty by yourself but seek assistance. *Move him only if absolutely necessary*, otherwise keep the casualty immobilized to prevent further damage to the neck/spine.

(2) *Expose the wound.* Remove the casualty's helmet (if necessary). In a nuclear, biological, and chemical (NBC) environment, the first aid provider must leave the casualty as much protection (such as protective mask, mission-oriented protective posture [MOPP] overgarments) as possible. What items of protective equipment can be removed is dependent upon the casualty's injuries (where on the body and what type), the MOPP level, integrity of protective equipment (such as tears in the garment or mask seal), availability of chemical protective shelters, and the tactical situation.

WARNING

DO NOT attempt to clean the wound or remove a protruding object.

NOTE

Always use the casualty's field dressing, not your own.

(3) *Apply a dressing to a wound of the forehead or back of head.* To apply a dressing to a wound of the forehead or back of the head—

(*a*) Remove the dressing from the wrapper.

(*b*) Grasp the tails of the dressing in both hands.

(*c*) Hold the dressing (white side down) directly over the wound. **DO NOT** touch the white (sterile) side of the dressing or allow anything except the wound to come in contact with it.

(*d*) Place it directly over the wound.

(*e*) Hold it in place with one hand. If the casualty is able, he may assist.

(*f*) Wrap the first tail horizontally around the head; ensure the tail covers the dressing (Figure 3-16).

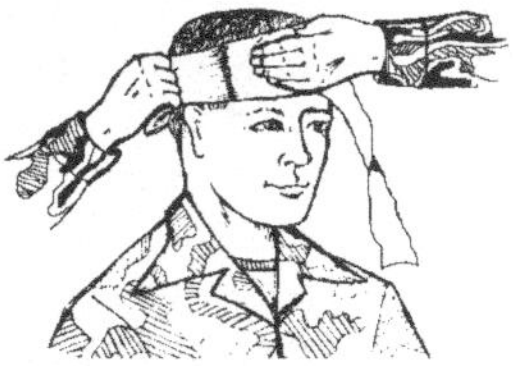

Figure 3-16. First tail of dressing wrapped horizontally around head.

(*g*) Hold the first tail in place and wrap the second tail in the opposite direction, covering the dressing (Figure 3-17).

Figure 3-17. Second tail wrapped in opposite direction.

(*h*) Tie a square knot and secure the tails at the side of the head, making sure they **DO NOT** cover the eyes or ears (Figure 3-18).

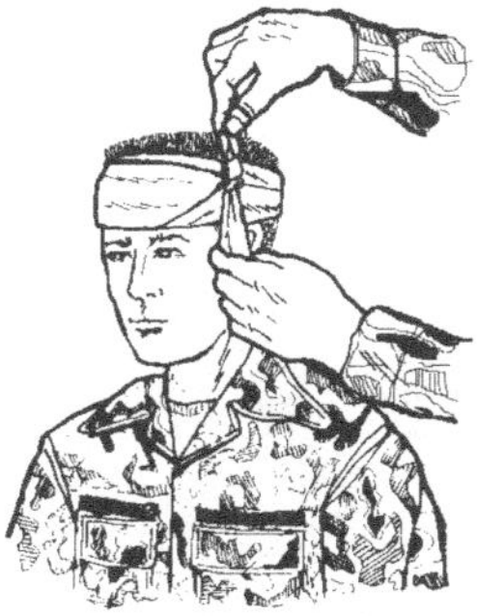

Figure 3-18. Tails tied in square knot at side of head.

(4) *Apply a dressing to a wound on top of the head.* To apply a dressing to a wound on top of the head—

(*a*) Remove the dressing from the wrapper.

(*b*) Grasp the tails of the dressing in both hands.

(*c*) Hold it (white side down) directly over the wound. **DO NOT** touch the white (sterile) side of the dressing or allow anything except the wound to come in contact with it.

(*d*) Place it over the wound (Figure 3-19).

Figure 3-19. Dressing placed over wound.

(*e*) Hold it in place with one hand. If the casualty is able, he may assist.

(*f*) Wrap one tail down under the chin (Figure 3-20), up in front of the ear, over the dressing, and in front of the other ear.

Figure 3-20. One tail of dressing wrapped under chin.

WARNING

Ensure the tails remain wide and close to the front of the chin to avoid choking the casualty.

(*g*) Wrap the remaining tail under the chin in the opposite direction and up the side of the face to meet the first tail (Figure 3-21).

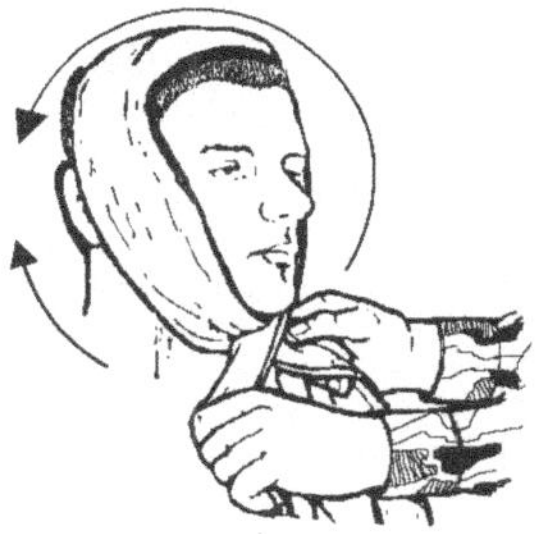

Figure 3-21. Remaining tail wrapped under chin in opposite direction.

(*h*) Cross the tails (Figure 3-22), bringing one around the forehead (above the eyebrows) and the other around the back of the head (at the base of the skull) to a point just above and in front of the opposite ear, and tie them using a square knot (Figure 3-23).

Figure 3-22. Tails of dressing crossed with one around forehead.

Figure 3-23. Tails tied in square knot (in front of and above ear).

(5) *Apply a triangular bandage to the head.* To apply a triangular bandage to the head—

(*a*) Turn the base (longest side) of the bandage up and center its base on the center of the forehead, letting the point (apex) fall on the back of the neck (Figure 3-24A).

(*b*) Take the ends behind the head and cross the ends over the apex.

(*c*) Take them over the forehead and tie them (Figure 3-24B).

(*d*) Tuck the apex behind the crossed part of the bandage or secure it with a safety pin, if available (Figure 3-24C).

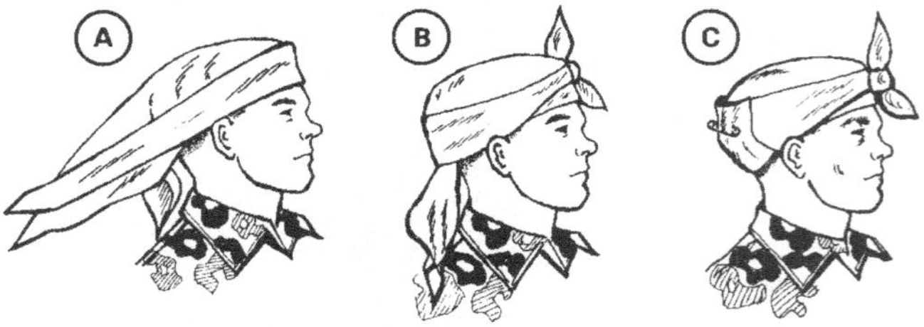

Figure 3-24. Triangular bandage applied to head (Illustrated A—C)

(6) *Apply a cravat bandage to the head.* To apply a cravat bandage to the head—

(*a*) Place the middle of the bandage over the dressing (Figure 3-25A).

(*b*) Cross the two ends of the bandage in opposite directions completely around the head (Figure 3-25B).

(*c*) Tie the ends over the dressing (Figure 3-25C).

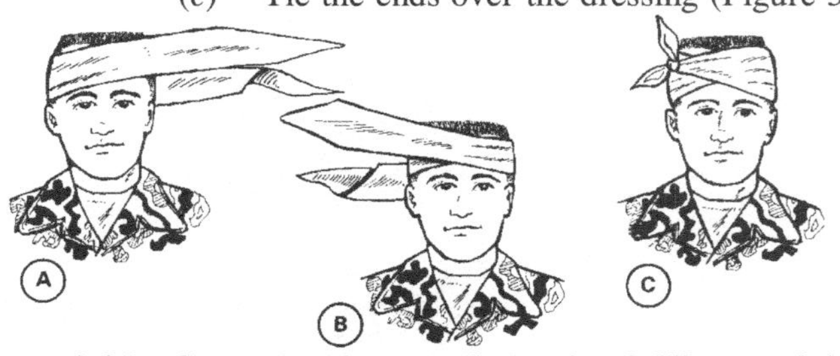

Figure 3-25. Cravat bandage applied to head (Illustrated A—C).

b. Eye Injuries. The eye is a vital sensory organ, and blindness is a severe physical handicap. Timely first aid of the eye may relieve pain and may also help to prevent shock, permanent eye injury, and possible loss of vision. Because the eye is very sensitive, any injury can be easily aggravated if it is improperly handled. Injuries of the eye may be quite severe. Cuts of the eyelids can appear to be very serious, but if the eyeball is not involved, a person's vision usually will not be damaged. However, lacerations (cuts) of the eyeball can cause permanent damage or loss of sight.

(1) *Lacerated/torn eyelids.* Lacerated eyelids may bleed heavily, but bleeding usually stops quickly. Cover the injured eye with a sterile dressing. **DO NOT** put pressure on the wound because you may injure the eyeball. Handle torn eyelids very carefully to prevent further injury. Place any detached pieces of the eyelid on a clean bandage or dressing and immediately send them with the casualty to the medical facility.

(2) *Lacerated eyeball (injury to the globe).* Lacerations or cuts to the eyeball may cause serious and permanent eye damage. Cover the injury with a loose sterile dressing. **DO NOT** put pressure on the eyeball because additional damage may occur. An important point to remember is that when one eyeball is injured, you should immobilize both eyes. This is done by applying a bandage to both eyes. Because the eyes move together, covering both will lessen the chances of further damage to the injured eye. (However, in hazardous surroundings, leave uninjured eye uncovered to enable casualty to see.)

CAUTION

DO NOT apply pressure when there is a possible laceration of the eyeball. The eyeball contains fluid. Pressure applied over the eye will force the fluid out, resulting in permanent injury. **APPLY PROTECTIVE DRESSING WITHOUT ADDED PRESSURE**.

(3) *Extruded eyeballs.* Service members may encounter casualties with severe eye injuries that include an extruded eyeball (eyeball out-of-socket). In such instances you should gently cover the extruded eye with a loose moistened dressing and also cover the unaffected eye. **DO NOT** bind or exert pressure on the injured eye while applying the dressing. Keep the casualty quiet, place him on his back, treat for shock, and evacuate him immediately.

(4) *Burns of the eyes.* Chemical burns, thermal (heat) burns, and light burns can affect the eyes.

(*a*) *Chemical burns*. Injuries from chemical burns require immediate first aid. Mainly acids or alkalies cause chemical burns. The first aid measures consist of flushing the eyes immediately with large amounts of water for at least 5 to 20 minutes, or as long as necessary to flush out the chemical and, once flushed, bandaging the eyes. If the burn is an acid burn, you should flush the eye for at least 5 to 10 minutes. If the burn is an alkali burn, you should flush the eye for at least 20 minutes. After the eye has been flushed evacuate the casualty immediately.

(*b*) *Thermal burns*. When an individual suffers burns of the face from a fire, the eyes will close quickly due to extreme heat. This reaction is a natural reflex to protect the eyeballs; however, the eyelids remain exposed and are frequently burned. If a casualty receives burns of the eyelids or face—

- **DO NOT** apply a dressing.
- **DO NOT** touch.
- **SEEK** medical assistance immediately.

(*c*) *Light burns*. Exposure to intense light can burn an individual. Infrared rays, eclipse light (if the casualty has looked directly at the sun), or laser burns cause injuries of the exposed eyeball. Ultraviolet rays from arc welding can cause a superficial burn to the surface of the eye. These injuries are generally not painful but may cause permanent damage to the eyes. Immediate first aid is usually not required. Loosely bandaging the eyes may make the casualty more comfortable and protect his eyes from further injury caused by exposure to other bright lights or sunlight.

CAUTION

With impaled objects or significant sized foreign bodies, both eyes are usually bandaged to help secure the foreign body in the injured eye. In a battlefield environment, leave the uninjured eye uncovered so that the casualty can see.

c. *Side-of-Head or Cheek Wound*. Facial injuries to the side of the head or the cheek may bleed profusely (Figure 3-26). Prompt action is necessary to ensure that the airway remains open and also to control the bleeding. It may be necessary to apply a dressing. To apply a dressing—

(1) Remove the dressing from its wrapper.

(2) Grasp the tails in both hands.

(3) Hold the dressing directly over the wound with the white side down and place it directly on the wound (Figure 3-27A). **DO NOT** touch the white (sterile) side of the dressing or allow anything except the wound to come in contact with it.

(4) Hold the dressing in place with one hand (the casualty may assist if able). Wrap the top tail over the top of the head and bring it down in front of the ear (on the side opposite the wound), under the chin (Figure 3-27B) and up over the dressing to a point just above the ear (on the wound side).

Figure 3-26. Side of head or cheek wound.

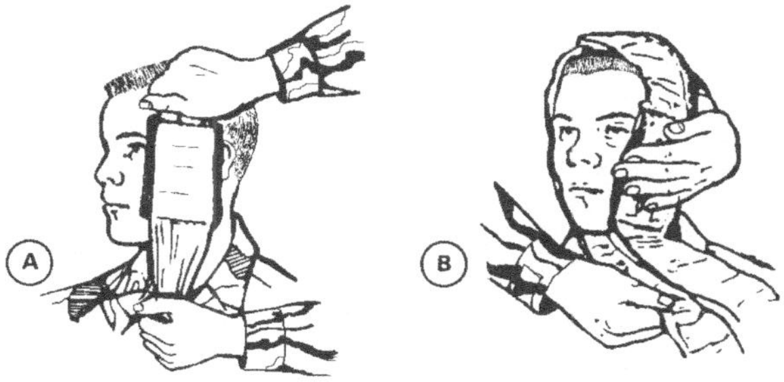

Figure 3-27. Dressing placed directly on wound. Top tail wrapped over top of head, down in front of ear, and under chin (Illustrated A—B).

NOTE

When possible, avoid covering the casualty's ear with the dressing, as this will decrease his ability to hear.

(5) Bring the second tail under the chin, up in front of the ear (on the side opposite the wound), and over the head to meet the other tail (on the wounded side) (Figure 3-28).

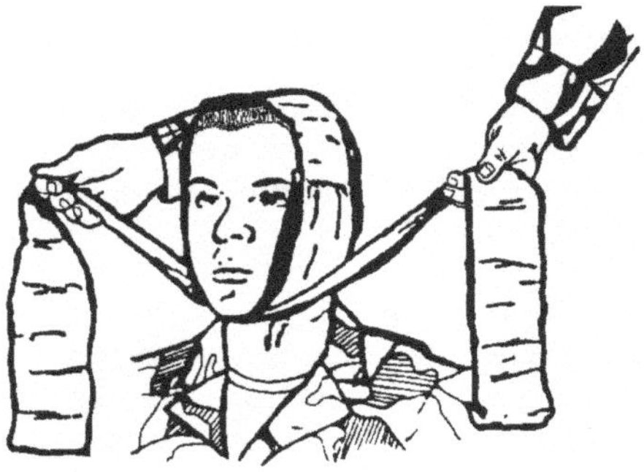

Figure 3-28. Bringing second tail under the chin.

(6) Cross the two tails (on the wound side) (Figure 3-29) and bring one end across the forehead (above the eyebrows) to a point just in front of the opposite ear (on the uninjured side).

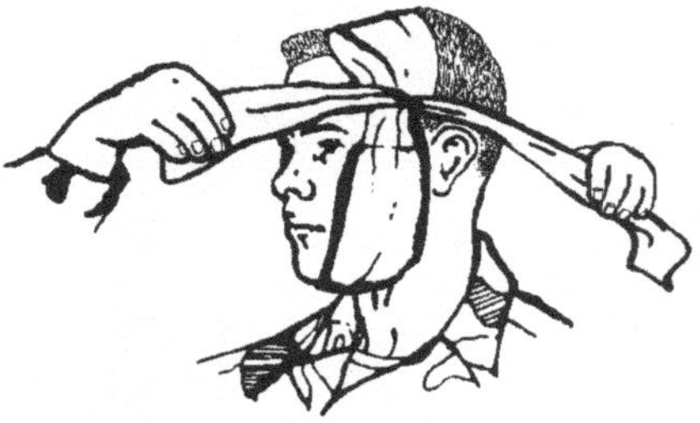

Figure 3-29. Crossing the tails on the side of the wound.

(7) Wrap the other tail around the back of the head (at the base of the skull), and tie the two ends just in front of the ear on the uninjured side with a square knot (Figure 3-30).

Figure 3-30. Tying the tails of the dressing in a square knot.

d. Ear Injuries. Lacerated (cut) or avulsed (torn) ear tissue may not, in itself, be a serious injury. Bleeding, or the drainage of fluids from the ear canal, however, may be a sign of a head injury, such as a skull fracture. **DO NOT** attempt to stop the flow from the inner ear canal nor put anything into the ear canal to block it. Instead, you should cover the ear lightly with a dressing. For minor cuts or wounds to the external ear, apply a cravat bandage as follows:

(1) Place the middle of the bandage over the ear (Figure 3-31A).

(2) Cross the ends, wrap them in opposite directions around the head, and tie them (Figures 3-31B and 3-31C).

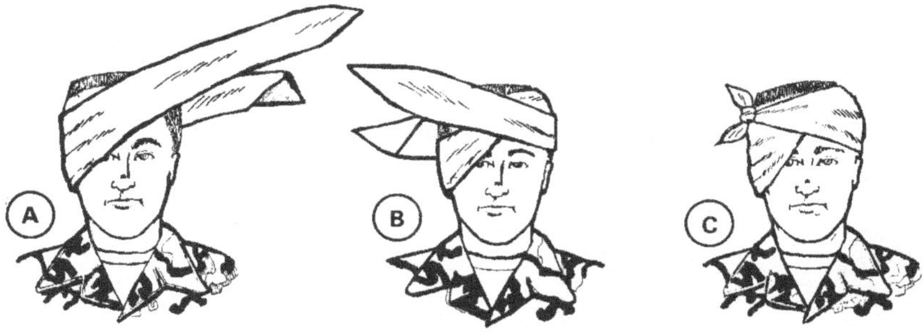

Figure 3-31. Applying cravat bandage to ear (Illustrated A—C).

(3) If possible, place some dressing material between the back of the ear and the side of the head to avoid crushing the ear against the head with the bandage.

e. Nose Injuries. Nose injuries generally produce bleeding. The bleeding may be controlled by placing an ice pack (if available) over the nose, or pinching the nostrils together. The bleeding may also be controlled by placing torn gauze (rolled) between the upper teeth and the lip.

CAUTION
DO NOT attempt to remove objects inhaled into the nose. An untrained person who removes such an object could worsen the casualty's condition and cause permanent injury.

f. Jaw Injuries. Before applying a bandage to a casualty's jaw, remove all loose or free-floating foreign material from the casualty's mouth.

If the casualty is unconscious, check for obstructions in the airway and remove if possible. If there is profuse bleeding in the oral cavity, the cavity may require loose packing with soft bandaging material (for example: Kerlix™ gauze) prior to applying a bandage. Care should be taken to avoid occluding the airway. When applying the bandage, allow the jaw enough freedom to permit passage of air and drainage from the mouth.

(1) *Apply bandages attached to field first aid dressing to the jaw.* After dressing the wound, apply the bandages using the same technique illustrated in Figure 3-32A—C.

NOTE

The dressing and bandaging procedure outlined for the jaw serves a twofold purpose. In addition to stopping the bleeding and protecting the wound, it also immobilizes a fractured jaw.

(2) *Apply a cravat bandage to the jaw.*

(*a*) Place the bandage under the chin and pull its ends upward. Adjust the bandage to make one end longer than the other (Figure 3-32A).

(*b*) Take the longer end over the top of the head to meet the short end at the temple and cross the ends over (Figure 3-32B).

(*c*) Take the ends in opposite directions to the other side of the head and tie them over the part of the bandage that was applied first (Figure 3-32C).

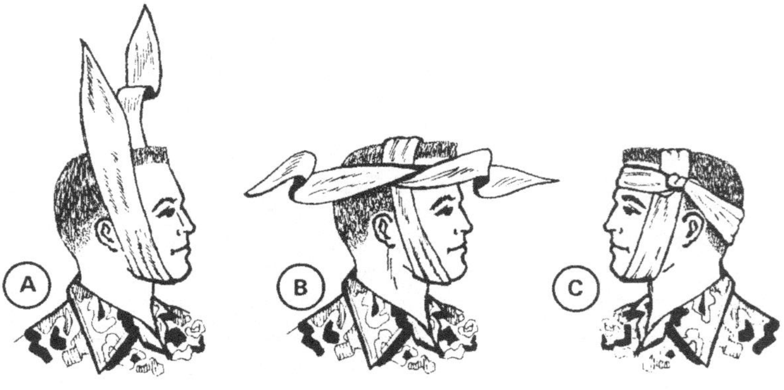

Figure 3-32. Applying a cravat bandage to jaw (Illustrated A—C).

NOTE

The cravat bandage technique is used to immobilize a fractured jaw or to maintain a sterile dressing that does not have tail bandages attached.

3-11. Shoulder Bandage

a. To apply bandages attached to the field first aid dressing—

(1) Take one bandage across the chest and the other across the back and under the arm opposite the injured shoulder.

(2) Tie the ends with a square knot (Figure 3-33).

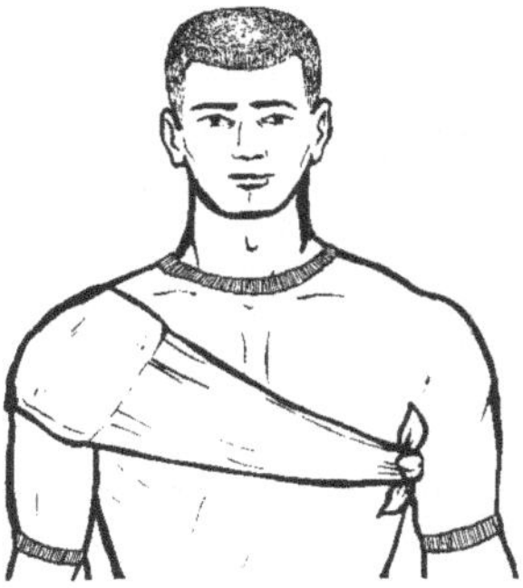

Figure 3-33. Shoulder bandage.

b. To apply a cravat bandage to the shoulder or armpit—

(1) Make an extended cravat bandage by using two triangular bandages (Figure 3-34A); place the end of the first triangular bandage along the base of the second one (Figure 3-34B).

(2) Fold the two bandages into a single extended bandage (Figure 3-34C).

(3) Fold the extended bandage into a single cravat bandage (Figure 3-34D). After folding, secure the thicker part (overlap) with two or more safety pins (Figure 3-34E).

(4) Place the middle of the cravat bandage under the armpit so that the front end is longer than the back end and safety pins are on the outside (Figure 3-34F).

(5) Cross the ends on top of the shoulder (Figure 3-34G).

(6) Take one of the bandage ends across the back and under the arm on the opposite side and the other end across the chest. Tie the ends (Figure 3-34H).

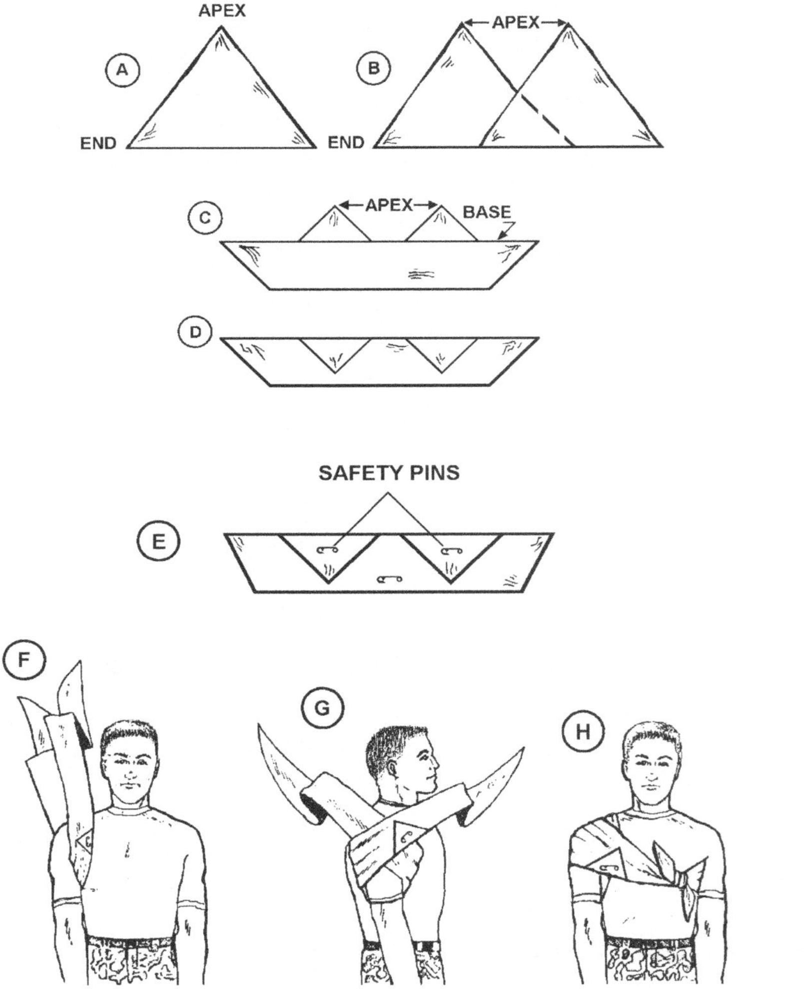

Figure 3-34. Extended cravat bandage applied to shoulder or armpit (Illustrated A—H).

Be sure to place sufficient wadding in the armpit. **DO NOT** tie the cravat bandage too tightly. Avoid compressing the major blood vessels in the armpit.

3-12. Elbow Bandage

To apply a cravat bandage to the elbow—

a. Bend the arm at the elbow and place the middle of the cravat at the point of the elbow bringing the ends upward (Figure 3-35A).

b. Bring the ends across, extending both downward (Figure 3-35B).

c. Take both ends around the arm and tie them with a square knot at the front of the elbow (Figure 3-35C).

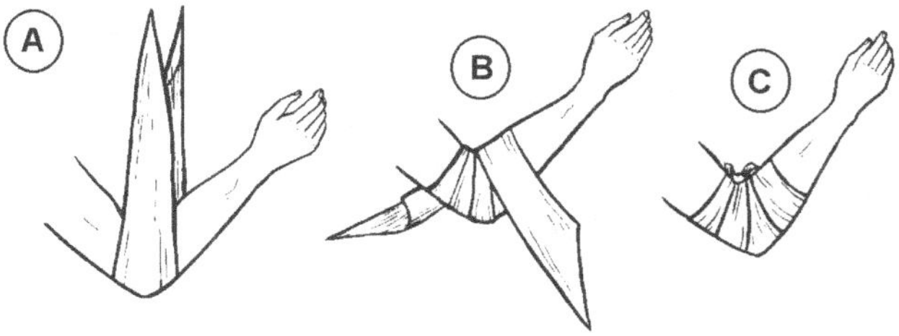

Figure 3-35. Elbow bandage (Illustrated A—C).

CAUTION

If an elbow fracture is suspected, **DO NOT** bend the elbow; bandage it in the position found.

3-13. Hand Bandage

a. To apply a triangular bandage to the hand—

(1) Place the hand in the middle of the triangular bandage with the wrist at the base of the bandage (Figure 3-36A). Ensure that the fingers are separated with absorbent material to prevent chafing and irritation of the skin.

(2) Place the apex over the fingers and tuck any excess material into the pleats on each side of the hand (Figure 3-36B).

(3) Cross the ends on top of the hand, take them around the wrist, and tie them (Figures 3-36C—E) with a square knot.

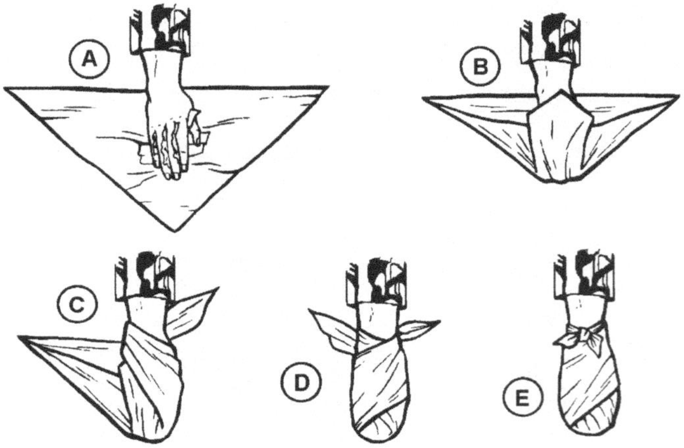

Figure 3-36. Triangular bandage applied to hand (Illustrated A—E).

b. To apply a cravat bandage to the palm of the hand—

(1) Lay the middle of the cravat over the palm of the hand with the ends hanging down on each side (Figure 3-37A).

(2) Take the end of the cravat at the little finger across the back of the hand, extending it upward over the base of the thumb; then bring it downward across the palm (Figure 3-37B).

(3) Take the thumb end across the back of the hand, over the palm, and through the hollow between the thumb and palm (Figure 3-37C).

(4) Take the ends to the back of the hand and cross them; then bring them up over the wrist and cross them again (Figure 3-37D).

(5) Bring both ends down and tie them with a square knot on top of the wrist (Figure 3-37E—F).

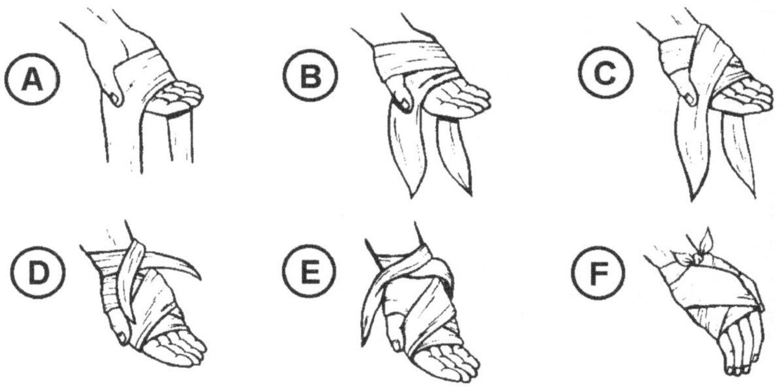

Figure 3-37. Cravat bandage applied to palm of hand (Illustrated A—F).

3-14. Leg (Upper and Lower) Bandage

To apply a cravat bandage to the leg—

a. Place the center of the cravat over the dressing (Figure 3-38A).

b. Take one end around and up the leg in a spiral motion and the other end around and down the leg in a spiral motion, overlapping part of each preceding turn (Figure 3-38B).

c. Bring both ends together and tie them (Figure 3-38C) with a square knot.

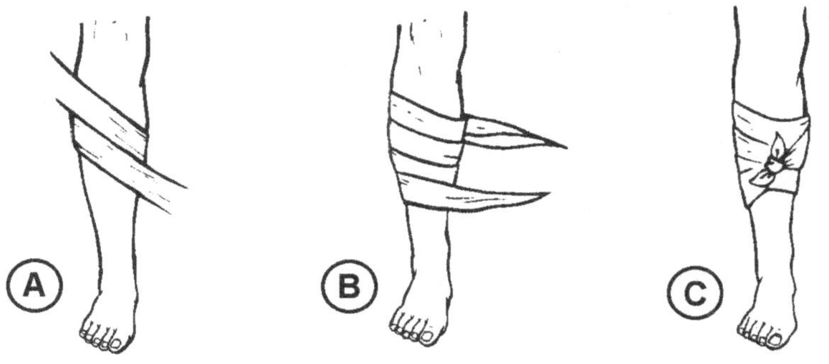

Figure 3-38. Cravat bandage applied to leg (Illustrated A—C).

3-15. Knee Bandage

To apply a cravat bandage to the knee as illustrated in Figure 3-39, use the same technique applied in bandaging the elbow.

> **CAUTION**
>
> If a fracture of the kneecap is suspected, **DO NOT** bend the knee; bandage it in the position found.

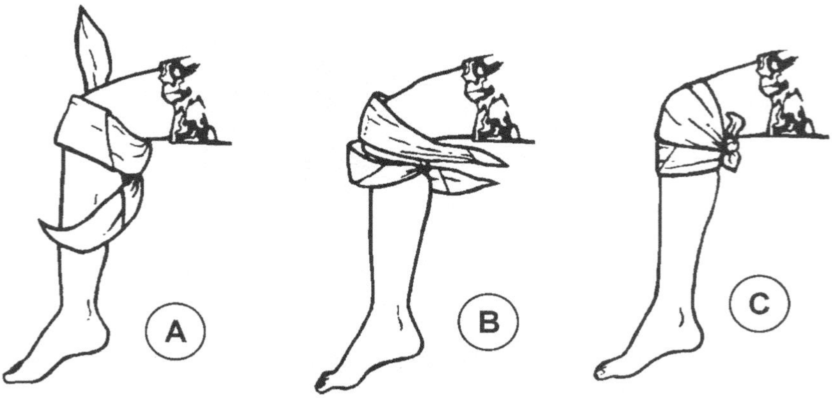

Figure 3-39. Cravat bandage applied to knee (Illustrated A—C).

3-16. Foot Bandage

To apply a triangular bandage to the foot—

a. Place the foot in the middle of the triangular bandage with the heel well forward of the base (Figure 3-40A). Ensure that the toes are separated by absorbent material to prevent chafing and irritation of the skin.

b. Place the apex over the top of the foot and tuck any excess material into the pleats on each side of the foot (Figure 3-40B).

c. Cross the ends on top of the foot, take them around the ankle, and tie them at the front of the ankle (Figure 3-40C—E).

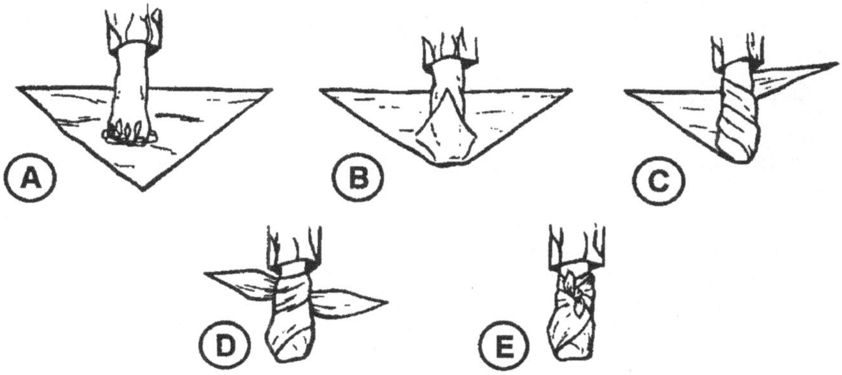

Figure 3-40. Triangular bandage applied to foot (Illustrated A—E).

CHAPTER 4

FIRST AID FOR FRACTURES

4-1. General

A fracture is any break in the continuity of a bone. Fractures can cause total disability or in some cases death by severing vital organs and/or arteries. On the other hand, they can most often be treated so there is a complete recovery. The potential for recovery depends greatly upon the first aid the individual receives before he is moved. First aid includes immobilizing the fractured part in addition to applying lifesaving measures when necessary. The basic splinting principle is to immobilize the joints above and below the fracture.

4-2. Kinds of Fractures

Figure 4-1 depicts types of fractures.

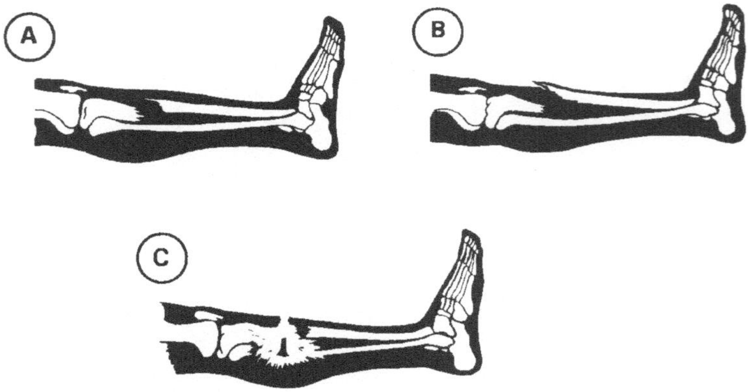

Figure 4-1. Types of fractures (Illustrated A—C).

a. *Closed Fracture* (Figure 4-1A). A closed fracture is a broken bone that does not break the overlying skin. The tissue beneath the skin may be damaged. A *dislocation* is when a joint, such as a knee, ankle, or shoulder, is not in the proper position. A *sprain* is when the connecting tissues of the joints have been torn. Dislocations and sprains (swelling, possible deformity, and discoloration) should be treated as closed fractures.

b. *Open Fracture* (Figure 4-1B and 4-1C). An open fracture is a broken bone that breaks (pierces) the overlying skin. The broken bone may come through the skin or a missile such as a bullet or shell fragment may go through the flesh and break the bone.

NOTE

An open fracture is contaminated and subject to infection.

4-3. Signs and Symptoms of Fractures

Indications of a fracture are deformity, tenderness, swelling, pain, inability to move the injured part, protruding bone, bleeding, or discolored skin at the injury site. A sharp pain when the service member attempts to move the part is also a sign of a fracture.

WARNING

DO NOT encourage the casualty to move the injured part in order to identify a fracture since such movement could cause further damage to surrounding tissues and promote shock. If you are not sure whether a bone is fractured, care for the injury as a fracture. At the site of the fracture, the bone ends are sharp and could cause vessel (artery and/or vein) damage.

4-4. Purposes of Immobilizing Fractures

A fracture is immobilized to prevent the sharp edges of the bone from moving and cutting tissue, muscle, blood vessels, and nerves. This reduces pain and helps prevent or control shock. In a closed fracture, immobilization keeps bone fragments from causing an open wound, which can become contaminated and subject to infection.

4-5. Splints, Padding, Bandages, Slings, and Swathes

a. Splints. Splints may be improvised from such items as boards, poles, sticks, tree limbs, or cardboard. If nothing is available for a splint, the chest wall can be used to immobilize a fractured arm and the uninjured leg can be used to immobilize (to some extent) the fractured leg.

b. Padding. Padding may be improvised from such items as a jacket, blanket, poncho, shelter half, or leafy vegetation.

c. Bandages. Bandages may be improvised from belts, rifle slings, kerchiefs, or strips torn from clothing or blankets. Narrow materials such as wire or cord should not be used to secure a splint in place. The application of wire and/or narrow material to an extremity could cause tissue damage and a tourniquet effect.

d. Slings. A sling is a bandage suspended from the neck to support an upper extremity. If a bandage is not available, a sling can be improvised by using the tail of a coat or shirt or pieces of cloth torn from such items as clothing and blankets. The triangular bandage is ideal for this purpose. Remember that the casualty's hand should be higher than his elbow, and the fingers should be showing at all times. The sling should be applied so that the supporting pressure is on the uninjured side.

e. Swathes. Swathes are any bands (pieces of cloth or load bearing equipment [LBE]) that are used to further immobilize a splinted fracture. Triangular and cravat bandages are often used and are called *swathe bandages*. The purpose of the swathe is to immobilize; therefore, the swathe bandage is placed above and/or below the fracture—not over it.

4-6. Procedures for Splinting Suspected Fractures

Before beginning first aid procedures for a fracture, gather whatever splinting materials are available. Ensure that splints are long enough to immobilize the joint above and below the suspected fracture. If possible, use at least four ties (two above and two below the fracture) to secure the splints. The ties should be square knots and should be tied away from the body on the splint. Distal pulses of the affected extremity should be checked before and after the application of the splint.

a. Evaluate the Casualty. Be prepared to perform any necessary lifesaving measures. Monitor the casualty for development of conditions that may require you to perform necessary lifesaving measures.

WARNING

Unless there is immediate life-threatening danger, such as a fire or an explosion, DO NOT move the casualty with a suspected back or neck injury. Improper movement may cause permanent paralysis or death.

WARNING

In a chemical environment, DO NOT remove any protective clothing. Apply the dressings and splints over the garments.

b. Locate the Site of the Suspected Fracture.

(1) Ask the casualty for the location of the injury.

- Does he have any pain?
- Where is it tender?
- Can he move the extremity?

NOTE

With the presence of an obvious deformity, do not make the casualty move extremity.

(2) Look for an unnatural position of the extremity.

(3) Look for a bone sticking out (protruding).

c. Prepare the Casualty for Splinting the Suspected Fracture.

(1) Reassure the casualty. Tell him that you will be providing first aid for him and that medical help is on the way.

(2) Loosen any tight or binding clothing.

(3) Remove all jewelry from the injured part and place it in the casualty's pocket. Tell the casualty you are doing this because if the jewelry is not removed and swelling occurs later, he may not be able to get it off and further bodily injury could result.

(4) Boots should not be removed from the casualty unless they are needed to stabilize a neck injury or there is actual bleeding from the foot.

d. Gather Splinting Materials. If standard splinting materials (splints, padding, and cravats) are not available, gather improvised materials. If splinting material is not available and the suspected fracture **CANNOT** be splinted, then swathes, or a combination of swathes and slings can be used to immobilize the extremity.

e. *Pad the Splints*. Pad the splints where they touch any bony part of the body, such as the elbow, wrist, knee, ankle, crotch, or armpit areas. Padding prevents excessive pressure on the area, which could lead to circulation problems.

f. *Check the Circulation Below the Site of the Injury*.

(1) Note any pale, white, or bluish-gray color of the skin, which may indicate impaired circulation. Circulation can also be checked by depressing the toe or fingernail beds and observing how quickly the color returns. A slower return of color to the injured side when compared with the uninjured side indicates a problem with circulation. The fingernail bed is the method to use to check the circulation in a dark-skinned casualty.

(2) Check the temperature of the injured extremity. Use your hand to compare the temperature of the injured side with the uninjured side. The body area below the injury may be colder to the touch indicating poor circulation.

(3) Question the casualty about the presence of numbness, tightness, cold, or tingling sensations.

WARNING

Casualties with fractures of the extremities may show impaired circulation, such as numbness, tingling, cold or pale to bluish skin tone. These casualties should be evacuated by medical personnel and treated as soon as possible. Prompt medical treatment may prevent possible loss of the limb.

WARNING

If it is an open fracture and the bone is protruding from the skin, *DO NOT ATTEMPT TO PUSH THE BONE BACK UNDER THE SKIN.* Apply a field dressing over the wound to protect the area.

g. Apply the Splint in Place.

(1) Splint the fracture in the position found. **DO NOT** attempt to reposition or straighten the injury. If it is an open fracture, stop the bleeding and protect the wound. Cover all wounds with field dressings before applying a splint. Remember to use the casualty's field dressing, not your own.

(2) Place one splint on each side of the fracture. Make sure that the splints reach, if possible, beyond the joints above and below the fracture.

(3) Tie the splints. Secure each splint in place above and below the fracture site with improvised (or actual) cravats. Improvised cravats, such as strips of cloth, belts, or whatever else you have, may be used. With minimal motion to the injured areas, place and tie the splints with the bandages. Push cravats through and under the natural body curvatures, and then gently position improvised cravats and tie in place. Use square knots. Tie all knots on the splint away from the casualty (Figure 4-2). **DO NOT** tie cravats directly over the suspected fracture site.

Figure 4-2. Square knots tied away from casualty.

h. Check the Splint for Tightness.

(1) **CHECK** to be sure that bandages are tight enough to securely hold splinting materials in place, but not so tight that circulation is impaired.

(2) **RECHECK** the circulation after application of the splint. Check the skin color and temperature. This is to ensure that the bandages holding the splint in place have not been tied too tightly. A fingertip check can be made by inserting the tip of the finger between the bandaged knot and the skin.

(3) **MAKE** any necessary adjustment without allowing the splint to become ineffective.

i. Apply a Sling. An improvised sling may be made from any available nonstretching piece of cloth, such as a battle dress uniform (BDU) shirt or trousers, poncho, or shelter half. Slings may also be improvised using the tail of a coat, belt, or a piece of cloth. Figure 4-3 depicts a shirttail used for support. A trousers belt or LBE may also be used for support (Figure 4-4). A sling should place the supporting pressure on the casualty's uninjured side. The supported arm should have the hand positioned slightly higher than the elbow showing the fingers.

Figure 4-3. Shirttail used for support.

Figure 4-4. Belt used for support.

(1) Insert the splinted arm in the center of the sling (Figure 4-5).

Figure 4-5. Arm inserted in center of improvised sling.

(2) Bring the ends of the sling up and tie them at the side (or hollow) of the neck on the uninjured side (Figure 4-6).

Figure 4-6. Ends of improvised sling tied to side of neck.

(3) Twist and tuck the corner of the sling at the elbow (Figure 4-7).

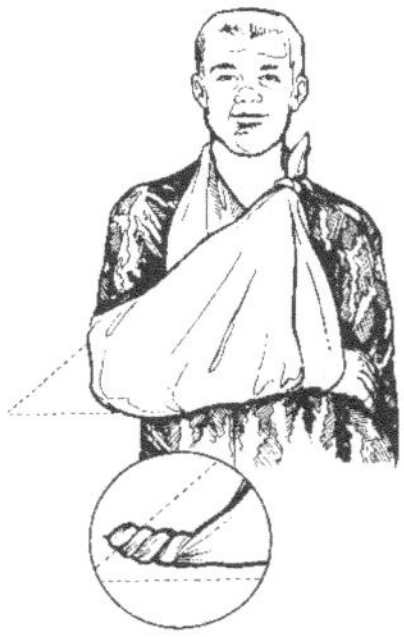

Figure 4-7. Corner of sling twisted and tucked at elbow.

j. Apply a Swathe. You may use any large piece of cloth, service member's belt, or pistol belt, to improvise a swathe.

WARNING

The swathe should not be placed directly on top of the injury, but positioned either above or below the fracture site.

(1) Apply swathes to the injured arm by wrapping the swathe over the injured arm, around the casualty's back, and under the arm on the uninjured side. Tie the ends on the uninjured side (Figure 4-8).

Figure 4-8. Arm immobilized with strip of clothing.

(2) A swathe is applied to an injured leg by wrapping the swathe around both legs and securing it on the uninjured side.

k. Seek Medical Assistance. Notify medical personnel, watch closely for development of life-threatening conditions and/or impaired circulation to the injured extremity. (Refer to Chapter 1 for additional information on life-threatening conditions.)

4-7. Upper Extremity Fractures

Figures 4-9 through 4-17 show how to apply slings, splints, and cravats (swathes) to immobilize and support fractures of the upper extremities. *Although the padding is not visible in some of the illustrations, it is always preferable to apply padding along the injured part for the length of the splint and especially where it touches any bony parts of the body.*

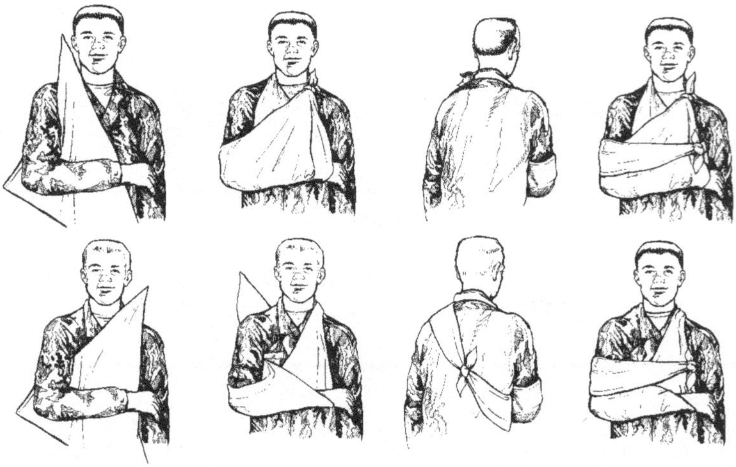

Figure 4-9. Application of triangular bandage to form sling (two methods).

Figure 4-10. Completing sling sequence by twisting and tucking the corner of the sling at the elbow.

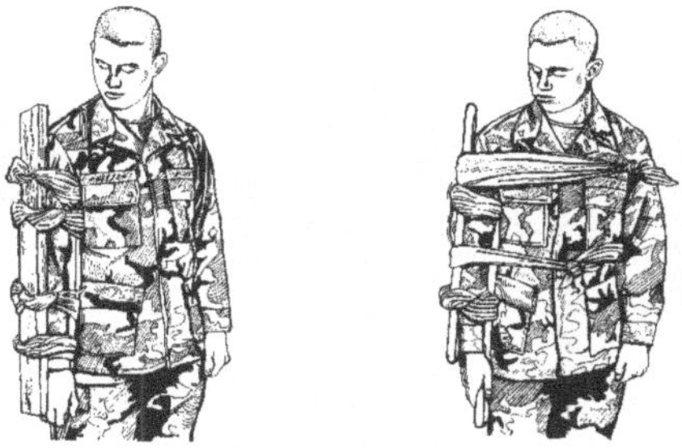

Figure 4-11. Board splints applied to fractured elbow when elbow is not bent (two methods).

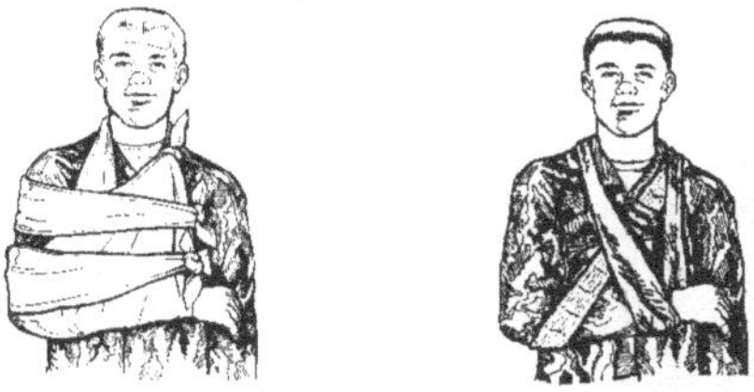

Figure 4-12. Chest wall used as splint for upper arm fracture when no splint is available.

Figure 4-13. Chest wall, sling, and cravat used to immobilize fractured elbow when elbow is bent.

Figure 4-14. Board splint applied to fractured forearm.

Figure 4-15. Fractured forearm or wrist splinted with sticks and supported with tail of shirt and strips of material.

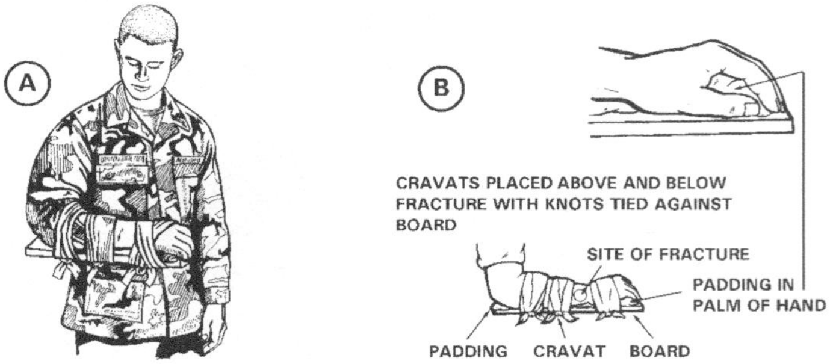

Figure 4-16. Board splint applied to fractured wrist and hand (Illustrated A—B).

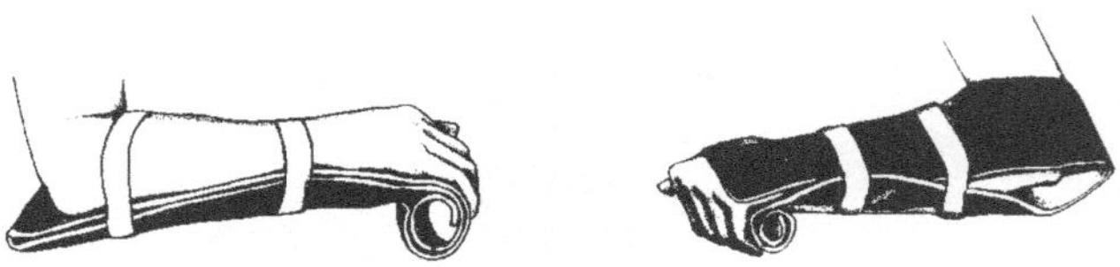

Figure 4-17. SAM® splint applied to fractured wrist or forearm.

4-8. Lower Extremity Fractures

Figures 4-18 through 4-24 show how to apply splints to immobilize fractures of the lower extremities. *Although padding is not visible in some of the illustrations, it is always preferable to apply padding along the injured part for the length of the splint and especially where it touches any bony parts of the body.*

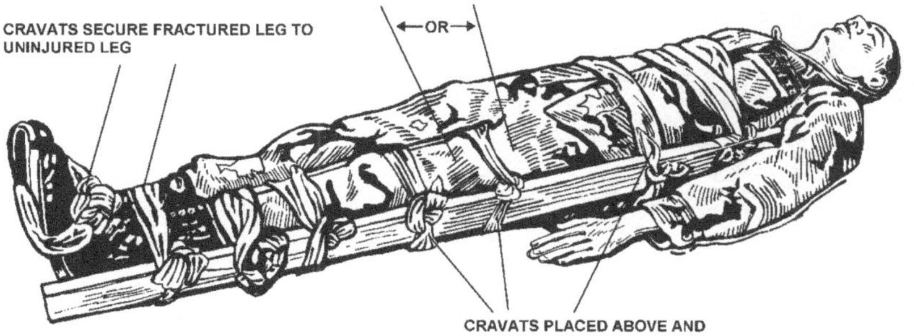

Figure 4-18. Board splints applied to fractured hip or thigh.

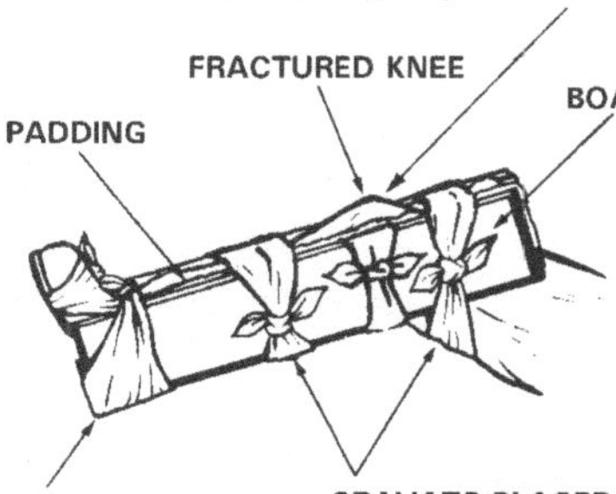

Figure 4-19. Board splint applied to fractured or dislocated knee.

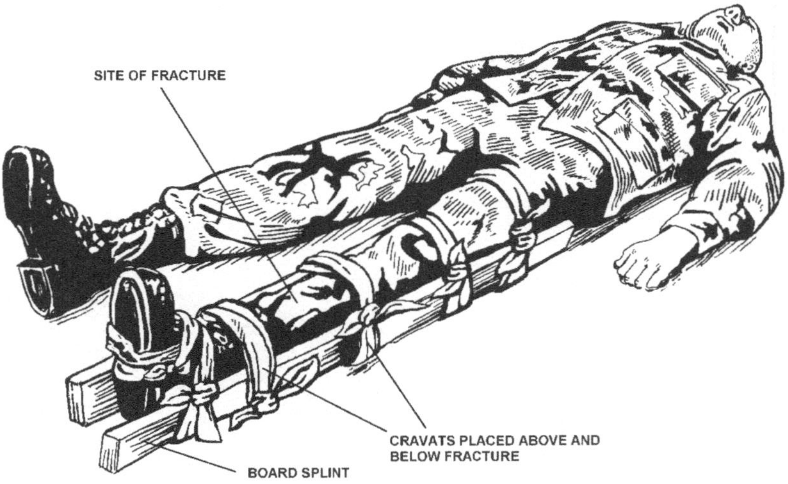

Figure 4-20. Board splints applied to fractured lower leg or ankle.

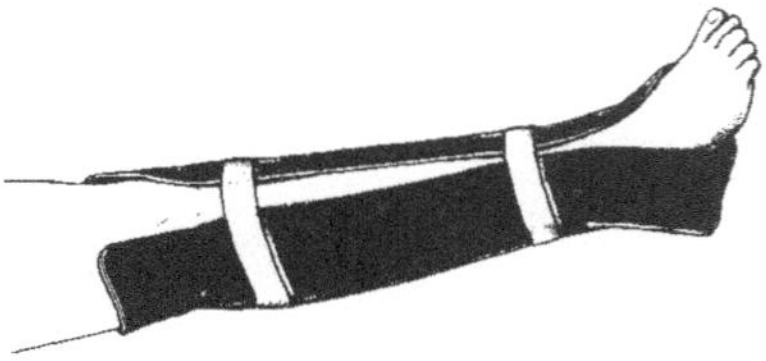

Figure 4-21. SAM® splint applied to fractured lower leg or ankle.

Figure 4-22. Improvised splints applied to fractured lower leg or ankle.

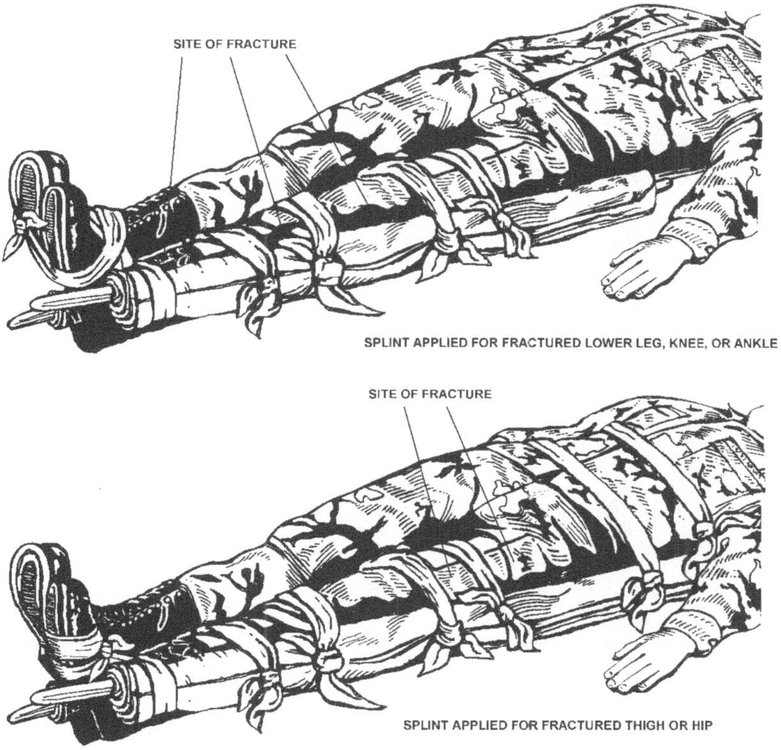

Figure 4-23. Poles rolled in a blanket and used as splints applied to fractured lower extremity.

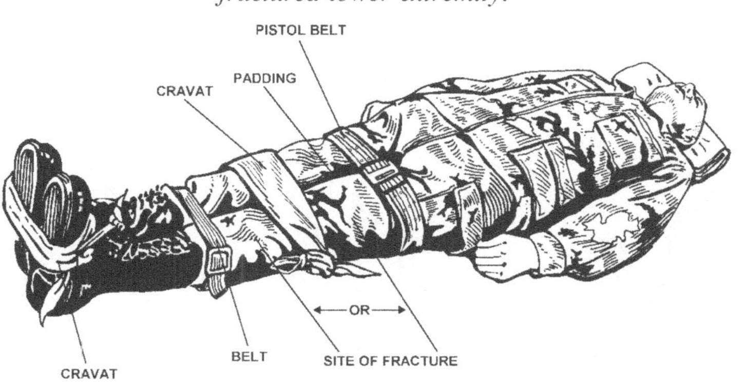

Figure 4-24. Uninjured leg used as splint for fractured leg (anatomical splint).

4-9. Jaw, Collarbone, and Shoulder Fractures

a. Apply a cravat to immobilize a fractured jaw as illustrated in Figure 4-25. Direct all bandaging support to the top of the casualty's head, not to the back of his neck. If incorrectly placed, the bandage will pull the casualty's jaw back and interfere with his breathing.

Figure 4-25. Fractured jaw immobilized.

WARNING

Casualties with lower jaw (mandible) fractures cannot be laid flat on their backs because facial muscles will relax and may cause an airway obstruction.

b. Apply two belts, a sling, and a cravat to immobilize a fractured collarbone, as illustrated in Figure 4-26.

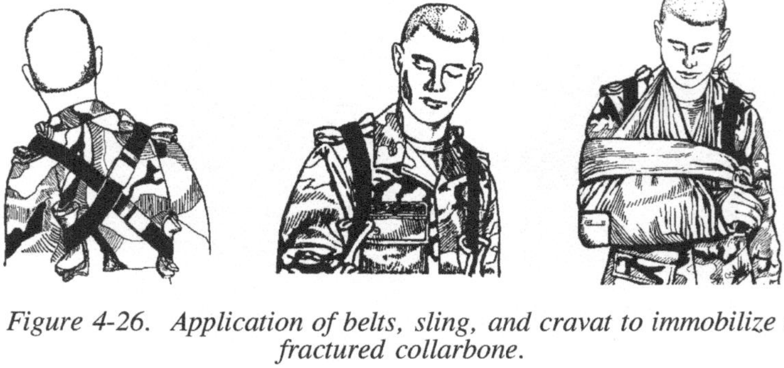

Figure 4-26. Application of belts, sling, and cravat to immobilize a fractured collarbone.

c. Apply a sling and a cravat to immobilize a fractured or dislocated shoulder, using the technique illustrated in Figure 4-27.

Figure 4-27. Application of sling and cravat to immobilize a fractured or dislocated shoulder.

4-10. Spinal Column Fractures

It is often impossible to be sure a casualty has a fractured spinal column. Be suspicious of any back injury, especially if the casualty has fallen or if his back has been sharply struck or bent. If a casualty has received such an injury and does not have feeling in his legs or cannot move them, you can be reasonably sure that he has a severe back injury, which should be managed as a fracture. Remember, that the possibility of a neck fracture or injury to the back should always be suspected, and it is often impossible to be sure if a casualty has a fractured spinal column. If the spine is fractured, bending it can cause the sharp bone fragments to bruise or cut the spinal cord and result in permanent paralysis or death (Figure 4-28A). The spinal column must maintain normal spinal position at the lower back (lumbar region) to help remove pressure from the spinal cord.

a. If the casualty is not to be transported until medical personnel arrive—

- Caution him not to move. Ask him if he is in pain or if he is unable to move any part of his body.

- Leave him in the position in which he is found. **DO NOT** move any part of his body, unless he is in imminent danger.

- If the casualty is lying face up, slip a blanket or other supporting material under the arch of his lower back to help support the spine in a normal position (Figure 4-28B). Take care not to place so much bulky padding as to cause potential damage by causing undo pressure on the spine. If he is lying face down, **DO NOT** put anything under any part of his body.

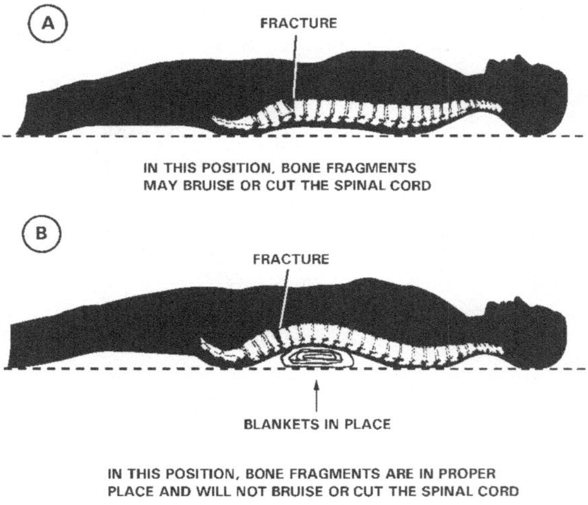

Figure 4-28. Spinal column must maintain a normal spine position.

b. If the casualty must be transported to a safe location before medical personnel arrive and if the casualty is in a—

- Face-up position, transport him by litter or use a firm substitute, such as a wide board or a door longer than his height. Loosely tie the casualty's wrists together over his waistline, using a cravat or a strip of cloth. Tie his feet together to prevent the accidental dropping or shifting of his legs. Lay a folded blanket across the litter where the arch of his back is to be placed. Using a four-man team (Figure 4-29), place the casualty on the litter without bending his spinal column or his neck.

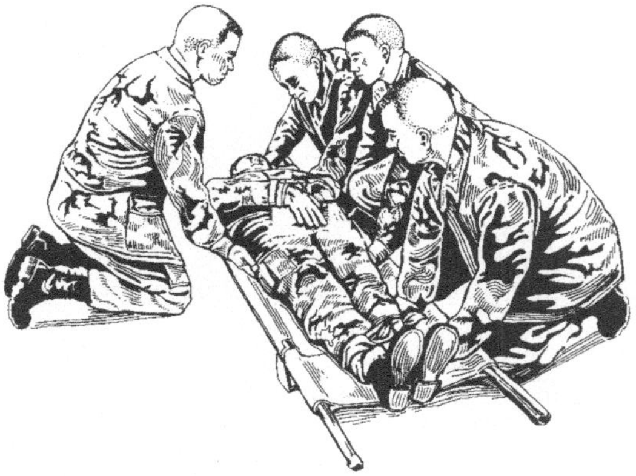

Figure 4-29. Placing face-up casualty with fractured back onto litter.

- The number two man positions himself at the casualty's head. His responsibility is to provide manual in-line (neutral) stabilization of the head and neck. The number three, and four men position themselves on one side of the casualty; all kneel on one knee along the side of the casualty. The number one man positions himself to the opposite side of the casualty (or can be on the same side of number three and four). The numbers two, three, and four men gently place their hands under the casualty. The number one man on the opposite side places his hands under the injured part to assist.

- When all four men are in position to lift, the number two man commands, "**PREPARE TO LIFT**" and then, "**LIFT**." All men, in unison, gently lift the casualty about 8 inches. Once the casualty is lifted, the number one man recovers and slides the litter under the casualty, ensuring that the blanket is in proper position. The number one man then returns to his original lift position (Figure 4-29).

- When the number two man commands, "**LOWER CASUALTY**," all men, in unison, gently lower the casualty onto the litter.

- Facedown position, he must be transported in this same position. The four-man team lifts him onto a regular or improvised litter, keeping the spinal column in a normal spinal position. If a regular litter is used, first place a folded blanket on the litter at the point where the chest will be placed.

4-11. Neck Fractures

A fractured neck is extremely dangerous. Bone fragments may bruise or cut the spinal cord just as they might in a fractured back.

a. If the casualty is not to be transported until medical personnel arrive—

- Caution him not to move. Moving may cause permanent injury or death.

- Leave the casualty in the position in which he is found. If his neck and head (cervical spine) are in an abnormal position, immediately immobilize the neck and head.

- Keep his head still, if the casualty is lying face up, raise his shoulders slightly, and slip a roll of cloth that has the bulk of a bath towel under his neck (Figure 4-31). The roll should be thick enough to arch his neck only slightly, leaving the back of his head on the ground. **DO NOT** bend his neck or head forward. **DO NOT** raise or twist his head. Immobilize the casualty's head (Figure 4-32). Do this by padding heavy objects (such as rocks or his boots filled with dirt, sand, gravel, or rock) and placing them on each side of his head. If it is necessary to use boots, after filling them, tie the top tightly or stuff with pieces of cloth to secure the contents.)

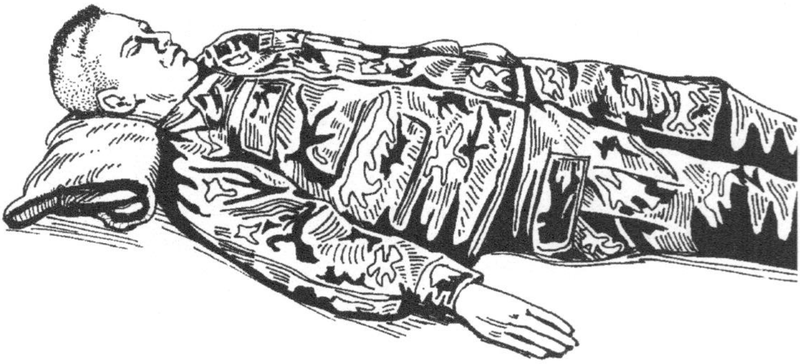

Figure 4-30. Casualty with roll of cloth (bulk) under neck.

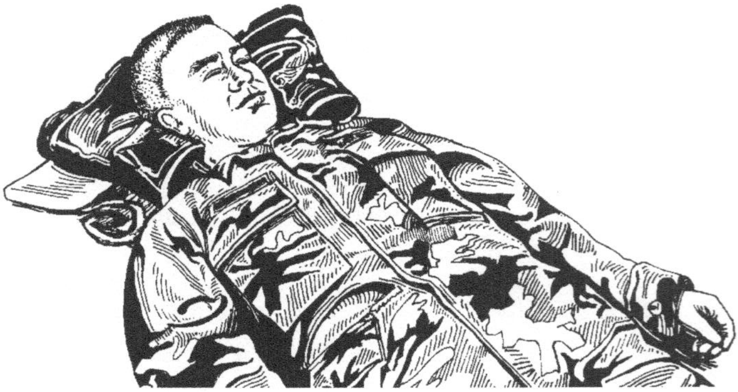

Figure 4-31. Immobilization of fractured neck.

- **DO NOT** move him if the casualty is lying face down. Immobilize the head and neck by padding heavy objects and placing them on each side of his head. DO NOT put a roll of cloth under the neck. **DO NOT** bend the neck or head, nor roll the casualty onto his back.

b. If the casualty must be prepared for transportation before medical personnel arrive—

- If the casualty has a fractured neck, at least two persons are needed because the casualty's head and trunk must be moved in unison. The two persons must work in close coordination (Figure 4-32) to avoid bending of the neck.

- A wide board is placed lengthwise beside the casualty. It should extend at least 4 inches beyond the casualty's head and feet (Figure 4-32A).

- If the casualty is lying face up, the number one man steadies the casualty's head and neck between his hands. At the same time, the number two man positions one foot and one knee against the board to prevent it from slipping. He then grasps the casualty underneath his shoulder and hip and gently slides him onto the board (Figure 4-32B).

- If the casualty is lying face down, the number one man steadies the casualty's head and neck between his hands, while the number two man gently rolls the casualty over onto the board (Figure 4-32C).

- The number one man continues to steady the casualty's head and neck. The number two man simultaneously raises the casualty's shoulders slightly, places padding under his neck, and immobilizes the casualty's head (Figures 4-32D—E).

- Any improvised supports are secured in position with a cravat or strip of cloth extended across the casualty's forehead and under the board (Figure 4-32D).

- The board is lifted onto a litter or blanket in order to transport the casualty (Figure 4-32E).

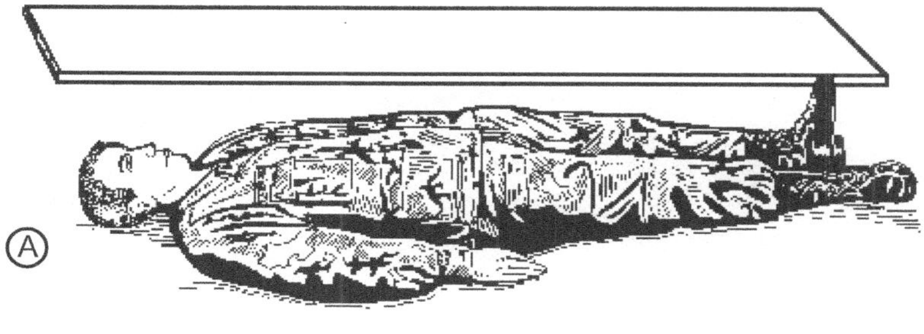

Figure 4-32. Preparing casualty with fractured neck for transportation (Illustrated A—E).

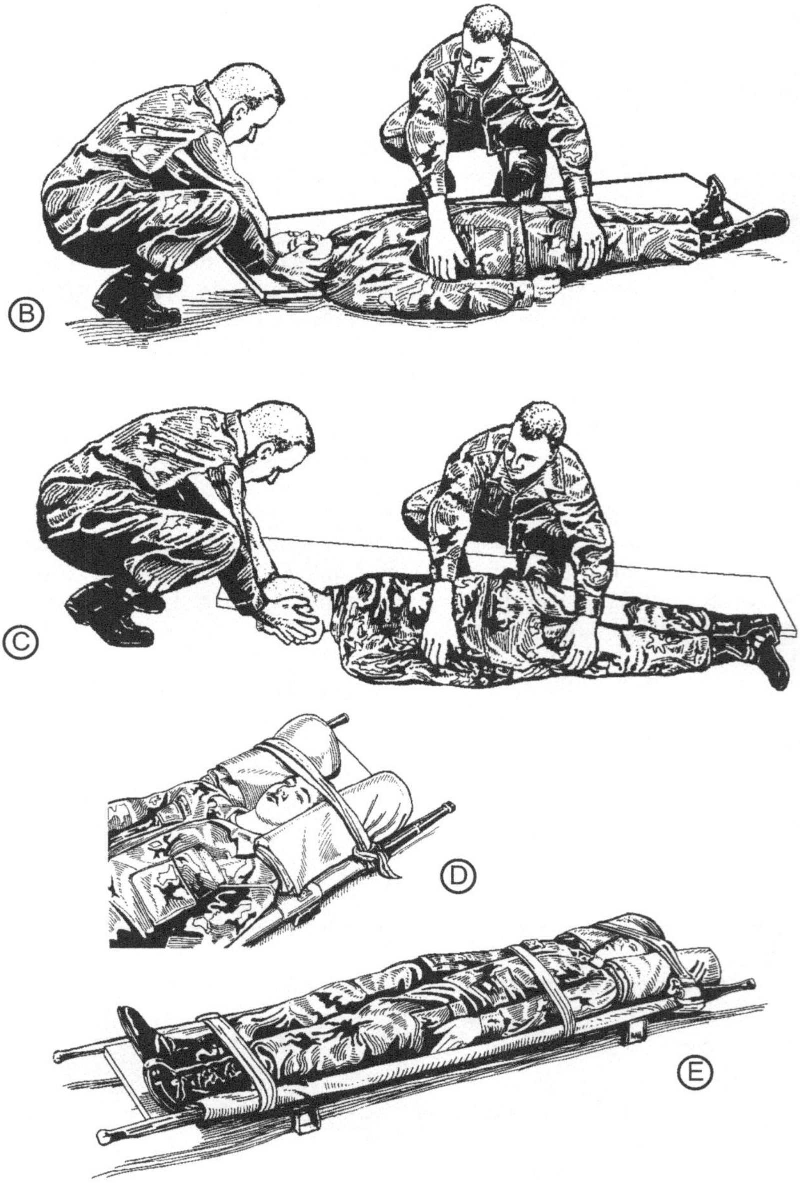

Figure 4-32. Preparing casualty with fractured neck for transportation (Illustrated A—E) (Continued).

CHAPTER 5

FIRST AID FOR CLIMATIC INJURIES

5-1. General

a. It is desirable, but not always possible, for an individual's body to become adjusted (acclimated) to an environment.

(1) The service members physical condition determines the amount of time their bodies need to adjust to the environment. Even those individuals in good physical condition need time before working or training in extremes of hot or cold weather. Climate-related injuries are usually preventable; prevention is both an individual and leadership responsibility.

(2) Several factors contribute to health and well-being in any environment—

- Diet.
- Sleep and rest.
- Exercise.
- Suitable clothing.

(3) Diet should be suited to an individual's needs in a particular climate. A special diet started for any purpose (such as weight reduction) should be done with appropriate medical supervision.

WARNING

Service members should use extreme caution when starting fad diets or taking over-the-counter herbal supplements. Medical records revealed that deaths and severe injuries occurred in individuals using dietary/ herbal supplements without medical monitoring.

NOTE

Weight loss and the use of weight loss supplements should be supervised by a trained health care provider.

(4) Specialized clothing and equipment (such as cold weather gear) for a specific environment should be obtained and used properly.

b. For information on the prevention of heat and cold injuries, refer to FM 21-10/Marine Corps Reference Publication (MCRP) 4-11.1D.

5-2. Heat Injuries

a. Heat injuries are environmental injuries. They may result when a service member—

- Is exposed to extreme heat, such as from the sun or from high temperatures.
- Does not wear proper clothing.
- Is in MOPP gear.
- Is inside closed spaces, such as inside an armored vehicle.
- Wears body armor.

b. Heat injury can be divided into three categories: heat cramps, heat exhaustion, and heatstroke.

c. Each service member must be able to recognize and give first aid for heat injuries.

WARNING

The heat casualty should be continually monitored for development of conditions which may require the performance of necessary basic lifesaving measures.

CAUTION

Do not use salt solutions in first aid procedures for heat injuries.

(1) Check the casualty for signs and symptoms of *cramping*.

- *Signs and symptoms.* Cramping is caused by an imbalance of chemicals (called electrolytes) in the body as a result of excessive sweating. This condition causes the casualty to exhibit:

- Cramping in the extremities (arms and legs).
- Abdominal (stomach) cramps.
- Excessive sweating.

NOTE

Thirst may or may not occur. Cramping can occur without the service member being thirsty.

- *First aid measures.*
  - Move the casualty to a cool, shady area or improvise shade if none is available.
  - Loosen his clothing (if not in a chemical environment).

NOTE

In a chemical environment, transport the heat casualty to a noncontaminated area as soon as the mission permits.

  - Have him slowly drink at least one canteen full of water. (The body absorbs cool water faster than warm or cold water; therefore, cool water is preferred if it is available.)
  - Seek medical assistance should cramps continue.

(2) Check the casualty for signs and symptoms of *heat exhaustion*.

- *Signs and symptoms.* Heat exhaustion is caused by loss of body fluids (dehydration) through sweating without adequate fluid replacement. It can occur in an otherwise fit individual who is involved in physical exertion in any hot environment especially if the service member is not acclimatized to that environment. These signs and symptoms are—
  - Excessive sweating with pale, moist, cool skin.
  - Headache.
  - Weakness.

- Dizziness.
- Loss of appetite.
- Cramping.
- Nausea (with or without vomiting).
- Urge to defecate.
- Chills (gooseflesh).
- Rapid breathing.
- Tingling of hands and/or feet.
- Confusion.

- *First aid measures.*
    - Move the casualty to a cool, shady area or improvise shade if none is available.
    - Loosen or remove his clothing and boots (unless in a chemical environment); pour water on him and fan him.
    - Have him slowly drink at least one canteen of water.
    - Elevate his legs.
    - If possible, the casualty should not participate in strenuous activity for the remainder of the day.
    - Monitor the casualty until the symptoms are gone, or medical assistance arrives.

(3) Check the casualty for signs and symptoms of *heatstroke*.

WARNING

Heatstroke is a medical emergency which may result in death if care is delayed.

- *Signs and symptoms.* A service member suffering from heatstroke has been exposed to high temperatures (such as direct sunlight) or been dressed in protective overgarments, which causes the body temperature to rise. Heatstroke occurs more rapidly in service members who are engaged in work or other physical activity in a high heat environment. Heatstroke is caused by a failure of the body's cooling mechanism which includes a decrease in the body's ability to produce sweat. The casualty's skin is red (flushed), hot, and dry. He may experience weakness, dizziness, confusion, headaches, seizures, nausea, stomach pains or cramps, and his respiration and pulse may be rapid and weak. Unconsciousness and collapse may occur suddenly.

- *First aid measures.* Cool casualty immediately by—

  - Moving him to a cool, shady area or improvising shade if none is available.

  - Loosening or removing his clothing (except in a chemical environment).

  - Spraying or pouring water on him; fanning him to permit the coolant effect of evaporation.

  - Massaging his extremities and skin, which increases the blood flow to those body areas, thus aiding the cooling process.

  - Elevating his legs.

  - Having him slowly drink at least one canteen full of water if he is conscious.

NOTE

Start cooling casualty immediately. Continue cooling while awaiting transportation and during transport to an MTF.

- *Medical assistance.* Seek medical assistance because the casualty should be transported to an MTF as soon as possible. Do not interrupt the cooling process or lifesaving measures to seek help; if someone else is present send them for help. The casualty should be continually monitored for development of conditions that may require the performance of necessary basic lifesaving measures.

d. *Table.* See Table 5-1 for further information.

Table 5-1. Heat Injuries

INJURIES	SIGNS AND SYMPTOMS	FIRST AID[1]
HEAT CRAMPS	THE CASUALTY EXPERIENCES MUSCLE CRAMPS OF THE ARMS, LEGS, AND/OR STOMACH. THE CASUALTY MAY ALSO HAVE EXCESSIVE SWEATING.	1. MOVE THE CASUALTY TO A COOL SHADY AREA OR IMPROVISE SHADE AND LOOSEN CLOTHING.[2] 2. HAVE HIM SLOWLY DRINK AT LEAST ONE CANTEEN FULL OF COOL WATER SLOWLY. 3. MONITOR THE CASUALTY AND GIVE HIM MORE WATER AS TOLERATED.
HEAT EXHAUSTION	THE CASUALTY EXPERIENCES HEAVY SWEATING WITH PALE, MOIST, COOL SKIN; HEADACHE, WEAKNESS, DIZZINESS, AND/OR LOSS OF APPETITE, HEAT CRAMPS, NAUSEA (WITH OR WITHOUT VOMITING), URGE TO DEFECATE, CHILLS (GOOSE-FLESH), RAPID BREATHING, CONFUSION, AND TINGLING OF THE HANDS AND/OR FEET.	1. MOVE THE CASUALTY TO A COOL, SHADY AREA OR IMPROVISE SHADE AND LOOSEN OR REMOVE HIS CLOTHING.[2] 2. POUR WATER ON HIM AND FAN HIM TO PERMIT THE COOLANT EFFECT OF EVAPORATION. 3. HAVE HIM SLOWLY DRINK AT LEAST ONE CANTEEN FULL OF COOL WATER. 4. ELEVATE THE CASUALTY'S LEGS. 5. SEEK MEDICAL ASSISTANCE IF SYMPTOMS CONTINUE; MONITOR UNTIL SYMPTOMS ARE GONE OR MEDICAL ASSISTANCE ARRIVES.
HEATSTROKE[3] (SUNSTROKE)	THE CASUALTY STOPS SWEATING (RED [FLUSHED] HOT, DRY SKIN). HE FIRST MAY EXPERIENCE HEADACHE, DIZZINESS, NAUSEA, FAST PULSE AND RESPIRATION, SEIZURES, AND MENTAL CONFUSION. HE MAY COLLAPSE	1. MOVE THE CASUALTY TO A COOL, SHADY AREA OR IMPROVISE SHADE AND LOOSEN OR REMOVE HIS CLOTHING, REMOVE THE OUTER GARMENTS AND PROTECTIVE

Table 5-1. Heat Injuries (Continued)

INJURIES	SIGNS AND SYMPTOMS	FIRST AID[1]
	AND SUDDENLY BECOME UNCONSCIOUS. **THIS IS A MEDICAL EMERGENCY.**	CLOTHING IF THE SITUATION PERMITS.[2] 2. START COOLING THE CASUALTY IMMEDIATELY. SPRAY OR POUR WATER ON HIM. FAN HIM. MASSAGE HIS EXTREMITIES AND SKIN. 3. ELEVATE HIS LEGS. 4. IF CONSCIOUS, HAVE HIM SLOWLY DRINK AT LEAST ONE CANTEEN FULL OF COOL WATER. 5. SEEK MEDICAL AID. CONTINUE COOLING WHILE AWAITING TRANSPORT AND CONTINUE FIRST AID WHILE EN ROUTE.

LEGEND:

1 THE FIRST AID PROCEDURE FOR HEAT RELATED INJURIES CAUSED BY WEARING INDIVIDUAL PROTECTIVE EQUIPMENT (IPE) IS TO MOVE THE CASUALTY TO A CLEAN AREA AND GIVE HIM WATER TO DRINK.
2 WHEN IN A CHEMICAL ENVIRONMENT, DO NOT LOOSEN OR REMOVE THE CASUALTY'S CLOTHING.
3 CAN BE FATAL IF NOT PROVIDED FIRST AID AND MEDICAL TREATMENT PROMPTLY.

5-3. Cold Injuries

Cold injuries are most likely to occur when conditions are moderately cold, but accompanied by wet or windy conditions. Cold injuries can usually be prevented. Well-disciplined and well-trained service members can be protected even in the most adverse circumstances. They and their leaders must know the hazards of exposure to the cold. They must know the importance of personal hygiene, exercise, care of the feet and hands, and the use of protective clothing.

a. Contributing Factors.

(1) Temperature, humidity, precipitation, and wind greatly increase likelihood of cold injuries, and the service members with wet clothing are at great risk of cold injuries. Riverine operations (river, swamp, and stream crossings) increase likelihood of cold injuries. Low temperatures and low relative humidity (dry cold) promote frostbite. Higher temperatures, together with moisture, promote immersion syndrome. Windchill accelerates the loss of body heat and may aggravate cold injuries.

(2) Relatively stationary activities such as being in an observation post or on guard duty increase the service member's vulnerability to cold injury. Also, a service member is more likely to receive a cold injury if he is—

- In contact with the ground (such as marching, performing guard duty, or engaging in other outside activities).
- Immobile for long periods (such as while riding in an unheated or open vehicle).
- Standing in water, such as in a foxhole.
- Out in the cold for days without being warmed.
- Deprived of an adequate diet and rest.
- Not able to take care of his personal hygiene.

(3) Physical fatigue contributes to apathy, which leads to inactivity, personal neglect, carelessness, and reduced heat production. In turn, these increase the risk of cold injury. Service members with prior cold injuries have a higher-than-normal risk of subsequent cold injury; not necessarily involving the body part previously injured.

(4) Depressed or unresponsive service members are also vulnerable because they are less active. These service members tend to be careless about precautionary measures, especially warming activities, when cold injury is a threat.

(5) Excessive use of alcohol or drugs leading to faulty judgment or unconsciousness in a cold environment increases the risk of becoming a cold injury casualty.

b. *Signs and Symptoms*. Once a service member becomes familiar with the factors that contribute to cold injury, he must learn to recognize cold injury signs and symptoms.

(1) Many service members suffer cold injury without realizing what is happening to them. They may be cold and generally uncomfortable. These service members often do not notice the injured part because it is already numb from the cold.

(2) Superficial cold injury usually can be detected by numbness or tingling sensations. These signs and symptoms often can be relieved simply by loosening boots or other clothing and by exercising to improve circulation. In more advanced cases involving deep cold injury, the service member often is not aware that there is a problem until the affected part feels like a stump or block of wood.

(3) Outward signs of cold injury include discoloration of the skin at the site of injury. In light-skinned persons, the skin first reddens and then becomes pale or waxy white. In dark-skinned persons, grayness in the skin is usually evident. An injured foot or hand feels cold to the touch. Swelling may be an indication of deep injury. Also note that blisters may occur after rewarming the affected parts. Service members should work in pairs (buddy teams) to check each other for signs of discoloration and other symptoms.

c. First Aid Measures. First aid for cold injuries depends on whether they are superficial or deep. Rewarming the affected part using body heat can adequately treat cases of superficial cold injury. (For example, this can be done by covering cheeks with hands, putting fingertips in armpits, or placing the casualty's feet under the clothing of a buddy [next to his belly].) The injured part should **NOT** be massaged, exposed to a fire or stove, rubbed with snow, slapped, chafed, or soaked in cold water. Walking on injured feet should be avoided. Deep cold injury (frostbite) is very serious and requires prompt first aid to avoid or to minimize the loss of parts or all of the fingers, toes, hands, or feet. The sequence for treating cold injuries depends on whether the condition is life-threatening. The first priority in managing cold injuries is to remove the casualty from the cold environment (such as building an improvised shelter). Other injuries the casualty may have are provided first aid simultaneously while waiting for transportation or evacuation. If the casualty is to be transported in a nonmedical vehicle, first aid measures should be continued en route to the MTF.

d. Conditions Caused by Cold. Conditions caused by cold include chilblain, immersion syndrome (immersion foot and trench foot), frostbite, snow blindness, dehydration, and hypothermia.

(1) *Chilblain.*

- *Signs and symptoms.* Chilblain is caused by repeated prolonged exposure of bare skin at temperatures from 60° Fahrenheit (F) to

32°F, or 20°F for acclimated, dry, unwashed skin. The area may be acutely swollen, red, tender, and hot with itchy skin. There may be no loss of skin tissue in untreated cases but continued exposure may lead to infected, ulcerated, or bleeding lesions.

- *First aid measures.* Within minutes, the area usually responds to locally applied body heat. Rewarm the affected part by applying firm steady pressure with your hands, or placing the affected part under your arms or against the stomach of a buddy. **DO NOT** rub or massage affected areas.

NOTE

Medical personnel should evaluate the injury, because signs and symptoms of tissue damage may be slow to appear.

(2) *Immersion syndrome (immersion foot and trench foot).* Immersion foot and trench foot are injuries that result from fairly long exposure of the feet to wet conditions at temperatures from approximately 32°F to 50°F. Inactive feet in damp or wet socks and boots, or tightly laced boots which impair circulation, are even more susceptible to injury. This injury can be very serious; it can lead to loss of toes or parts of the feet. If exposure of the feet has been prolonged and severe, the feet may swell so much that pressure closes the blood vessels and cuts off circulation. Should an immersion injury occur, dry the feet thoroughly and transport the casualty to an MTF by the fastest means possible.

- *Signs and symptoms.* At first, the parts of the affected foot are cold and painless, the pulse is weak, and numbness may be present. Second, the parts may feel hot, and burning and shooting pains may begin. In later stages, the skin is pale with a bluish cast and the pulse decreases. Other signs and symptoms that may follow are blistering, swelling, redness, heat, hemorrhaging (bleeding), and gangrene.

- *First aid measures.* First aid measures are required for all stages of immersion syndrome injury. Rewarm the injured part gradually by exposing it to warm air. Protect it from trauma and secondary infections. Dry, loose clothing or several layers of warm coverings are preferable to extreme heat. Under no circumstances should the injured part be exposed to an open fire. Elevate the injured part to relieve the swelling. Transport the casualty to an MTF as soon as possible. When the part is rewarmed, the casualty often feels a burning sensation and pain. Symptoms may persist for days or weeks even after rewarming.

NOTE

When providing first aid for immersion foot and trench foot—**DO NOT** massage the injured part. **DO NOT** moisten the skin. **DO NOT** apply heat or ice.

(3) *Frostbite.* Frostbite is the injury of tissue caused from exposure to cold, usually below 32°F depending on the windchill factor, duration of exposure, and adequacy of protection. Individuals with a history of cold injury are likely to suffer an additional cold injury. The body parts most easily frostbitten are the cheeks, nose, ears, chin, forehead, wrists, hands, and feet. Frostbite may involve only the skin (superficial), or it may extend to a depth below the skin (deep). Deep frostbite is very serious and requires prompt first aid to avoid or to minimize the loss of parts or all of the fingers, toes, hands, or feet.

- *Signs and symptoms.*
  - Loss of sensation (numb feeling) in any part of the body.
  - Sudden blanching (whitening) of the skin of the affected part, followed by a momentary tingling sensation.
  - Redness of skin in light-skinned service members; grayish coloring in dark-skinned service members.
  - Blisters.
  - Swelling or tender areas.
  - Loss of previous sensation of pain in affected area.
  - Pale, yellowish, waxy-looking skin.
  - Frozen tissue that feels solid (or wooden) to the touch.

CAUTION

Deep frostbite is a very serious injury and requires immediate first aid and subsequent medical treatment to avoid or minimize loss of body parts.

- *First aid measures.*

  - *Face, ears, and nose.* Cover the casualty's affected area with his and/or your bare hands until sensation and color return.

  - *Hands.* Open the casualty's field jacket and shirt. (In a chemical environment, do not loosen or remove the clothing and protective overgarments.) Place the affected hands under the casualty's armpits. Close the field jacket and shirt to prevent additional exposure.

  - *Feet.* Remove the casualty's boots and socks if he does not need to walk any further to receive additional treatment. (Thawing the casualty's feet and forcing him to walk on them will cause additional pain and injury.) Place the affected feet under clothing and against the body of another service member.

WARNING

DO NOT attempt to thaw the casualty's feet or other frozen areas if he will be required to walk or travel to an MTF for additional medical treatment. The possibility of additional injury from walking is less when the feet are frozen than when they are thawed. (However, if possible avoid walking.) Thawing in the field increases the possibilities of infection, gangrene, or other injury.

NOTE

Thawing may occur spontaneously during transportation to the MTF; this cannot be avoided since the body in general must be kept warm.

In all of the above areas, ensure that the casualty is kept warm and that he is covered (to avoid further injury). Seek medical treatment as soon as possible. Reassure the casualty, protect the affected area from further injury by covering it lightly with a blanket or any dry clothing, and seek shelter out of the wind. Remove or loosen constricting clothing (except in a contaminated environment) and increase insulation. Ensure the casualty exercises as much as possible, avoiding trauma to the injured part, and is prepared for pain when thawing occurs. Protect the frostbitten part from additional injury. **DO NOT—**

- Rub the injured part with snow or apply cold water soaks.

- Warm the part by massage or exposure to open fire because the frozen part may be burned due to the lack of feeling.

- Use ointments or other salves.

- Manipulate the part in any way to increase circulation.

- Use alcohol or tobacco because this reduces the body's resistance to cold.

NOTE

Remember, when freezing extends to a depth below the skin, it is a much more serious injury. Extra care is required to reduce or avoid the chances of losing all or part of the toes or feet. This also applies to the fingers and hands.

(4) *Snow blindness*. Snow blindness is the effect that glare from an ice field or snowfield has on the eyes. It is more likely to occur in hazy, cloudy weather than when the sun is shining. Glare from the sun will cause an individual to instinctively protect his eyes. However, in cloudy weather, he may be overconfident and expose his eyes longer than when the threat is more obvious. He may also neglect precautions such as the use of protective eyewear. Waiting until discomfort (pain) is felt before using protective eyewear is dangerous because a deep burn of the eyes may already have occurred.

- *Signs and symptoms*. Symptoms of snow blindness are a sensation of grit in the eyes with pain in and over the eyes, made worse by moving the eyeball. Other signs and symptoms are watering, redness, headache, and increased pain on exposure to light.

- *First aid measures*. First aid measures consist of blindfolding or covering the eyes with a dark cloth which stops painful eye movement. Complete rest is desirable. If further exposure to light is not preventable, the eyes should be protected with dark bandages or the darkest glasses available. Once unprotected exposure to sunlight stops, the condition usually heals in a few days without permanent damage. The casualty should be evacuated to the nearest MTF.

(5) *Dehydration*. Dehydration occurs when the body loses too much fluid. A certain amount of body fluid is lost through normal body processes. A normal daily intake of liquids replaces these losses. When individuals are engaged in any strenuous exercises or activities, fluid is lost

through sweating and this loss creates an imbalance of fluids in the body, and if not matched by rehydration it can contribute to dehydration. The danger of dehydration is as prevalent in cold regions as it is in hot regions. In hot weather, the individual is aware of his body losing fluids through sweat. In cold weather, however, it is extremely difficult to realize that this condition exists since sweating is not as apparent as in a hot environment. The danger of dehydration in cold weather operations is a serious problem. In cold climates, sweat evaporates so rapidly or is absorbed so thoroughly by layers of heavy clothing that it is rarely visible on the skin. Dehydration also occurs during cold weather operations because drinking is inconvenient. Dehydration will weaken or incapacitate a casualty for a few hours, or sometimes, several days. Because rest is an important part of the recovery process, casualties must take care that limited movement during their recuperative period does not enhance the risk of becoming a cold injury casualty.

- *Signs and symptoms.* The symptoms of cold weather dehydration are similar to those encountered in heat exhaustion. The mouth, tongue, and throat become parched and dry, and swallowing becomes difficult. The casualty may have nausea (with or without vomiting) along with extreme dizziness and fainting. The casualty may also feel generally tired and weak and may experience muscle cramps. Focusing the eyes may also become difficult.

- *First aid measures.* The casualty should be kept warm and his clothes should be loosened (if not in a chemical environment) to allow proper circulation. Shelter from wind and cold must be provided. Fluid replacement should begin immediately and the service member transported to an MTF as soon as possible.

(6) *Hypothermia (general cooling).* When exposed to prolonged cold weather a service member may become both mentally and physically numb, thus neglecting essential tasks or requiring more time and effort to achieve them. Under some conditions (particularly cold water immersion), even a service member in excellent physical condition may die in a matter of minutes. The destructive influence of cold on the body is called *hypothermia.* This means bodies lose heat faster than they can produce it. Hypothermia can occur from exposure to temperatures either above or below freezing, especially from immersion in cold water, wet-cold conditions, or from the effect of wind. Physical exhaustion and insufficient food intake may also increase the risk of hypothermia. General cooling of the entire body to a temperature below 95°F is caused by continued exposure to low or rapidly dropping temperatures, cold moisture, snow, or ice. Fatigue, poor physical condition, dehydration, faulty blood circulation, alcohol or other drug use, trauma, and immersion can cause hypothermia. Remember, cold

may affect the body systems slowly and almost without notice. Service members exposed to low temperatures for extended periods may suffer ill effects even if they are well protected by clothing.

- *Signs and symptoms*. As the body cools, there are several stages of progressive discomfort and impairment. A sign that is noticed immediately is shivering. Shivering is an attempt by the body to generate heat. The pulse is faint or very difficult to detect. People with temperatures around 90°F may be drowsy and mentally slow. Their ability to move may be hampered, stiff, and uncoordinated, but they may be able to function minimally. Their speech may be slurred. As the body temperature drops further, shock becomes evident as the person's eyes assume a glassy state, breathing becomes slow and shallow, and the pulse becomes weaker or absent. The person becomes very stiff and uncoordinated. Unconsciousness may follow quickly. As the body temperature drops even lower, the extremities freeze, and a deep (or core) body temperature (below 85°F) increases the risk of irregular heart action. This irregular heart action or heart standstill can result in sudden death.

- *First aid measures*. Except in cases of the most severe hypothermia (marked by coma or unconsciousness and a weak pulse), first aid measures for hypothermia are directed towards protecting the casualty from further loss of body heat. For the casualty who is conscious, first aid measures are directed at rewarming the body evenly and without delay. Provide heat by using a hot water bottle or field expedient or another service member's body heat.

CAUTION

DO NOT expose the casualty to an open fire, as he may become burned.

NOTE

When using a hot water bottle or field expedient (canteen filled with warm water), the bottle or canteen must be wrapped in cloth prior to placing it next to the casualty. This will reduce the chance of burning the casualty's skin.

Always call or send for help as soon as possible and protect the casualty immediately with dry clothing or a sleeping bag. Then, move him to a warm place. Evaluate other injuries and provide first aid as required. First aid measures can be performed while the casualty is waiting transportation or while he is en route. In the case of an accidental breakthrough into ice water, or other hypothermic accident, strip the casualty of wet clothing immediately and bundle him into a sleeping bag. Rescue breathing should be started at once if the casualty's breathing has stopped or is irregular or shallow. Warm liquids (**NOT HOT**) may be given gradually if the casualty is conscious. **DO NOT** force liquids on an unconscious or semiconscious casualty because he may choke. The casualty should be transported on a litter because the exertion of walking may aggravate circulation problems. Medical personnel should immediately treat any hypothermia casualty. Hypothermia is life threatening until normal body temperature has been restored. The first aid measures for a casualty with severe hypothermia are based upon the following principles: attempt to avoid further heat loss, handle the casualty gently, and transport the casualty as soon as possible to the nearest MTF. If at all possible, the casualty should be evacuated by medical personnel.

WARNING

Rewarming a severely hypothermic casualty is extremely dangerous in the field due to the possibility of such complications as rewarming, shock and disturbances in the rhythm of the heartbeat. These conditions require treatment by medical personnel.

NOTE

Resuscitation of casualties with hypothermic complications is difficult if not impossible to do outside of an MTF setting.

CAUTION

The casualty is unable to generate his own body heat. Therefore, merely placing him in a blanket or sleeping bag is not sufficient.

e. Table. See Table 5-2 for further information.

Table 5-2. Injuries Caused by Cold and Wet Conditions

INJURIES	SIGNS/SYMPTOMS	FIRST AID
CHILBLAIN	RED SWOLLEN, HOT, TENDER, ITCHING SKIN. CONTINUED EXPOSURE MAY LEAD TO INFECTED (ULCERATED OR BLEEDING) SKIN LESIONS.	1. AREA USUALLY RESPONDS TO LOCALLY APPLIED REWARMING (BODY HEAT). 2. DO NOT RUB OR MASSAGE AREA. 3. SEEK MEDICAL AID.
IMMERSION SYNDROME (IMMERSION FOOT/TRENCH FOOT)	AFFECTED PARTS ARE COLD, NUMB, AND PAINLESS. PARTS MAY THEN BE HOT, WITH BURNING AND SHOOTING PAINS. ADVANCED STAGE: SKIN PALE WITH BLUISH CAST; PULSE DECREASES; BLISTERING, SWELLING, HEAT, HEMORRHAGING, AND GANGRENE MAY FOLLOW.	1. GRADUAL REWARMING BY EXPOSURE TO WARM AIR. 2. DO NOT MASSAGE OR MOISTEN SKIN. 3. PROTECT AFFECTED PARTS FROM TRAUMA. 4. DRY FEET THOROUGHLY, AVOID WALKING. 5. SEEK MEDICAL AID.
FROSTBITE	LOSS OF SENSATION (NUMB FEELING) IN ANY PART OF THE BODY. SUDDEN BLANCHING (WHITENING) OF THE SKIN OF THE AFFECTED PART, FOLLOWED BY A MOMENTARY TINGLING SENSATION. REDNESS OF SKIN IN LIGHT-SKINNED SERVICE MEMBERS; GRAYISH COLORING IN DARK-SKINNED SERVICE MEMBERS. BLISTERS. SWELLING OR TENDER AREAS. LOSS OF PREVIOUS SENSATION OF PAIN IN THE AFFECTED AREA. PALE YELLOWISH, WAXY-LOOKING SKIN. FROZEN TISSUE THAT FEELS SOLID (WOODEN) TO THE TOUCH.	1. WARM THE AREA AT THE FIRST SIGN OF FROSTBITE, USING FIRM, STEADY PRESSURE OF THE HAND, UNDERARM, OR ABDOMEN. 2. FACE, EARS, NOSE: COVER AREA WITH HANDS (CASUALTY'S OWN OR BUDDY'S). 3. HANDS: OPEN FIELD JACKET AND PLACE CASUALTY'S HANDS AGAINST HIS BODY, THEN CLOSE THE JACKET TO PREVENT HEAT LOSS. 4. FEET: REMOVE THE CASUALTY'S BOOTS AND SOCKS AND PLACE HIS FEET AGAINST THE BODY OF ANOTHER SERVICE MEMBER.

Table 5-2. Injuries Caused by Cold and Wet Conditions (Continued)

INJURIES	SIGNS/SYMPTOMS	FIRST AID
		5. WARNING: DO NOT ATTEMPT TO THAW THE CASUALTY'S FEET OR OTHER FROZEN AREAS IF HE WILL BE REQUIRED TO WALK OR TRAVEL TO AN MTF FOR ADDITIONAL TREATMENT. THE POSSIBILITY OF INJURY FROM WALKING IS LESS WHEN THE FEET ARE FROZEN THAN WHEN THEY HAVE BEEN THAWED. (HOWEVER, IF POSSIBLE AVOID WALKING.) THAWING IN THE FIELD INCREASES THE POSSIBILITY OF INFECTION, GANGRENE, OR INJURY. 6. LOOSEN OR REMOVE CONSTRICTING CLOTHING AND REMOVE ANY JEWELRY. 7. INCREASE INSULATION (COVER WITH BLANKET OR OTHER DRY MATERIAL). ENSURE CASUALTY EXERCISES AS MUCH AS POSSIBLE, AVOIDING TRAUMA TO INJURED PART.
SNOW BLINDNESS	EYES MAY FEEL SCRATCHY. WATERING, REDNESS, HEADACHE, AND INCREASED PAIN WITH EXPOSURE TO LIGHT CAN OCCUR.	1. COVER THE EYES WITH A DARK CLOTH. 2. SEEK MEDICAL AID.
DEHYDRATION	SIMILAR TO HEAT EXHAUSTION (REFER TO TABLE 5-1).	1. KEEP WARM. 2. CASUALTY NEEDS FLUID REPLACEMENT, REST, AND PROMPT MEDICAL AID.
HYPOTHERMIA	CASUALTY IS COLD. SHIVERING. CORE TEMPERATURE IS LOW. CONSCIOUSNESS MAY BE	**MILD HYPOTHERMIA** 1. REWARM BODY

Table 5-2. Injuries Caused by Cold and Wet Conditions (Continued)

INJURIES	SIGNS/SYMPTOMS	FIRST AID
	ALTERED. UNCOORDINATED MOVEMENTS MAY OCCUR. SHOCK AND COMA MAY RESULT AS BODY TEMPERATURE DROPS.	EVENLY AND WITHOUT DELAY. (NEED TO PROVIDE HEAT SOURCE; CASUALTY'S BODY UNABLE TO GENERATE HEAT.) 2. KEEP DRY, PROTECT FROM THE ELEMENTS. 3. WARM (NOT HOT) LIQUIDS MAY BE GIVEN GRADUALLY (TO CONSCIOUS CASUALTIES ONLY). 4. BE PREPARED TO START BASIC LIFE SUPPORT MEASURES FOR THE CASUALTY. 5. SEEK MEDICAL TREATMENT IMMEDIATELY. **SEVERE HYPOTHERMIA** 1. STABILIZE THE TEMPERATURE. 2. ATTEMPT TO AVOID FURTHER HEAT LOSS. 3. HANDLE THE CASUALTY GENTLY. 4. EVACUATE TO THE NEAREST MTF AS SOON AS POSSIBLE. **5. WARNING: HYPOTHERMIA IS A MEDICAL EMERGENCY. PROMPT MEDICAL ATTENTION IS NECESSARY.**

CHAPTER 6

FIRST AID FOR BITES AND STINGS

6-1. General

Snakebites, insect bites, or stings can cause intense pain and/or swelling. If not treated promptly and correctly, they can cause serious illness or death. The severity of a snakebite depends upon: whether the snake is poisonous or nonpoisonous, the type of snake, the location of the bite, and the amount of venom injected. Bites from humans and other animals, such as dogs, cats, bats, raccoons, and rats, can cause severe bruises and infection and tears or lacerations of tissue. Awareness of the potential sources of injuries can reduce or prevent them from occurring. Knowledge and prompt application of first-aid measures can lessen the severity of injuries from bites and stings and keep the service member from becoming a serious casualty.

6-2. Types of Snakes

a. Nonpoisonous Snakes. There are approximately 130 different varieties of nonpoisonous snakes in the United States. They have oval-shaped heads and round eyes. Unlike poisonous snakes, discussed below, nonpoisonous snakes do not have fangs with which to inject venom. Figure 6-1 depicts the characteristics of a nonpoisonous snake.

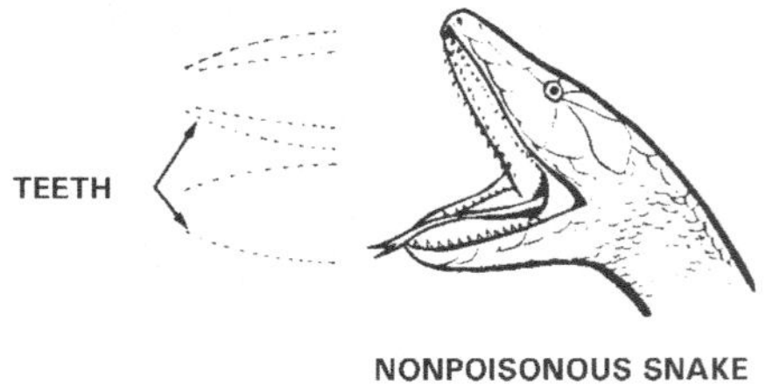

Figure 6-1. Characteristics of nonpoisonous snake.

b. Poisonous Snakes. Poisonous snakes are found throughout the world, primarily in tropical to moderate climates. Within the United States, there are four kinds: rattlesnakes, copperheads, water moccasins (cottonmouth), and coral snakes. Poisonous snakes in other parts of the world include sea snakes, the fer-de-lance, the bushmaster, and the tropical rattlesnake in tropical Central America; the Malayan pit viper in the tropical Far East; the cobra in Africa and Asia; the mamba (or black mamba) in central and southern Africa; and the krait in India and Southeast Asia. Refer to Figure 6-2 for characteristics of a poisonous pit viper.

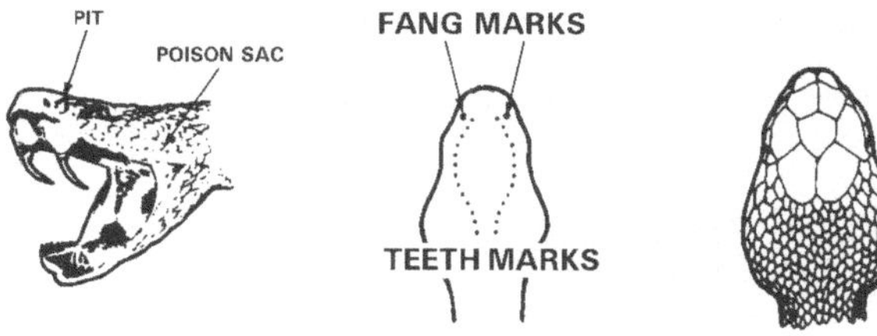

Figure 6-2. Characteristics of poisonous pit viper.

c. Pit Vipers (Poisonous). Figure 6-3 depicts a variety of poisonous snakes.

Figure 6-3. Poisonous snakes.

(1) Rattlesnakes, bushmasters, copperheads, fer-de-lance, Malayan pit vipers, and water moccasins (cottonmouth) are called pit vipers because of the small, deep pits between the nostrils and eyes on each side of the head (Figure 6-2). In addition to their long, hollow fangs, these snakes have other identifying features: thick bodies, slit-like pupils of the eyes, and flat, almost triangular-shaped heads. Color markings and other identifying characteristics, such as rattles or a noticeable white interior of the mouth (cottonmouth), also help distinguish these poisonous snakes. Further

identification is provided by examining the bite pattern of the wound for signs of fang entry. Occasionally there will be only one fang mark, as in the case of a bite on a finger or toe where there is no room for both fangs, or when the snake has broken off a fang.

(2) The casualty's condition provides the best information about the seriousness of the situation, or how much time has passed since the bite occurred. Pit viper bites are characterized by severe burning pain. Discoloration and swelling around the fang marks usually begins within 5 to 10 minutes after the bite. If only minimal swelling occurs within 30 minutes, the bite will almost certainly have been from a nonpoisonous snake or possibly from a poisonous snake which did not inject venom. The venom destroys blood cells, causing a general discoloration of the skin. Blisters and numbness in the affected area follow this reaction. Other signs, which can occur, are weakness, rapid pulse, nausea, shortness of breath, vomiting, and shock.

d. Corals, Cobras, Kraits, and Mambas. Corals (Figure 6-4), cobras (Figure 6-5), kraits, and mambas all belong to the same group even though they are found in different parts of the world. All four inject their venom through short, grooved fangs, leaving a characteristic bite pattern.

Figure 6-4. Coral snake.

(1) The small coral snake, found in the Southeastern US, is brightly colored with bands of red, yellow (or almost white), and black completely encircling the body. Other nonpoisonous snakes have the same coloring, but on the coral snake found in the US, the red ring always touches the yellow ring. To know the difference between a harmless snake and the coral snake found in the United States, remember the following:

"Red on yellow will kill a fellow,
Red on black, venom will lack."

Figure 6-5. Cobra snake.

(2) The venom of corals, cobras, kraits, and mambas produces symptoms different from those of pit vipers. Because there is only minimal pain and swelling, many people believe that the bite is not serious. Delayed reactions in the nervous system normally occur between 1 to 7 hours after the bite. Symptoms include blurred vision, drooping eyelids, slurred speech, drowsiness, and increased salivation and sweating. Nausea, vomiting, shock, respiratory difficulty, paralysis, convulsions, and coma will usually develop if the bite is not treated promptly.

e. Sea Snakes. Sea snakes (Figure 6-6) are found in the warm water areas of the Pacific and Indian oceans, along the coasts, and at the mouths of some larger rivers. Their venom is **VERY** poisonous, but their fangs are only 1/4 inch long. The first aid outlined for land snakes also applies to sea snakes.

Figure 6-6. Sea snake.

6-3. Snakebites

a. Poisonous snakes **DO NOT** always inject venom when they bite or strike a person. However, all snakes may carry tetanus (lockjaw); anyone bitten by a snake, whether poisonous or nonpoisonous, should immediately seek medical attention.

- Poison is injected from the venom sacs through grooved or hollow fangs. Depending on the species, these fangs are either long or short. Pit vipers have long hollow fangs. These fangs are folded against the roof of the mouth and extend when the snake strikes. This allows them to strike quickly and then withdraw. Cobras, coral snakes, kraits, mambas, and sea snakes have short, grooved fangs. These snakes are less effective in their attempts to bite, since they must chew after striking to inject enough venom (poison) to be effective. Figure 6-7 depicts the characteristics of a poisonous snakebite.

- In the event you are bitten, attempt to identify and/or kill the snake. Take it to medical personnel for inspection/identification. This provides valuable information to medical personnel who deal with snakebites. **TREAT ALL SNAKEBITES AS POISONOUS**.

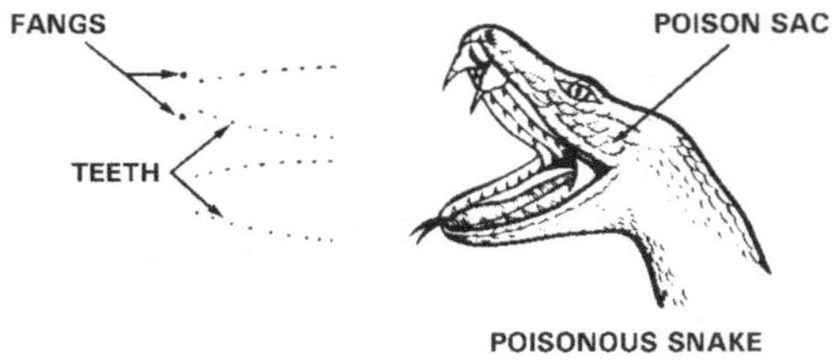

Figure 6-7. Characteristics of poisonous snakebite.

b. The venoms of different snakes cause different effects. Pit viper venom (hemotoxin [blood toxin]) destroys tissue and blood cells. Cobras, adders, and coral snakes inject powerful venom (neurotoxin [nerve toxin]) which affect the central nervous system, causing respiratory paralysis. Water moccasins and sea snakes have venom that is both hemotoxic and neurotoxic.

c. The identification of poisonous snakes is very important since medical treatment will be different for each type of venom. *Unless it can be positively identified, the snake should be killed and saved.* When this is not possible or when doing so is a serious threat to others, identification may sometimes be difficult since many venomous snakes resemble harmless varieties. When dealing with snakebite problems in foreign countries, seek advice, professional or otherwise, which may help identify species in the particular area of operations.

d. Get the casualty to an MTF as soon as possible and with minimum movement. Until evacuation or treatment is possible, have the casualty lie quietly and not move any more than necessary. If the casualty has been bitten on an extremity, **DO NOT** elevate the limb; keep the extremity level with the body. Keep the casualty comfortable and reassure him. If the casualty is alone when bitten, he should go to the medical facility himself rather than wait for someone to find him. Unless the snake has been positively identified, attempt to kill it and send it with the casualty. Be sure that retrieving the snake does not endanger anyone or delay transporting the casualty.

(1) If the bite is on an arm or leg, place a constricting band (narrow cravat [swathe], or narrow gauze bandage) one to two fingerbreadths above and below the bite (Figure 6-8). If the bite is on the hand or foot, place a single band above the wrist or ankle. The band should be tight enough to stop the flow of blood near the skin, but not tight enough to interfere with circulation. In other words, it should not have a tourniquet-like affect. If no swelling is seen, place the bands about 1 inch from either side of the bite. If swelling is present, put the bands on the unswollen part at the edge of the swelling. If the swelling extends beyond the band, move the band to the new edge of the swelling. (If possible, leave the old band on, place a new one at the new edge of the swelling, and then remove and save the old one in case the process has to be repeated.)

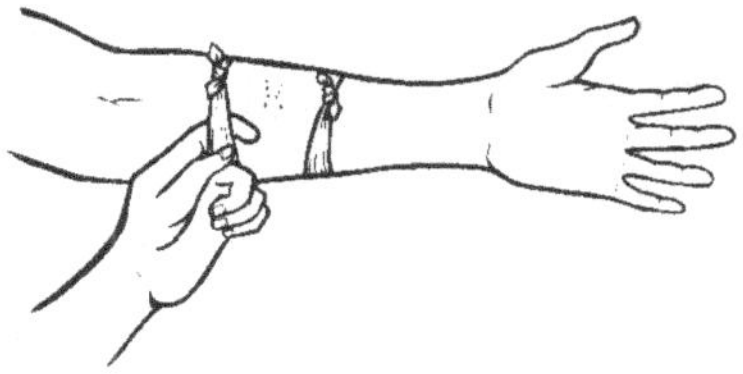

Figure 6-8. Constricting band.

CAUTION

DO NOT attempt to cut open the bite nor suck out the venom. If the venom should seep through any damaged or lacerated tissues in your mouth, you could immediately lose consciousness or even die.

(2) If the bite is located on an arm or leg, immobilize it at a level below the heart. **DO NOT** elevate an arm or leg even with or above the level of the heart.

> **CAUTION**
>
> When a splint is used to immobilize the arm or leg, take **EXTREME** care to ensure the splinting is done properly and does not bind. Watch it closely and adjust it if any changes in swelling occur.

(3) When possible, clean the area of the bite with soap and water. **DO NOT** use ointments of any kind.

(4) **NEVER** give the casualty food, alcohol, stimulants (coffee or tea), drugs, or tobacco.

(5) Remove rings, watches, or other jewelry from the affected limb.

6-4. Human or Animal Bites

Human or other land animal bites may cause lacerations or bruises. In addition to damaging tissue, bites always present the possibility of infection.

a. Human Bites. Human bites that break the skin may become seriously infected since the mouth is heavily contaminated with bacteria. Medical personnel **MUST** treat all human bites.

b. Animal Bites. Land animal bites can result in both infection and disease. Tetanus, rabies, and various types of fevers can follow an untreated animal bite. Because of these possible complications, the animal causing the bite should, if possible, be captured or killed (without damaging its head) so that it can be tested for disease.

c. First Aid.

(1) Cleanse the wound thoroughly with soap.

(2) Flush it well with water.

(3) Cover it with a sterile dressing.

(4) Immobilize the injured arm or leg, if appropriate.

(5) Transport the casualty immediately to an MTF.

NOTE

If unable to capture or kill the animal, provide medical personnel with any information that will help identify it.

6-5. Marine (Sea) Animals

With the exception of sharks and barracuda, most marine animals will not deliberately attack. The most frequent injuries from marine animals are wounds by biting, stinging, or puncturing. Wounds inflicted by marine animals can be very painful, but are rarely fatal.

a. Sharks, Barracuda, and Alligators. Wounds from these marine animals can involve major trauma as a result of bites and lacerations. Bites from large marine animals are potentially the most life threatening of all injuries from marine animals. Major wounds from these animals can be treated by controlling the bleeding, preventing shock, giving basic life support, splinting the injury, and by securing prompt medical aid.

b. Turtles, Moray Eels, and Corals. These animals normally inflict minor wounds. Treat by cleansing the wound(s) thoroughly and by splinting if necessary.

c. Jellyfish, Portuguese Man-of-War, Anemones, and Others. This group of marine animals inflict injury by means of stinging cells in their tentacles. Contact with the tentacles produces burning pain with a rash and small hemorrhages on the skin. Shock, muscular cramping, nausea, vomiting, and respiratory distress may also occur. Gently remove the clinging tentacles with a towel and wash or treat the area. Use diluted ammonia or alcohol, meat tenderizer, and talcum powder. If symptoms become severe or persist, seek medical assistance.

d. Spiny Fish, Urchins, Stingrays, and Cone Shells. These animals inject their venom by puncturing the skin with their spines. General signs and symptoms include swelling, nausea, vomiting, generalized cramps, diarrhea, muscular paralysis, and shock. Deaths are rare. Treatment consists of soaking the wounds in hot water (when available) for 30 to 60 minutes. This inactivates the heat sensitive toxin. In addition, further first aid measures (controlling bleeding, applying a dressing, and so forth) should be carried out as necessary.

CAUTION

Be careful not to scald the casualty with water that is too hot because the pain of the wound will mask the normal reaction to heat.

6-6. Insect (Arthropod) Bites and Stings

An insect bite or sting can cause great pain, allergic reaction, inflammation, and infection. If not treated correctly, some bites/stings may cause serious illness or even death. When an allergic reaction is not involved, first aid is a simple process. In any case, medical personnel should examine the casualty at the earliest possible time. It is important to properly identify the spider, bee, or creature that caused the bite/sting, especially in cases of allergic reaction.

a. Types of Insects. The insects found throughout the world that can produce a bite or sting are too numerous to mention in detail. Commonly encountered stinging or biting insects include brown recluse spiders (Figure 6-9), black widow spiders (Figure 6-10), tarantulas (Figure 6-11), scorpions (Figure 6-12), urticating caterpillars, bees, wasps, centipedes, conenose beetles (kissing bugs), ants, and wheel bugs. Upon being reassigned, especially to overseas areas, take the time to become acquainted with the types of insects to avoid.

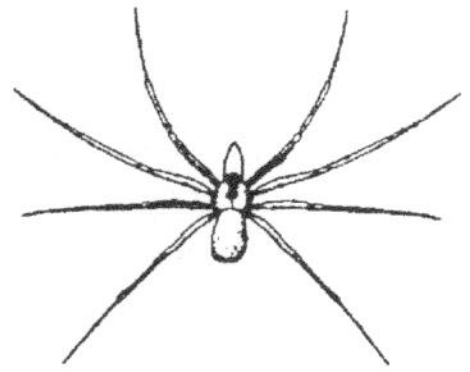

Figure 6-9. Brown recluse spider.

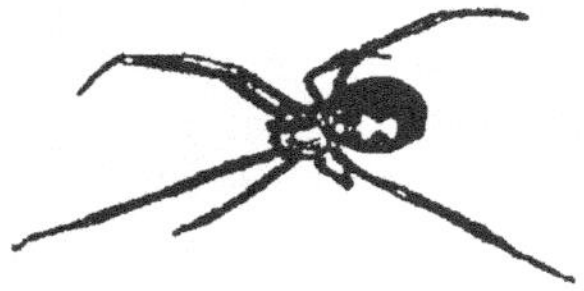

Figure 6-10. Black widow spider.

Figure 6-11. Tarantula.

Figure 6-12. Scorpion.

b. Signs and Symptoms. Discussed in paragraphs (1) and (2) below are the most common effects of insect bites/stings. They can occur alone or in combination with the others.

(1) *Less serious.* Commonly seen signs/symptoms are pain, irritation, swelling, heat, redness, and itching. Hives or wheals (raised areas of the skin that itch) may occur. These are the least severe of the allergic reactions that commonly occur from insect bites/stings. They are usually dangerous only if they affect the air passages (mouth, throat, nose, and so forth), which could interfere with breathing. The bites/stings of bees, wasps, ants, mosquitoes, fleas, and ticks are usually not serious and normally produce mild and localized symptoms. A tarantula's bite is usually no worse than that of a bee sting. Scorpions are rare and their stings (except for a specific species found only in the Southwest desert) are painful but usually not dangerous.

(2) *Serious.* Emergency allergic or hypersensitive reactions sometimes result from the stings of bees, wasps, and ants. Many people are allergic to the venom of these particular insects. Bites or stings from these insects may produce more serious reactions, to include generalized itching and hives, weakness, anxiety, headache, breathing difficulties, nausea, vomiting, and diarrhea. Very serious allergic reactions (called *anaphylactic shock*) can lead to complete collapse, shock, and even death. Spider bites (particularly from the black widow and brown recluse spiders) can also be

serious. Venom from the black widow spider affects the nervous system. This venom can cause muscle cramps, a rigid, nontender abdomen, breathing difficulties, sweating, nausea, and vomiting. The brown recluse spider generally produces local rather than system-wide problems; however, local tissue damage around the bite can be severe and can lead to an ulcer and even gangrene.

c. *First Aid.* There are certain principles that apply regardless of what caused the bite/sting. Some of these are—

- If there is a stinger present (for example, from a bee), remove the stinger by scraping the skin's surface with a fingernail or knife. **DO NOT** squeeze the sac attached to the stinger because it may inject more venom.

- Wash the area of the bite/sting with soap and water (alcohol or an antiseptic may also be used) to help reduce the chances of an infection and remove traces of venom.

- Remove jewelry from bitten extremities because swelling may occur.

- In most cases of insect bites the reaction will be mild and localized; use ice or cold compresses (if available) on the site of the bite/sting. This will help reduce swelling, ease the pain, and slow the absorption of venom. Meat tenderizer (to neutralize the venom) or calamine lotion (to reduce itching) may be applied locally. If necessary, seek medical assistance.

- In more serious reactions (severe and rapid swelling, allergic symptoms, and so forth) treat the bite/sting like you would treat a snakebite; that is, apply constricting bands above and below the site.

- Be prepared to perform basic life-support measures, such as rescue breathing.

- Reassure the casualty and keep him calm.

- In serious reactions, attempt to capture the insect for positive identification; however, be careful not to become a casualty yourself.

- If the reaction to the bite/sting appears serious, seek medical assistance.

WARNING

Insect bites/stings may cause *anaphylactic shock* (a shock caused by a severe allergic reaction). This is a life-threatening event and a TRUE MEDICAL EMERGENCY. Be prepared to perform the basic life-support measures and to immediately transport the casualty to an MTF.

NOTE

Be aware that some allergic or hypersensitive individuals may carry identification or emergency insect bite treatment kits. If the casualty is having an allergic reaction and has such a kit, administer the medication in the kit according to the instructions which accompany the kit.

d. Supplemental Information. For additional information concerning biting insects, see FM 21-10.

6-7. First Aid for Bites and Stings

See the table below for information on bites and stings.

Table 6-1. First Aid Measures for Bites and Stings

TYPES	FIRST AID MEASURES
SNAKEBITE	1. MOVE CASUALTY AWAY FROM THE SNAKE. 2. REMOVE JEWELRY FROM THE AFFECTED AREA, IF APPLICABLE. 3. REASSURE CASUALTY AND KEEP HIM QUIET. 4. APPLY CONSTRICTING BAND, 1-2 FINGERBREADTHS FROM THE BITE. YOU SHOULD BE ABLE TO INSERT A FINGER BETWEEN THE BAND AND THE SKIN. *a. ARM OR LEG BITE.* PLACE ONE BAND ABOVE AND ONE BAND BELOW THE BITE SITE. *b. HAND OR FOOT BITE.* PLACE ONE BAND ABOVE THE WRIST OR ANKLE.

Table 6-1. First Aid Measures for Bites and Stings

TYPES	FIRST AID MEASURES
	5. IMMOBILIZE THE AFFECTED PART IN A POSITION BELOW THE LEVEL OF THE HEART. 6. KILL THE SNAKE (IF POSSIBLE, WITHOUT DAMAGING ITS HEAD OR ENDANGERING YOURSELF) AND SEND IT WITH THE CASUALTY. 7. SEEK MEDICAL ASSISTANCE IMMEDIATELY.
BROWN RECLUSE SPIDER OR BLACK WIDOW SPIDER BITE	1. KEEP CASUALTY QUIET. 2. REMOVE ALL JEWELRY FROM AFFECTED PART, IF APPLICABLE. 3. WASH THE AREA. 4. APPLY ICE OR FREEZE PACK, IF AVAILABLE. 5. SEEK MEDICAL ASSISTANCE.
TARANTULA BITE OR SCORPION STING OR ANT BITE	1. WASH THE AREA. 2. REMOVE ALL JEWELRY FROM AFFECTED PART, IF APPLICABLE. 3. APPLY ICE OR FREEZE PACK, IF AVAILABLE. 4. APPLY BAKING SODA, CALAMINE LOTION, OR MEAT TENDERIZER (IF AVAILABLE) TO BITE SITE TO RELIEVE PAIN AND ITCHING. 5. IF THE SITE OF THE BITE IS ON THE FACE, NECK (POSSIBLE AIRWAY PROBLEMS), OR GENITAL AREA, OR IF LOCAL REACTION SEEMS SEVERE, OR IF THE STING IS BY THE DANGEROUS TYPE OF SCORPION FOUND IN THE SOUTHWEST UNITED STATES DESERT, KEEP THE CASUALTY AS QUIET AS POSSIBLE. SEEK MEDICAL ASSISTANCE.
BEE STING	1. IF THE STINGER IS PRESENT, REMOVE BY SCRAPING WITH A KNIFE OR FINGERNAIL. DO NOT SQUEEZE VENOM SAC ON STINGER; MORE VENOM MAY BE INJECTED. 2. REMOVE ALL JEWELRY FROM AFFECTED PART, IF APPLICABLE.

Table 6-1. First Aid Measures for Bites and Stings

TYPES	FIRST AID MEASURES
	3. WASH THE AREA. 4. APPLY ICE OR FREEZE PACK, IF AVAILABLE. 5. IF ALLERGIC SIGNS OR SYMPTOMS APPEAR, BE PREPARED TO PERFORM BASIC LIFE SUPPORT MEASURES. SEEK IMMEDIATE MEDICAL ASSISTANCE.

CHAPTER 7

FIRST AID IN A NUCLEAR, BIOLOGICAL, AND CHEMICAL ENVIRONMENT

7-1. General

American forces have not been exposed to NBC weapons/agents on the battlefield since World War I. In future conflicts and wars we can expect the use of such agents. Nuclear, biological, and chemical weapons will rapidly degrade unit effectiveness by forcing troops to wear protective clothing and by creating confusion and fear. Through training in protective procedures and first aid, units can maintain their effectiveness on the integrated battlefield.

7-2. First Aid Materials

You may be issued the following materials to protect, decontaminate, and use as first aid for NBC exposure. You must know how to use the items; some items are described in *a* through *d* below. It is equally important that you know when to use them.

a. Nerve Agent Pyridostigmine Pretreatment (*NAPP*). You may be issued a blister pack of pretreatment tablets when your commander directs. The NAPP is a pretreatment; it is not an antidote. It improves the effectiveness of the nerve agent antidote. When ordered to take the pretreatment you must take one tablet every 8 hours, mission permitting. This must be taken prior to exposure to nerve agents, since it may take several hours to develop adequate blood levels.

NOTE

Commanders must follow investigational new drug protocols for use of the NAPP.

b. M291 Skin Decontaminating Kit. The M291 Skin Decontaminating Kit (Figure 7-1) contains six packets of XE-555 decontaminant resin.

WARNING

For external use only. May be slightly irritating to the eyes. Keep decontaminating powder out of eyes. Use water to wash toxic agent out of eyes.

c. *Nerve Agent Antidote Kit, MARK I.* Each service member is issued three MARK Is for use in first aid for nerve agent poisoning (Figure 7-2 and paragraph 7-6).

Figure 7-1. M291 Skin Decontamination Kit.

d. *Antidote Treatment, Nerve Agent, Autoinjector.* A new nerve agent antidote injection device, Antidote Treatment, Nerve Agent, Autoinjector (ATNAA) is currently under development that will replace the MARK I. The ATNAA is a multichambered device with the atropine and pralidoxime chloride in separate chambers. Both antidotes will be administered through a single needle.

7-3. Classification of Chemical and Biological Agents

a. Chemical agents are classified according to the primary physiological effects they produce, such as blistering, choking, vomiting, and incapacitating agents.

b. Biological warfare agents are classified according to the effect they have on man. The effects include their ability to incapacitate and cause death. Most biological warfare agents are delivered as aerosols that effect the respiratory tract; some can be delivered by releasing infected insects, by contaminating food and water, and by injection (injecting material in individuals by terrorist, not mass exposure). These agents are found in living organisms such as fungi, bacteria, and viruses.

WARNING

Swallowing water or food contaminated with nerve, blister, and other chemical agents and with some biological agents can be fatal. NEVER consume water or food that is suspected of being contaminated until it has been tested and found safe for consumption by medical personnel.

7-4. Conditions for Masking Without Order or Alarm

a. Once an attack with a chemical or biological agent is detected or suspected, or information is available that such an agent is about to be used, you must **STOP BREATHING** and mask immediately. **DO NOT WAIT** to receive an order or alarm under the following circumstances:

- Your position is hit by artillery missiles, rockets that produce vapors, smoke, and mists, and aerial sprays.
- Smoke or vapor cloud from an unknown source is present or approaching.
- A suspicious odor, liquid, or solid is present.
- A chemical or biological warfare agent attack is occurring.
- You are entering an area known or suspected of being contaminated.
- When casualties are being received from an area where chemical or biological agents have reportedly been used.
- You have one or more of the following symptoms:
  - An unexplained runny nose.
  - A sudden unexplained headache.
  - A feeling of choking or tightness in the chest or throat.
  - Dimness of vision.
  - Irritation of the eyes.
  - Difficulty in or increased rate of breathing without obvious reasons.
  - Sudden feeling of depression.
  - Dread, anxiety, or restlessness.
  - Dizziness or light-headedness.

- Slurred speech.
- Unexplained laughter or unusual behavior is noted in others.
- Numerous unexplained ill personnel.
- Service members suddenly collapsing without evident cause.
- Animals or birds exhibiting unusual behavior or suddenly dying.

b. For further information on protection and masking procedures, refer to FM 3-4, FM 4-02.7, FM 8-284, and FM 8-285.

7-5. First Aid for a Chemical Attack

Your field protective mask gives protection against biological and chemical warfare agents as well as radiological fallout. With practice you can mask in 9 seconds or less, or put on your mask with hood within 15 seconds.

a. Stop breathing. Don on your mask, seal it properly, and clear and check it; then resume breathing. Give the alarm, and continue the mission. Keep your mask on until the "all clear" signal has been given.

NOTE

Keep your mask on until the area is no longer hazardous and you are told to unmask.

b. If symptoms of nerve agent poisoning (paragraph 7-7) appear, immediately give yourself one MARK I or ATNAA.

CAUTION

Do not inject a nerve agent antidote until you are sure you need it.

c. If your eyes and face become contaminated, you must immediately try to get under cover. You need shelter to prevent further contamination while performing decontamination procedures on your face. If no overhead cover is available, put your poncho over your head before beginning the decontamination process. Then you put on the remaining protective clothing. If vomiting occurs, the mask should be lifted momentarily and drained—with your eyes closed and while holding your breath—then replaced, cleared, and sealed.

d. If nerve agents are used, mission permitting, watch for persons needing nerve agent antidotes and immediately follow procedures outlined in paragraph 7-8*b* or *c*.

e. Decontaminate your skin immediately and clothing and equipment as soon as the mission permits.

7-6. Background Information on Nerve Agents

a. Nerve agents are among the deadliest of chemical agents. Nerve agents enter the body by inhalation, by ingestion, and through the skin. Depending on the route of entry and the amount, nerve agents can produce injury or death within minutes. Nerve agents can achieve their effects with small amounts. Nerve agents are absorbed rapidly, and the effects are felt immediately upon entry into the body. You will be issued three MARK Is or three ATNAAs and one Convulsant Antidote for Nerve Agent (CANA). Each MARK I consists of one atropine autoinjector and one pralidoxime chloride (2 PAM Cl) autoinjector (Figure 7-2A). Each ATNAA consist of a multichambered autoinjector with the atropine and pralidoxime chloride in separate chambers (Figure 7-2C). The CANA is a single autoinjector with flanges (Figure 7-2B). Procedures for use of both the MARK I and ATNAA are described below. You will use either the MARK I or the ATNAA in self-aid and buddy aid as issued.

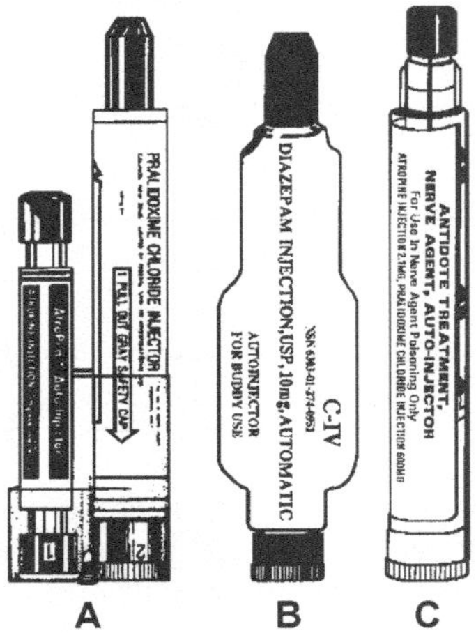

A B C

Figure 7-2. Nerve Agent Antidote Kit, MARK I, CANA, and ATNAA.

b. When you have the signs and symptoms of nerve agent poisoning, you should immediately put on the protective mask and then inject yourself with one set of the MARK I or ATNAA. Do not administer the CANA. You should inject yourself in the outer (lateral) thigh muscle (Figure 7-3) or if you are thin, in the upper outer (lateral) part of the buttocks (Figure 7-4).

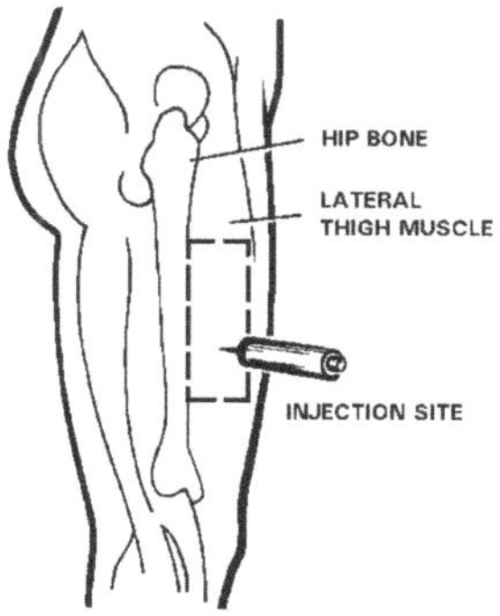

Figure 7-3. Thigh injection site.

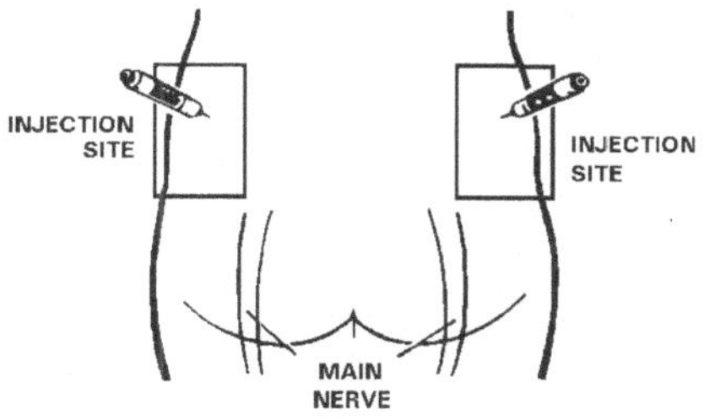

Figure 7-4. Buttocks injection site.

c. Also, you may come upon an unconscious chemical agent casualty who will be unable to care for himself and who will require first aid. You should be able to successfully—

(1) Mask him if he is unmasked.

(2) Inject him, if necessary, with all of **HIS** autoinjectors.

(3) Decontaminate his skin.

(4) Seek medical assistance.

7-7. Signs and Symptoms of Nerve Agent Poisoning

The symptoms of nerve agent poisoning are grouped as **MILD**—those that you recognize and for which you can perform self-aid, and **SEVERE**—those which require buddy aid.

a. MILD Signs and Symptoms.

- Unexplained runny nose.
- Unexplained sudden headache.
- Sudden drooling.
- Difficulty seeing (dimness of vision and miosis).
- Tightness in the chest or difficulty in breathing.
- Localized sweating and muscular twitching in the area of contaminated skin.
- Stomach cramps.
- Nausea.
- Tachycardia followed by bradycardia. (*Tachycardia* is an abnormally rapid heartbeat with a heart rate of over 100 beats per minute. *Bradycardia* is a slow heart rate of less than 60 beats per minute.)

b. SEVERE Signs and Symptoms.

- Strange or confused behavior.
- Wheezing, dyspnea (difficulty in breathing), and coughing.
- Severely pinpointed pupils.
- Red eyes with tearing.
- Vomiting.
- Severe muscular twitching and general weakness.
- Involuntary urination and defecation.

- Convulsions.
- Unconsciousness.
- Respiratory failure.
- Bradycardia.

7-8. First Aid for Nerve Agent Poisoning

First aid for nerve agent poisoning consists of administering the MARK I or ATNAA and CANA.

a. Injection Site. The injection site for administering the antidotes is normally in the outer thigh muscle. The thigh injection site is the area about a hand's width above the knee to a hand's width below the hip joint (Figure 7-3). It is important that the injection be given into a large muscle area. If the individual is thinly built, then the injections should be administered into the upper outer quarter (quadrant) of the buttock (Figure 7-4). Injecting in the buttocks of a thinly built individual avoids injury to the thighbone.

b. Self-Administer MARK I. If you experience any or all of the nerve agent **MILD** symptoms (paragraph 7-7*a*), you must **IMMEDIATELY** put on your protective mask and self-administer one MARK I (Figure 7-2A). Follow the procedure given in Table 7-1. The MARK I is carried in your protective mask carrier, pocket of the MOPP overgarment, or other location as specified in your unit tactical standing operating procedure (TSOP). (In cold weather, the MARK I should be stored in an inside pocket of your clothing to protect the antidote from freezing. A frozen MARK I cannot be immediately used to provide you with antidote, when needed. (However, the MARK I can still be used after complete thawing.)

Table 7-1. Self Aid for Nerve Agent Poisoning

MARK I*	ATNAA*
STEP 1. OBTAIN ONE MARK I.**	STEP 1. OBTAIN ONE ATNAA.**
STEP 2. CHECK INJECTION SITE.	STEP 2. CHECK INJECTION SITE.
STEP 3. HOLD MARK I AT EYE LEVEL WITH NONDOMINANT HAND WITH THE LARGE INJECTOR ON TOP (FIGURE 7-5A).	STEP 3. HOLD ATNAA WITH DOMINANT HAND (FIGURE 7-12A).

Table 7-1. Self Aid for Nerve Agent Poisoning (Continued)

MARK I*	ATNAA*
STEP 4. GRASP SMALL INJECTOR (ATROPINE) (FIGURE 7-5B) AND REMOVE FROM CLIP (FIGURE 7-5C).	STEP 4. GRASP SAFETY CAP WITH NONDOMINANT HAND AND REMOVE FROM INJECTOR (FIGURE 7-12B).
STEP 5. CLEAR HARD OBJECTS FROM INJECTION SITE.	STEP 5. CLEAR HARD OBJECTS FROM INJECTION SITE.
STEP 6. INJECT ATROPINE AT INJECTION SITE APPLYING EVEN PRESSURE TO THE INJECTOR (FIGURE 7-6 OR 7-7). HOLD IN PLACE FOR 10 SECONDS.	STEP 6. INJECT ATNAA AT INJECTION SITE APPLYING EVEN PRESSURE TO THE INJECTOR (FIGURE 7-14 OR 7-15). HOLD IN PLACE FOR 10 SECONDS.
STEP 7. HOLD USED INJECTOR WITH NONDOMINANT HAND.	STEP 7. BEND NEEDLE OF USED INJECTOR BY PRESSING ON A HARD SURFACE TO FORM A HOOK.
STEP 8. GRASP THE LARGE (2 PAM CI) INJECTOR (FIGURE 7-8B) AND PULL IT FROM CLIP (FIGURE 7-8C). DROP CLIP TO GROUND.	STEP 8. ATTACH USED INJECTOR TO BLOUSE POCKET FLAP OF BDO/JSLIST (FIGURE 7-16).
STEP 9. INJECT 2 PAM CI AT INJECTION SITE APPLYING EVEN PRESSURE TO THE INJECTOR (FIGURE 7-6 OR 7-7). HOLD IN PLACE FOR 10 SECONDS.	STEP 9. MASSAGE INJECTION SITE, MISSION PERMITTING.
STEP 10. BEND THE NEEDLES OF ALL USED INJECTORS BY PRESSING ON A HARD SURFACE TO FORM A HOOK.	
STEP 11. ATTACH ALL USED INJECTORS TO BLOUSE POCKET FLAP OF BDO/JSLIST (FIGURE 7-9).	
STEP 12. MASSAGE INJECTION SITE, MISSION PERMITTING.	

* USE STEPS LISTED FOR TYPE OF ANTIDOTE DEVICE ISSUED.

** ONLY ADMINISTER ONE MARK I OR ATNAA AS SELF-AID. DO NOT SELF-ADMINISTER CANA.

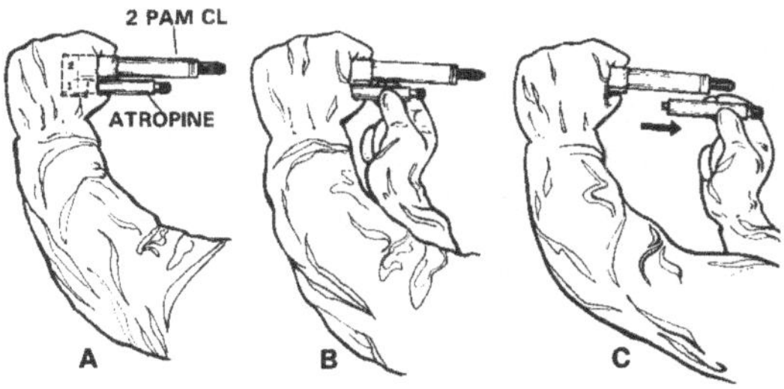

Figure 7-5. Removing the atropine autoinjector from the MARK I clip.

CAUTION

DO NOT cover or hold the needle end with your hand, thumb, or fingers—you might accidentally inject yourself. An accidental injection into the hand **WILL NOT** deliver an effective dose of the antidote, especially if the needle goes through the hand.

Figure 7-6. Thigh injection site for self-aid.

NOTE

If you are thinly built, inject yourself into the upper outer quadrant of the buttock (Figure 7-7). There is a nerve that crosses the buttocks; hitting this nerve can cause paralysis. Therefore, you must only inject into the *upper outer quadrant* of the buttock.

Figure 7-7. Buttocks injection site for self-aid.

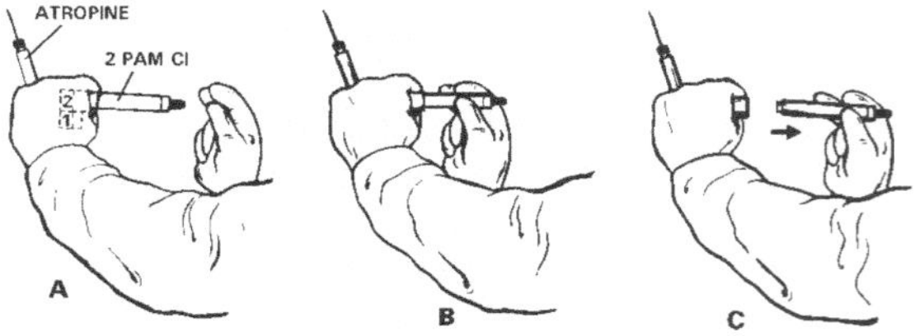

Figure 7-8. Removing the 2 PAM Cl autoinjector from the MARK I clip.

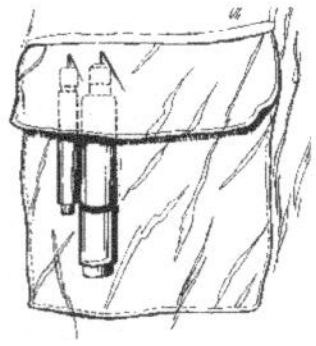

Figure 7-9. One set of used MARK I autoinjectors attached to pocket flap.

NOTES

1. **DO NOT** give yourself another set of injections. If you are able to walk without assistance, know who you are, and where you are, you **WILL NOT** need the second set of injections. (If not needed, giving yourself a second set of MARK I injections or ATNAA may create a nerve agent antidote overdose, which could cause incapacitation [inability to perform mission or defend yourself].)

2. If you continue to have symptoms of nerve agent poisoning, seek someone else (a buddy) to check your symptoms and administer the additional sets of injections, if required.

c. *Buddy Evaluation and Buddy Aid.* Service members may seek assistance after self-aid (self-administering one MARK I or ATNAA) or may become incapacitated after self-aid. A buddy must evaluate the individual to determine if additional antidotes are required to counter the effects of the nerve agent. Also, service members may experience **SEVERE** symptoms of nerve agent poisoning (paragraph 7-7*b*); they will not be able to treat themselves. In either case, other service members must perform buddy aid as quickly as possible. Before initiating buddy aid, determine if one set of MARK I autoinjectors has already been used so that no more than three sets of the antidote are administered. Buddy aid also includes administering the CANA with the third MARK I or ATNAA to prevent convulsions. Follow the procedures indicated in Table 7-2.

WARNING

Squat, DO NOT kneel, when masking the casualty or administering the nerve agent antidote to the casualty. Kneeling may force the chemical agent into or through your protective clothing.

CAUTION

DO NOT use your own MARK I, ATNAA, or CANA on a casualty. If you use your own, you may not have any antidote if needed for self-aid.

WARNING

DO NOT inject into areas close to the hip, knee, or thigh-bone.

Table 7-2. Buddy Aid/Combat Lifesaver Aid for Nerve Agent Casualty.

MARK I*	ATNAA*	CANA**
STEP 1. MASK THE CASUALTY AND POSITION HIM ON HIS SIDE (SWIMMER'S POSITION).	STEP 1. MASK THE CASUALTY AND POSITION HIM ON HIS SIDE (SWIMMER'S POSITION).	STEP 1. OBTAIN BUDDY'S CANA.
STEP 2. POSITION YOURSELF NEAR THE CASUALTY'S THIGH.	STEP 2. POSITION YOURSELF NEAR THE CASUALTY'S THIGH.	STEP 2. CHECK INJECTION SITE.

Table 7-2. Buddy Aid/Combat Lifesaver Aid for Nerve Agent Casualty (Continued).

MARK I*	ATNAA*	CANA**
STEP 3. OBTAIN BUDDY'S THREE OR REMAINING MARK Is.	STEP 3. OBTAIN BUDDY'S THREE OR REMAINING ATNAAs.	STEP 3. HOLD CANA IN A CLOSED FIST WITH DOMINANT HAND (FIGURE 7-12A).
STEP 4. CHECK INJECTION SITE.	STEP 4. CHECK INJECTION SITE.	STEP 4. GRASP SAFETY CAP WITH NONDOMINANT HAND AND REMOVE FROM INJECTOR (FIGURE 7-12B).
STEP 5. HOLD MARK I WITH NONDOMINANT HAND (FIGURE 7-5A).	STEP 5. HOLD ATNAA IN A CLOSED FIST WITH DOMINANT HAND (FIGURE 7-12A).	STEP 5. CLEAR HARD OBJECTS FROM INJECTION SITE.
STEP 6. GRASP SMALL INJECTOR (ATROPINE) AND REMOVE FROM CLIP (FIGURE 7-5B).	STEP 6. GRASP SAFETY CAP WITH NONDOMINANT HAND AND REMOVE FROM INJECTOR (FIGURE 7-12B).	STEP 6. INJECT CANA AT INJECTION SITE BY APPLYING EVEN PRESSURE TO THE INJECTOR, NOT A JABBING MOTION (FIGURE 7-14 OR 7-15). HOLD IN PLACE FOR 10 SECONDS.
STEP 7. CLEAR HARD OBJECTS FROM INJECTION SITE.	STEP 7. CLEAR HARD OBJECTS FROM INJECTION SITE.	STEP 7. BEND NEEDLE OF INJECTOR BY PRESSING ON A HARD SURFACE TO FORM A HOOK.
STEP 8. INJECT ATROPINE AT INJECTION SITE BY APPLYING EVEN PRESSURE TO THE INJECTOR, NOT A JABBING MOTION (FIGURE 7-10 OR 7-11). HOLD IN PLACE FOR 10 SECONDS.	STEP 8. INJECT ATNAA AT INJECTION SITE BY APPLYING EVEN PRESSURE TO THE INJECTOR, NOT A JABBING MOTION (FIGURE 7-14 OR 7-15). HOLD IN PLACE FOR 10 SECONDS.	STEP 8. ATTACH USED INJECTOR TO BLOUSE POCKET FLAP OF BDO/JSLIST (FIGURE 7-16).
STEP 9. HOLD USED INJECTOR BETWEEN LITTLE FINGER AND RING FINGER OF NONDOMINANT HAND (FIGURE 7-5A).	STEP 9. BEND NEEDLE OF INJECTOR BY PRESSING ON A HARD SURFACE TO FORM A HOOK.	STEP 9. MASSAGE INJECTION SITE, MISSION PERMITTING.
STEP 10. PULL LARGE INJECTOR (2 PAM CI) FROM CLIP (FIGURE 7-5C). DROP CLIP TO GROUND.	STEP 10. ATTACH ALL USED INJECTORS TO BLOUSE POCKET FLAP OF BDO/JSLIST (FIGURE 7-16).	

Table 7-2. Buddy Aid/Combat Lifesaver Aid for Nerve Agent Casualty (Continued).

MARK I*	ATNAA*	CANA**
STEP 11. INJECT 2 PAM CI AT INJECTION SITE BY APPLYING EVEN PRESSURE TO THE INJECTOR, NOT A JAB-BING MOTION (FIGURE 7-10 OR 7-11). HOLD IN PLACE FOR 10 SECONDS.	STEP 11. MASSAGE INJECTION SITE, MISSION PERMITTING.	
STEP 12. REPEAT STEPS ABOVE FOR REMAINING MARK Is.		
STEP 13. BEND THE NEEDLES OF ALL USED INJECTORS BY PRESSING ON A HARD SURFACE TO FORM A HOOK.		
STEP 14. ATTACH ALL USED INJECTORS TO BLOUSE POCKET FLAP OF BDO/JSLIST (FIGURE 7-13).		
STEP 15. MASSAGE INJECTION SITE, MISSION PERMITTING.		

* USE STEPS LISTED FOR TYPE OF ANTIDOTE DEVICE ISSUED.
** CANA IS USED IN BUDDY AID/CLS AID ONLY. DO NOT USE IN SELF-AID.

NOTE

If the casualty is thinly built, inject the antidote into the buttock. Only inject the antidote into the upper outer portion of the casualty's buttock (Figure 7-11). This avoids hitting the nerve that crosses the buttocks (Figure 7-4). Hitting this nerve can cause paralysis.

Figure 7-10. Injecting the casualty's thigh (Mark I or CANA).

Figure 7-11. Injecting the casualty's buttocks (Mark I or CANA).

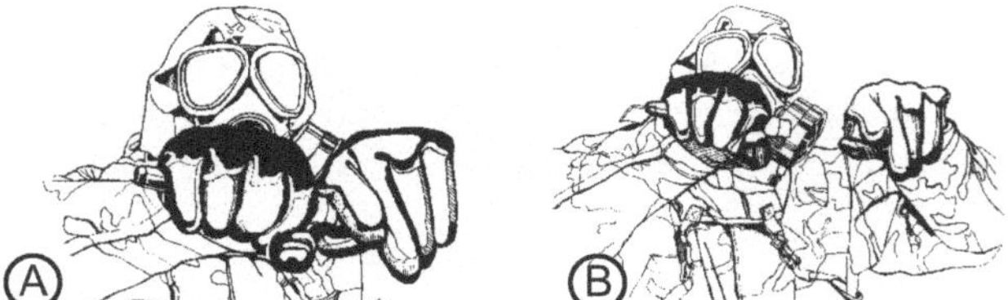

Figure 7-12. Preparing CANA or ATNAA for injection.

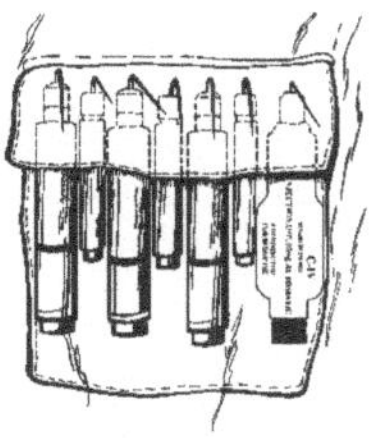

Figure 7-13. Three sets of used MARK I autoinjectors and one CANA autoinjector attached to pocket flap.

d. Self-Administer Antidote Treatment Nerve Agent Autoinjector. If you experience any or all of the nerve agent **MILD** symptoms (paragraph 7-7*b*), you must **IMMEDIATELY** self-administer one ATNAA following the procedure given Table 7-1.

NOTE

If you are thinly-built, inject yourself into the upper outer quarter (quadrant) of the buttock (Figure 7-15). There is a nerve that crosses the buttocks; hitting this nerve can cause paralysis. Therefore, you must only inject into the upper outer quarter (quadrant) of the buttocks.

Figure 7-14. Self-administration of ATNAA (thigh).

Figure 7-15. Self-administration of ATNAA (buttock).

NOTE

If you continue to have symptoms of nerve agent poisoning, seek someone else (a buddy) to check your symptoms and administer your remaining sets of injections, if required.

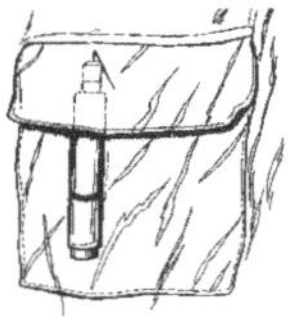

Figure 7-16. Used ATNAA attached to clothing.

e. Buddy Assistance. Service members may seek assistance after self-aid (self-administering one ATNAA) or may become incapacitated after self-aid. A buddy must evaluate the individual to determine if additional antidotes are required to counter the effects of the nerve agent. Also, service members may experience **SEVERE** symptoms of nerve agent poisoning (paragraph 7-7*b*); they will not be able to treat themselves. In either case, other service members must perform buddy aid as quickly as possible. Before initiating buddy aid, determine if one ATNAA has already been used so that no more than three ATNAA are administered. Buddy aid also includes administering the CANA with the third ATNAA to prevent convulsions. Follow the procedures indicated in Table 7-2.

WARNING

Squat, DO NOT kneel, when masking the casualty or administering the nerve agent antidotes to the casualty. Kneeling may force any chemical agent on your overgarment into or through your protective clothing.

Figure 7-17. Buddy injecting casualty's outer thigh (ATNAA or CANA).

NOTE

If the casualty is thinly built, inject the antidote into the buttocks (Figure 7-18). Only inject the antidote into the upper outer portion of the casualty's buttocks. This avoids hitting the nerve that crosses the buttocks (Figure 7-4). Hitting this nerve can cause paralysis.

WARNING

DO NOT inject into areas close to the hip, knee, or thighbone.

Figure 7-18. Buddy injecting casualty's buttocks (ATNAA or CANA).

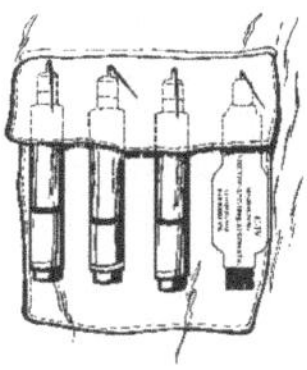

Figure 7-19. Three used ATNAAs and one CANA autoinjector attached to clothing.

f. Combat Lifesaver.

(1) The combat lifesaver must check to verify if the individual has received three sets of MARK I or ATNAAs. If not, the combat lifesaver performs first aid as described for buddy aid above. If the

individual has received the initial three sets of MARK I, then the combat lifesaver may administer additional atropine injections at approximately 15 minute intervals until atropinization is achieved (that is a heart rate above 90 beats per minute, reduced bronchial secretions, and reduced salivations). Administer additional atropine at intervals of 30 minutes to 4 hours to maintain atropinization or until the casualty is placed under the care of medical personnel. Check the heart rate by lifting the casualty's mask hood and feeling for a pulse at the carotid artery. Request medical assistance as soon as the tactical situation permits.

(2) The CLS should administer additional CANA to casualties suffering convulsions. Administer a second, and if needed, a third CANA at 5 to 10 minute intervals for a maximum of three injections (30 milligrams diazepam). Follow the steps and procedures described in buddy aid for administering the CANA. **DO NOT** give more than two additional injections for a total of three (one self-aid plus two by the CLS).

7-9. Blister Agents

Blister agents (vesicants) include mustard (H and HD), nitrogen mustards (HN), lewisite (L), and other arsenicals, mixtures of mustards and arsenicals, and phosgene oxime (CX). Blister agents may act on the eyes, mucous membranes, lungs, and skin. They burn and blister the skin or any other body parts they contact. Even relatively low doses may cause serious injury. Blister agents damage the respiratory tract (nose, sinuses, and windpipe) when inhaled and cause vomiting and diarrhea when absorbed. Lewisite and CX cause immediate pain on contact. However, mustard agents are deceptive as there is little or no pain at the time of exposure. Thus, in some cases, signs of injury may not appear for several hours after exposure.

a. Protective Measures. Your protective mask with hood and protective overgarment provide protection against blister agents. If it is known or suspected that blister agents are being used, **STOP BREATHING**, put on your mask and your protective overgarment.

CAUTION

Large drops of liquid vesicants on the protective overgarment ensemble may penetrate it if allowed to stand for an extended period. Remove large drops as soon as possible.

b. Signs and Symptoms of Blister Agent Poisoning.

(1) Immediate and intense pain upon contact with L, LH (lewisite and mustard) mixture, and CX. No initial pain upon contact with mustard.

(2) Inflammation and blisters (burns) resulting in tissue destruction. The severity of a chemical burn is directly related to the concentration of the agent and the duration of contact with the skin. The longer the agent is in contact with the tissue, the more serious the injury will be.

(3) Vomiting and diarrhea. Exposure to high concentrations of vesicants may cause vomiting or diarrhea.

(4) Death. The blister agent vapors absorbed during ordinary field exposure will probably not cause enough internal body (systemic) damage to result in death. However, death may occur from prolonged exposure to high concentrations of vapor or from extensive liquid contamination over wide areas of the skin, particularly *when decontamination is neglected or delayed.*

c. First Aid Measures.

(1) Use your M291 Skin Decontaminating Kit to decontaminate your skin and use water to flush contaminated eyes. Decontamination of vesicants must be done immediately (within 1 minute is best).

(2) If blisters form, cover them loosely with a field dressing and secure the dressing.

CAUTION

Blisters are actually burns. **DO NOT** attempt to decontaminate the skin where blisters have formed, as the agent has already been absorbed.

(3) If you receive blisters over a wide area of the body, you are considered seriously burned. Seek medical assistance immediately.

(4) If vomiting occurs, the mask should be lifted momentarily and drained—while the eyes are closed and the breath is held—and replaced, cleared, and sealed.

(5) Remember, if vomiting or diarrhea occurs after having been exposed to blister agents, seek medical assistance immediately.

7-10. Choking Agents (Lung-Damaging Agents)

Chemical agents that attack lung tissue, primarily causing fluid buildup (pulmonary edema), are classified as choking agents (lung-damaging agents). This group includes phosgene (CG), diphosgene (DP), chlorine (Cl), and chloropicrin (PS). Of these four agents, CG is the most dangerous and is more likely to be employed by the enemy in future conflict.

a. Protective Measures. Your protective mask gives adequate protection against choking agents.

b. Signs and Symptoms. During and immediately after exposure to choking agents (depending on agent concentration and length of exposure), you may experience some or all of the following signs and symptoms:

- Tears (lacrimation).
- Coughing.
- Choking.
- Tightness of chest.
- Nausea and vomiting.
- Headaches.

c. Self-Aid.

(1) The protective mask should be put on immediately when any of the conditions described in *b* above exist. Another indication of a CG attack is an odor like newly mown hay; however, **DO NOT** rely upon odor as indication of a chemical attack.

(2) If some CG is inhaled, normal combat duties should be continued unless there is difficulty in breathing, nausea, vomiting, or more than the usual shortness of breath during exertion. If any of the above symptoms occur and the mission permits, remain at quiet rest until medical evacuation is accomplished.

d. Death. With ordinary field exposure to choking agents, death will probably not occur. *However, prolonged exposure to high concentrations of the vapor and neglect or delay in masking can be fatal.*

7-11. Cyanogen (Blood) Agents

Cyanogen agents interfere with proper oxygen utilization in the body. Hydrogen cyanide (AC) and cyanogen chloride (CK) are the primary agents in this group.

a. Protective Measures. Your protective mask with a fresh filter gives adequate protection against field concentrations of cyanogen agent vapor. The protective overgarments, as well as the mask, are needed when exposed to liquid AC.

b. Signs and Symptoms. During and immediately after exposure to cyanogen agents (depending on agent concentration and length of exposure), you may experience some or all of the following signs and symptoms:

- Tearing (lacrimation).
- Eye, nose, and throat irritation.
- Sudden stimulation of breathing (unable to hold breath).
- Nausea.
- Coughing.
- Tightness of chest.
- Headache.
- Light-headedness (dizziness).
- Unconsciousness.

c. First Aid.

(1) *Hydrogen cyanide.* During any chemical attack, if you get a sudden stimulation of breath or detect an odor like bitter almonds, **PUT ON YOUR MASK IMMEDIATELY**. Speed is absolutely essential since this agent acts so rapidly that within a few seconds its effects will make it impossible for service members to put on their mask by themselves. Stop breathing until the mask is on, if at all possible. This may be very difficult since the agent strongly stimulates respiration.

(2) *Cyanogen chloride.* Put your mask on immediately if you experience any irritation of the eyes, nose, or throat. Service members

who are unable to mask should be masked by the nearest service member (buddy).

d. Medical Assistance. If you suspect that you have been exposed to blood agents, seek medical assistance immediately.

7-12. Incapacitating Agents

An incapacitating agent is a chemical agent which produces temporary, disabling conditions which persist for hours to days after exposure. Unlike riot control agents, which usually are momentary or fleeting in action, incapacitating agents have a persistent effect. It is likely that smoke-producing munitions or aerosols will disseminate such agents, thus making breathing their means of entry into the body. The protective mask is, therefore, essential.

a. There are no specific first aid measures to relieve the symptoms of incapacitating agents. Supportive first aid and physical restraint may be indicated. If the casualty is stuporous or comatose, be sure that respiration is unobstructed; then turn him on his side in case vomiting should occur. Complete cleansing of the skin with soap and water should be done as soon as possible; or, the M291 Skin Decontaminating Kit can be used if washing is impossible. Remove weapons and other potentially harmful items from service members who are suspected of having these symptoms. Harmful items include cigarettes, matches, medications, and small items that might be swallowed accidentally. Delirious (confused) persons have been known to attempt to eat items bearing only a superficial resemblance to food.

b. Incapacitating agents (anticholinergic drugs BZ type) may produce alarming dryness and coating of the lips and tongue; however, there is usually no danger of immediate dehydration. Fluids should be given sparingly, if at all, because of the danger of vomiting and because of the likelihood of temporary urinary retention due to paralysis of bladder muscles.

c. If the body temperature is elevated and mucous membranes are dry, immediate and vigorous cooling (as for heatstroke) is indicated. Methods that can be used to cool the skin are spraying with cool water and air circulation (fanning); applying alcohol soaked cloths and air circulation; and providing maximum exposure to air in a shaded area, along with maximum air circulation. Such cases are usually a result of anticholinergic poisoning. Rapid evacuation should be accomplished since medical treatment with the appropriate medication may be lifesaving.

CAUTION

DO NOT use **ice** for cooling the skin.

d. Reassurance and a firm, but friendly, attitude by individuals providing first aid will be beneficial if the casualty appears to comprehend what is being said. Conversation is a waste of time if the service member is incoherent or cannot understand what is being said. In such cases, the less said, the better it is—these casualties will benefit more from prompt and vigorous restraint and evacuation to an MTF.

7-13. Incendiaries

Incendiaries can be grouped as WP, thickened gasoline, metal, and oil and metal. You must learn to protect yourself against these incendiaries.

a. White phosphorus is used primarily as a smoke producer but can be used for its incendiary effect to ignite field expedients and combustible materials. The burns from WP are usually multiple, deep, and variable in size. When particles of WP get on the skin or clothing, they continue to burn until deprived of air. They also have a tendency to stick to a surface and must be brushed off or picked out.

(1) If burning particles of WP strike and stick to your clothing, quickly take off the contaminated clothing before the WP burns through to the skin.

(2) If burning WP strikes your skin, smother the flame with water, a wet cloth, or mud.

NOTE

Since WP is soluble in oil, **DO NOT** use grease, oily ointments, or eye ointments to smother the flame.

(3) Keep the WP particles covered with a wet material to exclude air until you can remove them or have them removed from your skin.

(4) Remove the WP particles from the skin by brushing them with a wet cloth and by picking them out with a knife, bayonet, stick, or other available object.

(5) Seek medical assistance when the mission permits.

b. Thickened fuel mixtures (napalm) have a tendency to cling to clothing and body surfaces, thereby producing prolonged exposure and severe burns. The first aid for these burns is the same as for other heat burns. The heat and irritating gases given off by these combustible mixtures may cause lung damage, which must be treated by medical personnel.

c. Metal incendiaries pose special problems. Thermite particles on the skin should be immediately cooled with water and then removed. The first aid for these burns is the same as for other heat burns. Particles of magnesium on the skin burn quickly and deeply. Like other metal incendiaries, they must be removed. Ordinarily, medical personnel should do the complete removal of these particles as soon as possible. Immediate medical treatment is required.

d. Oil and metal incendiaries have much the same effect on contact with the skin and clothing as those discussed (*b* and *c* above). First aid measures for burns are discussed in Chapter 3.

7-14. Biological Agents and First Aid

a. Biological attacks can result in combat ineffectiveness by introducing disease-causing organisms into a troop population.

b. Once a disease is identified, first aid or medical treatment is initiated, depending on the seriousness of the disease. First aid measures are concerned with observable symptoms of the disease such as diarrhea or vomiting.

7-15. Toxins

Toxins are alleged to have been used in past conflicts. Witnesses and victims have described the agent as toxic rain (or yellow rain) because it was reported to have been released from aircraft as a yellow powder or liquid that covered ground, structures, vegetation, and people.

a. Signs and Symptoms. The occurrence of the symptoms from toxins may appear in a period of a few minutes to several hours depending on the particular toxin, the service member's susceptibility, and the amount of toxin inhaled, ingested, or deposited on the skin. Symptoms from toxins usually involve the central nervous system but are often preceded by less prominent symptoms, such as nausea, vomiting, diarrhea, cramps, or stomach irritation and burning sensation. Typical neurological symptoms often develop rapidly in severe cases; for example, visual disturbances, inability to swallow, speech difficulty, lack of muscle coordination, and sensory abnormalities (numbness of mouth, throat, or extremities). Yellow rain (mycotoxins) also may have hemorrhagic symptoms, which could include any or all of the following:

- Dizziness.
- Severe itching or tingling of the skin.
- Formation of multiple, small, hard blisters.
- Coughing up blood.
- Shock (which could result in death).

b. *Self-Aid.* Upon recognition of an attack employing toxins, you must immediately take the following actions:

(1) Stop breathing, put on your protective mask with hood, and then resume breathing. Next, put on your protective clothing.

(2) Should severe itching of the face become unbearable, quickly—

- Loosen the cap on your canteen.
- Take and hold a deep breath and lift your mask.
- While holding your breath, close your eyes and flush your face with generous amounts of water.

CAUTION

DO NOT rub or scratch your eyes. Try not to let the water run onto your clothing or protective overgarment.

- Put your protective mask back on, seat it properly, clear it, and check it for a seal; then resume breathing.
- Decontaminate your skin by bathing with soap and water as soon as the mission permits.

- Change clothing and decontaminate your protective mask using soap and water. Replace the filters if directed.

(3) If vomiting occurs, the mask should be lifted momentarily and drained—while the eyes are closed and the breath is held—and replaced, cleared, and sealed.

c. *Medical Assistance.* If you suspect that you have been exposed to toxins, you should seek medical assistance immediately.

7-16. Nuclear Detonation

a. Three types of injuries may result from a nuclear detonation. These are thermal, blast, and radiation injuries. Many times the casualty will have a combination of these types of injuries. First aid for thermal and blast injuries is provided based on observable injuries, such as burns, hemorrhage, or fractures.

b. The signs and symptoms of radiation illness in the initial phase include the rapid onset of nausea, vomiting, and malaise (tiredness). The only first aid procedure for radiological casualties is decontamination.

CHAPTER 8

FIRST AID FOR PSYCHOLOGICAL REACTIONS

8-1. General

Psychological first aid is as natural and reasonable as physical first aid and is just as familiar. When you were hurt as a child, the understanding attitude of your parents did as much as the psychological effect of a bandage. Later, your disappointment or grief was eased by supportive words from a friend. Certainly, taking a walk and talking things out with a friend are familiar ways of dealing with an emotional crisis. The same natural feelings that make us want to help a person who is injured make us want to give a helping hand to a buddy who is upset. *Psychological first aid* really means nothing more complicated than assisting people with emotional distress whether it results from physical injury, disease, or excessive stress. Emotional distress is not always as visible as a wound or a broken bone. However, overexcitement, severe fear, excessive worry, deep depression, misdirected irritability, and anger are signs that stress has reached the point of interfering with effective coping. The more noticeable the symptoms become, the more urgent the need for you to be of help and the more important it is for you to know *how* to help.

8-2. Importance of Psychological First Aid

You must know how to give psychological first aid to be able to help yourself, your buddies, and your unit in order to keep performing the mission. Psychological first aid measures are simple and easy to understand. Your decision of what to do depends upon your ability to observe the service member and understand his needs. Making the best use of resources requires ingenuity on your part. A stress reaction resulting in poor judgment can cause injury or even death to yourself or others on the battlefield. It can be even more dangerous if other persons are affected by the judgment of an emotionally upset service member. If it is detected early enough, the affected service member stands a good chance of remaining in his unit as an effective member. If it is not detected early and if the service member becomes more emotionally upset, he may become a threat to himself and to others.

8-3. Situations Requiring Psychological First Aid

- Psychological first aid (buddy aid) is most needed at the first sign that a service member cannot perform the mission because of emotional distress. Stress is inevitable in combat, in hostage and terrorist situations, and in civilian disasters such as floods, hurricanes, or industrial accidents. Most emotional reactions to such situations are temporary, and the service member can still carry on with encouragement. Painful or disruptive symptoms may last for minutes, hours, or days. However, if the stress symptoms are seriously disabling, they may be psychologically contagious and endanger not only the emotionally upset service member but also the entire unit.

- Sometimes people continue to function well during a disastrous event, but suffer from emotional scars which impair their job performance or quality of life at a later time. Painful memories and dreams may recur for months and years and still be considered a normal reaction. However, if the memories are so painful that the person must avoid all situations which arouse them, becomes socially withdrawn, or shows symptoms of anxiety, depression, or substance abuse, he needs treatment. Experience with police, firemen, emergency medical technicians, and others who deal with disasters has proved that the routine application of psychological first aid to all the participants, including those who have functioned well, greatly reduces the likelihood of future serious post-traumatic stress disorders (PTSDs).

8-4. Interrelationship of Psychological and Physical First Aid

Psychological first aid should go hand in hand with physical first aid. The discovery of a physical injury or cause for an inability to function does not rule out the possibility of a psychological injury (or vice versa). The person suffering from pain, shock, fear of serious injury, or fear of death does not respond well to joking, indifference, or fearful-tearful attention. Fear and anxiety may take as high a toll of the service member's strength as does the loss of blood.

8-5. Goals of Psychological First Aid

The goals of psychological first aid are to—

- Be supportive; assist the service member in dealing with his stress reaction.

- Prevent, and if necessary control, behavior harmful to himself and to others.

- Return the service member to duty as soon as possible after dealing with the stress reaction.

8-6. Respect for Others' Feelings

a. Accept the service member you are trying to help without censorship or ridicule. Respect his right to his own feelings. Even though your feelings, beliefs, and behavior are different, DO NOT blame or make light of him for the way he feels or acts. Your purpose is to help him in this tough situation, not to be his critic. A person DOES NOT WANT to be upset and worried. When he seeks help, he needs and expects consideration of his fears, not abrupt dismissal or ridicule.

b. Realize that people are the products of a wide variety of factors. All people DO NOT react the same way to the same situations. Each individual has complex needs and motivations, both conscious and unconscious, that are uniquely his own. Often the one thing that finally causes the person to become overloaded by a stressful situation is not the stressor itself, but some other problem.

8-7. Emotional and Physical Disability

a. Accept emotional disability as being just as real as physical disability. If a service member's ankle is seriously sprained in a fall, no one expects him to run right away. A service member's emotions may be temporarily strained by the overwhelming stress of battle or other traumatic incident. DO NOT demand that he pull himself together immediately and carry on without a break. Some individuals can pull themselves together immediately, but others cannot. The service member whose emotional stability has been disrupted has a disability just as real as the service member who has sprained his ankle. There is an unfortunate tendency in many people to regard as real only what they can see, such as a wound or bleeding. Some people tend to assume that damage involving a person's mind and emotions is just imagined, that he is not really sick or injured, and that he could overcome his trouble by using his will power.

b. The terms *it's all in your head, snap out of it,* and *get control of yourself* are often used by people who believe they are being helpful. Actually, these terms are expressions of hostility because they show lack of understanding. They only emphasize weakness and inadequacy. Such terms are of no use in psychological first aid.

c. Every physically injured person has some emotional reaction to the fact that he is injured.

(1) It is normal for an injured person to feel upset. The more severe the injury, the more insecure and fearful he becomes, especially

if the injury is to a body part which is highly valued. For example, an injury to the eyes or the genitals, even though relatively minor, is likely to be extremely upsetting. An injury to some other part of the body may be especially disturbing to an individual for his own particular reason. For example, an injury of the hand may be a terrifying blow to a surgeon or an injury to the eye of a pilot.

(2) An injured service member always feels less secure, more anxious, and more afraid not only because of what has happened to him but because of what he imagines may happen as a result of his injury. This fear and insecurity may cause him to be irritable, uncooperative, or unreasonable. As you help him, always keep in mind that such behavior has little or nothing to do with you personally. He needs your patience, reassurance, encouragement, and support.

8-8. Combat and Other Operational Stress Reactions

Stress reaction is a temporary emotional disorder or inability to function, experienced by a previously normal service member as a reaction to the overwhelming or cumulative stress of combat. Stress reaction gets better with reassurance, rest, physical replenishment, and activities that restore confidence. All service members are likely to feel stress reaction under conditions of intense and/or prolonged stress. They may even become stress reaction casualties, unable to perform their mission for hours or days. Other combat and operational stress reactions (COSRs) may result in negative behavior, but are not termed *stress reaction,* as they need more intensive treatment. These negative COSRs may result in misconduct stress behaviors such as drug and alcohol abuse, criminal acts, looting, desertion, and self-inflicted wounds. These harmful COSRs can often be prevented by good psychological first aid. Service members who commit misconduct stress behaviors may require disciplinary action rather than medical treatment.

8-9. Reactions to Stress

Most service members react to stressful incidents after the situation has passed. All service members feel some fear. This fear may be greater than they have experienced at any other time, or they may be more aware of their fear. In such a situation, they should not be surprised if they feel shaky or become sweaty, nauseated, or confused. These reactions are normal and are not a cause for concern. However, some reactions, either short- or long-term, will cause problems if left unchecked. See paragraph 8-13 for more information.

a. Emotional Reactions.

(1) The most obvious combat stress reaction (CSR) is inefficient performance. This can be demonstrated by—

- Slow thinking (or reaction time).
- Difficulty recognizing priorities and seeing what needs to be done.
- Difficulty getting started.
- Indecisiveness and having trouble focusing attention.
- Tendency to do familiar tasks and be preoccupied with familiar details. (This can reach the point where the person is very passive, such as just sitting or wandering about not knowing what to do.)

(2) A less common reaction may be uncontrolled emotional outbursts; this can be demonstrated by crying, screaming, or laughing. Some service members will react in the opposite way. They will be very withdrawn and silent and try to isolate themselves from everyone. These service members should be encouraged to remain with their assigned unit. Uncontrolled reactions may appear by themselves or in any combination (the person may be crying uncontrollably one minute and then laughing the next). In this state, the person is restless and cannot keep still. He may run about, apparently without purpose. Inside, he feels a great rage or fear and his physical acts may show this. In his anger he may indiscriminately strike out at others.

b. Loss of Adaptability.

(1) In a desperate attempt to get away from the danger, which has overwhelmed him, a service member may panic and become confused. His mental ability may be so impaired he cannot think clearly or even follow simple commands. His judgment may be faulty and he may not be aware of his actions, such as standing up in his fighting position during an attack.

(2) In other cases, overwhelming stress may produce symptoms that are often associated with head injuries. For example, the service member may appear dazed or be found wandering around aimlessly. He may appear confused and disoriented and may seem to have a complete or partial loss of memory. In such cases, especially when no eyewitnesses can provide evidence that the service member has NOT suffered a head injury, it is necessary for him to be rapidly medically evacuated. **DO NOT** allow the

service member to expose himself to further personal danger until the cause of the problem has been determined.

c. *Sleep Disturbance and Repetition of Dreams.* A person who has been overwhelmed by stress often has difficulty sleeping. The service member may experience nightmares related to the stressors. Remember that nightmares, in themselves, are not considered abnormal when they occur soon after a period of intensive stress. As time passes, the nightmares usually become less frequent and less intense. In extreme cases, a service member, even when awake, may think repeatedly of the incident, feel as though it is happening again, and act out parts of his stress over and over again. For some persons, this repetitious reexperiencing of the stressful event may be necessary for eventual recovery; therefore, it should not be discouraged or viewed as abnormal. For the person reexperiencing the event, such reaction may be disruptive. The service member needs to be encouraged to *ventilate* about the incident. Ventilation is a technique where the service member is given the opportunity to talk extensively, often repetitiously about the experience.

8-10. Severe Stress or Stress Reaction

You do not need specialized training to recognize severe stress or stress reaction that will cause problems for the service member, the unit, or the mission. Reactions that are less severe, however, are more difficult to detect. To determine whether a person needs help, you must observe him to see whether he is doing something meaningful, performing his duties, taking care of himself, behaving in an unusual fashion, or acting out of character.

8-11. Application of Psychological First Aid

The emotionally disturbed service member has built a barrier against fear. He does this for his own protection, although he is probably not aware that he is doing it. If he finds that he does not have to be afraid and that there are normal, understandable things about him, he will feel safer in dropping this barrier. Persistent efforts to make him realize that you want to understand him will be reassuring, especially if you remain calm. Nothing can cause an emotionally disturbed person to become even more fearful than feeling that others are afraid of him. Try to remain calm. Familiar things, such as a cup of coffee, the use of his name, attention to a minor wound, being given a simple job to do, or the sight of familiar people and activities, will add to his ability to overcome his fear. He may not respond well if you get excited, angry, or abrupt.

a. Ventilation. After the service member becomes calmer, he is likely to have dreams about the stressful event. He also may think about it when he is awake or even repeat his personal reaction to the event. One benefit of this natural pattern is that it helps him master the stress by going over it just as one masters the initial fear of parachuting from an aircraft by doing it over and over again. Eventually, it is difficult to remember how frightening the event was initially. In giving first aid to the emotionally disturbed service member, you should let him follow this natural pattern. Encourage him to talk. Be a good listener. Let him tell, in his own words, what actually happened. If home front problems or worries have contributed to the stress, it will help him to talk about them. Your patient listening will prove to him that you are interested in him, and by describing his personal problem, he can work at mastering his fear. If he becomes overwhelmed in the telling, suggest a cup of coffee or a break. Whatever you do, assure him that you will listen again as soon as he is ready. Do try to help put the service member's perception of what happened back into realistic perspective; but DO NOT argue about it.

b. Activity.

(1) A person who is emotionally disturbed as the result of a combat action is a casualty of anxiety and fear. He is disabled because he has become temporarily overwhelmed by his anxiety. A good way to control fear is through activity. Almost all service members, for example, experience a considerable sense of anxiety and fear while they are poised, awaiting the opening of a big offensive; but this is normally relieved, and they actually feel better once they begin to move into action. They take pride in effective performance and pleasure in knowing that they are good service members, perhaps being completely unaware that overcoming their initial fear was their first major accomplishment.

(2) Useful activity is very beneficial to the emotionally disturbed service member who is not physically incapacitated. After you help a service member get over his initial fear, help him to regain some self-confidence. Make him realize his job is continuing by finding him something useful to do. Encourage him to be active. Get him to help load trucks, clean up debris, or dig fighting positions. If possible, get him back to his usual duty. Seek out his strong points and help him apply them. Avoid having him just sit around. You may have to provide direction by telling him what to do and where to do it. The instructions should be clear and simple and should be repeated. A person who has panicked is likely to argue. Respect his feelings, but point out more immediate, obtainable, and demanding needs. Channel his excessive energy and, above all, DO NOT argue. If you cannot get him interested in doing more profitable work, it may be necessary to enlist aid in controlling his overactivity before it spreads to the group and

results in more panic. Prevent the spread of such infectious feelings by restraining and segregating if necessary.

(3) Involvement in activity helps a service member in three ways; he—

- Forgets himself.
- Has an outlet for his excessive tensions.
- Proves to himself he can do something useful.

c. Rest. There are times, particularly in combat, when physical exhaustion is a principal cause for emotional reactions. A unit sleep plan should be established and implemented. When possible, service members should be given a safe and relatively comfortable area in which to sleep. Examples would be an area away from heavy traffic, noise, and congestion or a place that is clean and dry and protected from environmental conditions. The more uninterrupted sleep a service member gets the better he will be able to function in the tactical environment.

d. Hygiene. Field hygiene is an important ingredient in a service member's morale. A service member who is dirty and unkempt will not function as well as a service member who has had the opportunity to bathe and put on clean, dry clothing. During combat, unit leaders should stress the importance of personal hygiene. Good personal hygiene not only improves morale, it also is a preventive measure against disease and nonbattle injury (DNBI).

e. Group Activity. You have probably already noticed that a person works, faces danger, and handles serious problems better if he is a member of a closely-knit group. Each service member in the team supports the other team members. Esprit de corps is built because the service members have the same interests, goals, and mission, and as a result they are more productive; furthermore, they are less worried because everyone is involved. It is this spirit that takes a strategic hill in battle. It is so powerful that it is one of the most effective tools you have in your *psychological first aid bag*. Getting the service member back into the team or squad activities will reestablish his sense of belonging and security and will go far toward making him a useful member of the unit.

8-12. Reactions and Limitations

Up to this point the discussion has been primarily about the feelings of the emotionally distressed service member. What about your feelings toward him? Whatever the situation, you will have emotional reactions (conscious or unconscious) toward this service member. Your reactions can either help or hinder your ability to help him. When you are tired or worried, you may very easily become impatient with him if he is unusually slow or exaggerates. You may even feel resentful toward him. At times when many physically wounded lie about you, it will be especially natural for you to resent disabilities that you cannot see. Physical wounds can be seen and easily accepted. Emotional reactions are more difficult to accept as injuries. On the other hand, will you tend to be overly sympathetic? Excessive sympathy for an incapacitated person can be as harmful as negative feelings in your relationship with him. He needs strong help, but not your sorrow. To overwhelm him with pity will make him feel even more inadequate. You must expect your buddy to recover, to be able to return to duty, and to become a useful service member again. This expectation should be displayed in your behavior and attitude as well as in what you say. If he can see your calmness, confidence, and competence, he will be reassured and will feel a sense of greater security.

8-13. Stress Reactions

See Tables 8-1, 8-2, and 8-3 for more information.

Table 8-1. Mild Stress Reaction

PHYSICAL SIGNS*	EMOTIONAL SIGNS*
1. TREMBLING, TEARFUL 2. JUMPINESS, NERVOUSNESS 3. COLD SWEAT, DRY MOUTH 4. POUNDING HEART, DIZZINESS 5. INSOMNIA, NIGHTMARES 6. NAUSEA, VOMITING, DIARRHEA 7. FATIGUE 8. THOUSAND-YARD STARE 9. DIFFICULTY THINKING, SPEAKING, AND COMMUNICATING	1. ANXIETY, INDECISIVENESS 2. IRRITABLE, COMPLAINING 3. FORGETFUL, UNABLE TO CONCENTRATE 4. EASILY STARTLED BY NOISE, MOVEMENT 5. GRIEF, TEARFUL 6. ANGER, BEGINNING TO LOSE CONFIDENCE IN SELF AND UNIT
SELF- AND BUDDY AID	
1. CONTINUE MISSION PERFORMANCE, FOCUS ON IMMEDIATE MISSION. 2. EXPECT SERVICE MEMBER TO PERFORM ASSIGNED DUTIES. 3. REMAIN CALM AT ALL TIMES; BE DIRECTIVE AND IN CONTROL. 4. LET SERVICE MEMBER KNOW HIS REACTION IS NORMAL, AND THAT THERE IS NOTHING SERIOUSLY WRONG WITH HIM. 5. KEEP SERVICE MEMBER INFORMED OF THE SITUATION, OBJECTIVES, EXPECTATIONS, AND SUPPORT. CONTROL RUMORS. 6. BUILD SERVICE MEMBER'S CONFIDENCE, TALK ABOUT SUCCEEDING. 7. KEEP SERVICE MEMBER PRODUCTIVE (WHEN NOT RESTING) THROUGH RECREATIONAL ACTIVITIES, EQUIPMENT MAINTENANCE.	

8. ENSURE SERVICE MEMBER MAINTAINS GOOD PERSONAL HYGIENE.
9. ENSURE SERVICE MEMBER EATS, DRINKS, AND SLEEPS AS SOON AS POSSIBLE.
10. LET SERVICE MEMBER TALK ABOUT HIS FEELINGS. DO NOT "PUT DOWN" HIS FEELINGS OF GRIEF OR WORRY. GIVE PRACTICAL ADVICE AND PUT EMOTIONS INTO PERSPECTIVE.

* MOST OR ALL OF THESE SIGNS ARE PRESENT IN MILD STRESS REACTION. THEY CAN BE PRESENT IN ANY NORMAL SERVICE MEMBER IN COMBAT YET HE CAN STILL DO HIS JOB.

Table 8-2. More Serious Stress Reaction

PHYSICAL SIGNS*	EMOTIONAL SIGNS*
1. CONSTANTLY MOVES AROUND 2. FLINCHING OR DUCKING AT SUDDEN SOUNDS 3. SHAKING, TREMBLING (WHOLE BODY OR ARMS) 4. CANNOT USE PART OF BODY, NO PHYSICAL REASON (HAND, ARM, LEGS) 5. CANNOT SEE, HEAR, OR FEEL (PARTIAL OR COMPLETE LOSS) 6. PHYSICAL EXHAUSTION, CRYING 7. FREEZING UNDER FIRE, OR TOTAL IMMOBILITY 8. VACANT STARES, STAGGERS, SWAYS WHEN STANDS 9. PANIC RUNNING UNDER FIRE	1. RAPID AND/OR INAPPROPRIATE TALKING 2. ARGUMENTATIVE, RECKLESS MOVEMENTS/ACTIONS 3. INATTENTIVE TO PERSONAL HYGIENE 4. INDIFFERENT TO DANGER 5. MEMORY LOSS 6. SEVERE STUTTERING, MUMBLING, OR CANNOT SPEAK AT ALL 7. INSOMNIA, NIGHTMARES 8. SEEING OR HEARING THINGS THAT DO NOT EXIST 9. RAPID EMOTIONAL SHIFTS 10. SOCIAL WITHDRAWAL 11. APATHETIC 12. HYSTERICAL OUTBURSTS 13. FRANTIC OR STRANGE BEHAVIOR

TREATMENT PROCEDURES**

1. IF A SERVICE MEMBER'S BEHAVIOR ENDANGERS THE MISSION, SELF, OR OTHERS, DO WHATEVER IS NECESSARY TO CONTROL HIM.
2. IF THE SERVICE MEMBER IS UPSET, CALMLY TALK HIM INTO COOPERATING.
3. IF CONCERNED ABOUT THE SERVICE MEMBER'S RELIABILITY:
   - UNLOAD HIS WEAPON.
   - TAKE WEAPON IF SERIOUSLY CONCERNED.
   - PHYSICALLY RESTRAIN HIM ONLY WHEN NECESSARY FOR SAFETY OR TRANSPORTATION.
4. REASSURE EVERYONE THAT THE SIGNS ARE PROBABLY JUST STRESS REACTION AND WILL QUICKLY IMPROVE.
5. IF STRESS REACTION SIGNS CONTINUE:
   - GET THE SERVICE MEMBER TO A SAFER PLACE.
   - DO NOT LEAVE THE SERVICE MEMBER ALONE, KEEP SOMEONE HE KNOWS WITH HIM.
   - NOTIFY SENIOR NONCOMMISSIONED OFFICER (NCO) OR OFFICER.
   - HAVE THE SERVICE MEMBER EXAMINED BY MEDICAL PERSONNEL.

Table 8-2. More Serious Stress Reaction (Continued)

TREATMENT PROCEDURES**

6. GIVE THE SERVICE MEMBER EASY TASKS TO DO WHEN NOT SLEEPING, EATING, OR RESTING.
7. ASSURE THE SERVICE MEMBER HE WILL RETURN TO FULL DUTY IN 24 HOURS; AND, RETURN HIM TO NORMAL DUTIES AS SOON AS HE IS READY.

* THESE SIGNS ARE PRESENT IN ADDITION TO THE SIGNS OF MILD STRESS REACTION.

** DO THESE PROCEDURES IN ADDITION TO THE SELF- AND BUDDY AID CARE.

Table 8-3. Preventive Measures to Combat Stress Reaction

1. WELCOME NEW MEMBERS INTO YOUR TEAM, GET TO KNOW THEM QUICKLY. IF YOU ARE NEW, BE ACTIVE IN MAKING FRIENDS.
2. BE PHYSICALLY FIT (STRENGTH, ENDURANCE, AND AGILITY).
3. KNOW AND PRACTICE LIFESAVING SELF- AND BUDDY AID.
4. PRACTICE RAPID RELAXATION TECHNIQUES (FM 22-51).
5. HELP EACH OTHER OUT WHEN THINGS ARE TOUGH AT HOME OR IN THE UNIT.
6. KEEP INFORMED; ASK YOUR LEADER QUESTIONS, IGNORE RUMORS.
7. WORK TOGETHER TO GIVE EVERYONE FOOD, WATER, SHELTER, HYGIENE, AND SANITATION.
8. SLEEP WHEN MISSION AND SAFETY PERMIT; LET EVERYONE GET TIME TO SLEEP.
   - SLEEP ONLY IN SAFE PLACES AND BY STANDING OPERATING PROCEDURE (SOP).
   - IF POSSIBLE, SLEEP 6 TO 9 HOURS PER DAY.
   - TRY TO GET AT LEAST 4 HOURS SLEEP PER DAY.
   - GET GOOD SLEEP BEFORE GOING ON SUSTAINED OPERATIONS.
   - CATNAP WHEN YOU CAN, BUT ALLOW TIME TO WAKE UP FULLY.
   - CATCH UP ON SLEEP AFTER GOING WITHOUT.

APPENDIX A

FIRST AID CASE AND KITS, DRESSINGS, AND BANDAGES

A-1. First Aid Case with Field Dressings and Bandages

Every service member is issued a first aid case (Figure A-1A) with a field first aid dressing encased in a plastic wrapper (Figure A-1B). He carries it at all times for his use. The field first aid dressing is a standard sterile (germ-free) compress or pad with bandages attached (Figure A-1C). This dressing is used to cover the wound, to protect against further contamination, and to stop bleeding (pressure dressing). When a service member administers first aid to another person, he must remember to use the wounded person's dressing; he may need his own later. The service member must check his first aid case regularly and replace any used or missing dressing. The field first aid dressing may normally be obtained from his unit supply.

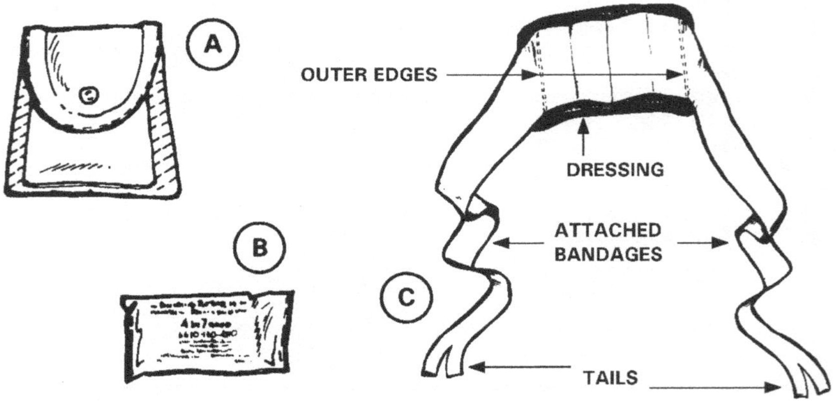

Figure A-1. Field first aid case and dressing (Illustrated A—C).

A-2. General Purpose First Aid Kits

General-purpose first aid kits are listed in the common table of allowances (CTA) 8-100. The operators, crew, and passengers carry these kits on Army vehicles, aircraft, and boats for use. Individuals designated by unit TSOP to be responsible for these kits are required to check them regularly and replace all items used. The general-purpose kit and its contents can be obtained through the unit supply system.

NOTE

Periodically check the dressings (for holes or tears in the packaging) and the medicines (for expiration date) that are in the first aid kits. If necessary, replace defective or outdated items.

A-3. Dressings

Dressings are sterile pads or compresses used to cover wounds. They usually are made of gauze or cotton wrapped in gauze (Figure A-1C). In addition to the standard field first aid dressing, other dressings such as sterile gauze compresses and small sterile compresses on adhesive strips may be available under CTA 8-100.

A-4. Standard Bandages

a. Standard bandages are made of gauze or muslin and are used over a sterile dressing to secure the dressing in place, to close off its edge from dirt and germs, and to create pressure on the wound and control bleeding. A bandage can also support an injured part or secure a splint.

b. Tailed bandages may be attached to the dressing as indicated on the field first aid dressing (Figure A-1C).

A-5. Triangular and Cravat (Swathe) Bandages

a. Triangular and cravat (or swathe) bandages (Figure A-2) are fashioned from a triangular piece of muslin (37 by 37 by 52 inches) provided in the general-purpose first aid kit. If it is folded into a strip, it is called a cravat. Two safety pins are packaged with each bandage. These bandages are valuable in an emergency since they are easily applied.

b. To improvise a triangular bandage, cut a square of available material, slightly larger than 3 feet by 3 feet, and *fold it diagonally*. If two bandages are needed, cut the material along the diagonal fold.

c. A cravat can be improvised from such common items as T-shirts, other shirts, bed linens, trouser legs, scarfs, or any other item made of pliable and durable material that can be folded, torn, or cut to the desired size.

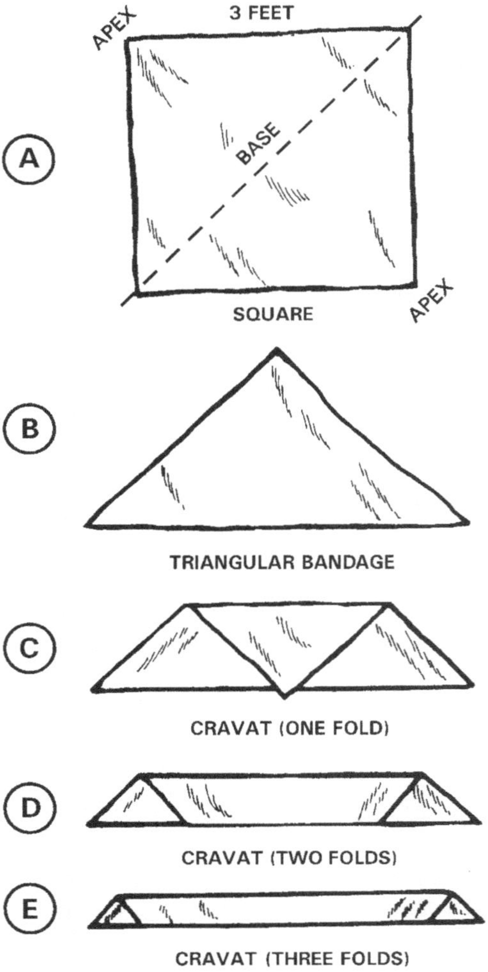

Figure A-2. Triangular and cravat bandages (Illustrated A—E).

APPENDIX B

RESCUE AND TRANSPORTATION PROCEDURES

B-1. General

A basic principle of first aid is to evaluate the casualty's injuries and administer first aid before moving him. However, adverse situations or conditions may jeopardize the lives of both the rescuer and the casualty if this is done. It may be necessary *first to rescue* the casualty before first aid can be effectively or safely given. The life and/or the well-being of the casualty will depend as much upon the manner in which he is *rescued and transported,* as it will upon the first aid and medical treatment he receives. Rescue actions must be done quickly and safely. Careless or rough handling of the casualty during rescue operations can aggravate his injuries.

B-2. Principles of Rescue Operations

a. When faced with the necessity of rescuing a casualty who is threatened by hostile action, fire, water, or any other immediate hazard, DO NOT take action without first determining the extent of the hazard and your ability to handle the situation. DO NOT become a casualty.

b. The rescuer must evaluate the situation and analyze the factors involved. This evaluation involves three major steps:

- Identify the task.
- Evaluate circumstances of the rescue.
- Plan the action.

B-3. Considerations

a. First determine if a rescue attempt is actually needed. It is a waste of time, equipment, and personnel to rescue someone not in need of rescuing. It is also a waste to look for someone who is not lost or needlessly risk the lives of the rescuer(s). In planning a rescue, attempt to obtain the following information:

- Who, what, where, when, why, and how the situation happened?

- How many casualties are involved and the nature of their injuries?

- What is the tactical situation?

- What are the terrain features and the location of the casualties?

- Will there be adequate assistance available to aid in the rescue/evacuation?

- Can first aid and/or medical treatment be provided at the scene; will the casualties require movement to a safer location?

- What specialized equipment will be required for the rescue operation?

- Is the rescue area contaminated? Will decontamination equipment and materiel be required for casualties, rescue personnel, and rescue equipment?

- How much time is available?

b. The time element can play a significant role in how the rescue is attempted. If the casualties are in imminent danger of losing their lives (such as near a burning vehicle or in a burning building) the time available will be relatively short and will sometimes cause a rescuer to compromise planning stages and/or the first aid which can be given. However, if the casualty is in a relatively secure area and his physical condition is strong, more deliberate planning can take place. A realistic estimate of time available must be made as quickly as possible to determine action time remaining. The key elements are the casualty's physical and mental condition, the tactical situation, and the environment.

B-4. Plan of Action

a. The casualty's ability to endure is of primary importance in estimating the time available. Age, physical condition, and extent of wounds and/or injuries will differ from casualty to casualty. Therefore, to determine the time available, you will have to consider—

- Endurance time of the casualty.

- Extent of injuries.

- Type of situation.
- Personnel and/or equipment availability.
- Weather.
- Terrain (natural and man-made).
- Environment (contaminated or uncontaminated).

b. In respect to terrain, you must consider altitude and visibility. In some cases, the casualty may be of assistance because he knows more about the particular terrain or situation than you do. Maximum use of secure/reliable trails or roads is essential.

c. When taking weather into account, ensure that blankets and/or rain gear are available. Even a mild rain can complicate a normally simple rescue. In high altitudes and/or extreme cold and gusting winds, the time available is critically shortened. Be prepared to provide shelter and warmth for the casualty as well as the rescuers.

B-5. Proper Handling of Casualties

a. You may have saved the casualty's life through the application of appropriate first aid measures. However, his life can be lost through rough handling or careless transportation procedures. Before you attempt to move the casualty—

- Evaluate the type and extent of his injuries.
- Ensure that dressings over wounds are adequately reinforced.
- Ensure that fractured bones are properly immobilized and supported to prevent them from cutting through muscle, blood vessels, and skin.

b. Based upon your evaluation of the type and extent of the casualty's injury and your knowledge of the various manual carries, you must select the best possible method of manual transportation. If the casualty is conscious, tell him how he is to be transported. This will help allay his fear of movement and gain his cooperation and confidence.

c. Buddy aid for chemical agent casualties includes those actions required to prevent an incapacitated casualty from receiving additional injury from the effects of chemical hazards. If a casualty is physically unable to decontaminate himself or administer the proper chemical agent antidote, the casualty's buddy assists him and assumes responsibility for his care. Buddy-aid includes—

- Administering the proper chemical agent antidote.
- Decontaminating the incapacitated casualty's exposed skin.
- Ensuring that his protective ensemble remains correctly emplaced.
- Maintaining respiration.
- Controlling bleeding.
- Providing other standard first aid measures
- Transporting the casualty out of the contaminated area.

B-6. Positioning the Casualty

The first step in any manual carry is to position the casualty to be lifted. If he is conscious, he should be told how he is to be positioned and transported. This helps lessen his fear of movement and to gain his cooperation. It may be necessary to roll the casualty onto his abdomen, or his back, depending upon the position in which he is lying and the particular carry to be used.

a. To roll a casualty onto his abdomen, kneel at the casualty's uninjured side.

(1) Place his arms above his head; cross his ankle which is farther from you over the one that is closer to you.

(2) Place your hands on the shoulder which is farther from you; place your other hand in the area of his hip or thigh (Figure B-1).

(3) Roll him gently toward you onto his abdomen (Figure B-2).

b. To roll a casualty onto his back, follow the same procedure described in *a* above, except gently roll the casualty onto his back, rather than onto his abdomen.

Figure B-1. Positioning the casualty.

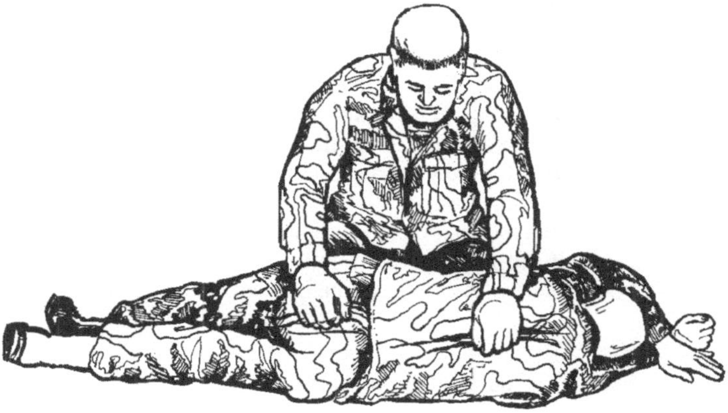

Figure B-2. Rolling casualty onto his abdomen.

B-7. Medical Evacuation and Transportation of Casualties

a. Medical evacuation of the sick and wounded (with en route medical care) is the responsibility of medical personnel who have been provided special training and equipment. Therefore, unless a good reason for you to transport a casualty arises, wait for some means of medical evacuation to be provided. When the situation is urgent and you are unable to obtain medical assistance or know that no medical evacuation assets are available, you will have to transport the casualty. For this reason, you must know how to transport him without increasing the seriousness of his condition.

b. Transporting a casualty by litter (FM 8-10-6) is safer and more comfortable for him than by manual means; it is also easier for you.

Manual transportation, however, may be the only feasible method because of the terrain or the combat situation; or it may be necessary to save a life. In these situations, the casualty should be transferred to a litter as soon as one can be made available or improvised.

B-8. Manual Carries

Casualties carried by manual means must be carefully and correctly handled, otherwise their injuries may become more serious or possibly fatal. Situation permitting, transport of a casualty should be organized and unhurried. Each movement should be performed as deliberately and gently as possible. Casualties should not be moved before the type and extent of injuries are evaluated and the required first aid is administered. The exception to this occurs when the situation dictates immediate movement for safety purposes (for example, it may be necessary to remove a casualty from a burning vehicle); that is, the situation dictates that the urgency of casualty movement outweighs the need to administer first aid. Manual carries are tiring for the bearers and involve the risk of increasing the severity of the casualty's injury. In some instances, however, they are essential to save the casualty's life. Although manual carries are accomplished by one or two bearers, the two-man carries are used whenever possible. They provide more comfort to the casualty, are less likely to aggravate his injuries, and are also less tiring for the bearers. The distance a casualty can be carried depends on many factors, such as—

- Nature of the casualty's injuries.
- Strength and endurance of the bearer(s).
- Weight of the casualty.
- Obstacles encountered during transport (natural or manmade).
- Type of terrain.

a. One-man Carries. These carries should be used when only one bearer is available to transport the casualty.

(1) The *fireman's carry* (Figure B-3) is one of the easiest ways for one individual to carry another. After an unconscious or disabled casualty has been properly positioned, he is raised from the ground, then supported and placed in the carrying position.

(*a*) After rolling the casualty onto his abdomen, straddle him. Extend your hands under his chest and lock them together.

(*b*) Lift the casualty to his knees as you move backward.

(*c*) Continue to move backward, thus straightening the casualty's legs and locking his knees.

(*d*) Walk forward, bringing the casualty to a standing position; tilt him slightly backward to prevent his knees from buckling.

(*e*) As you maintain constant support of the casualty with one arm, free your other arm, quickly grasp his wrist, and raise his arm high. Instantly pass your head under his raised arm, releasing it as you pass under it.

(*f*) Move swiftly to face the casualty and secure your arms around his waist. Immediately place your foot between his feet and spread them apart (approximately 6 to 8 inches).

(*g*) Grasp the casualty's wrist and raise his arm high over your head.

(*h*) Bend down and pull the casualty's arm over and down on your shoulder, bringing his body across your shoulders. At the same time, pass your arm between his legs.

(*i*) Grasp the casualty's wrist with one hand, and place your other hand on your knee for support.

(*j*) Rise with the casualty positioned correctly. Your other hand is free for use.

Figure B-3. Fireman's carry (Illustrated A—J).

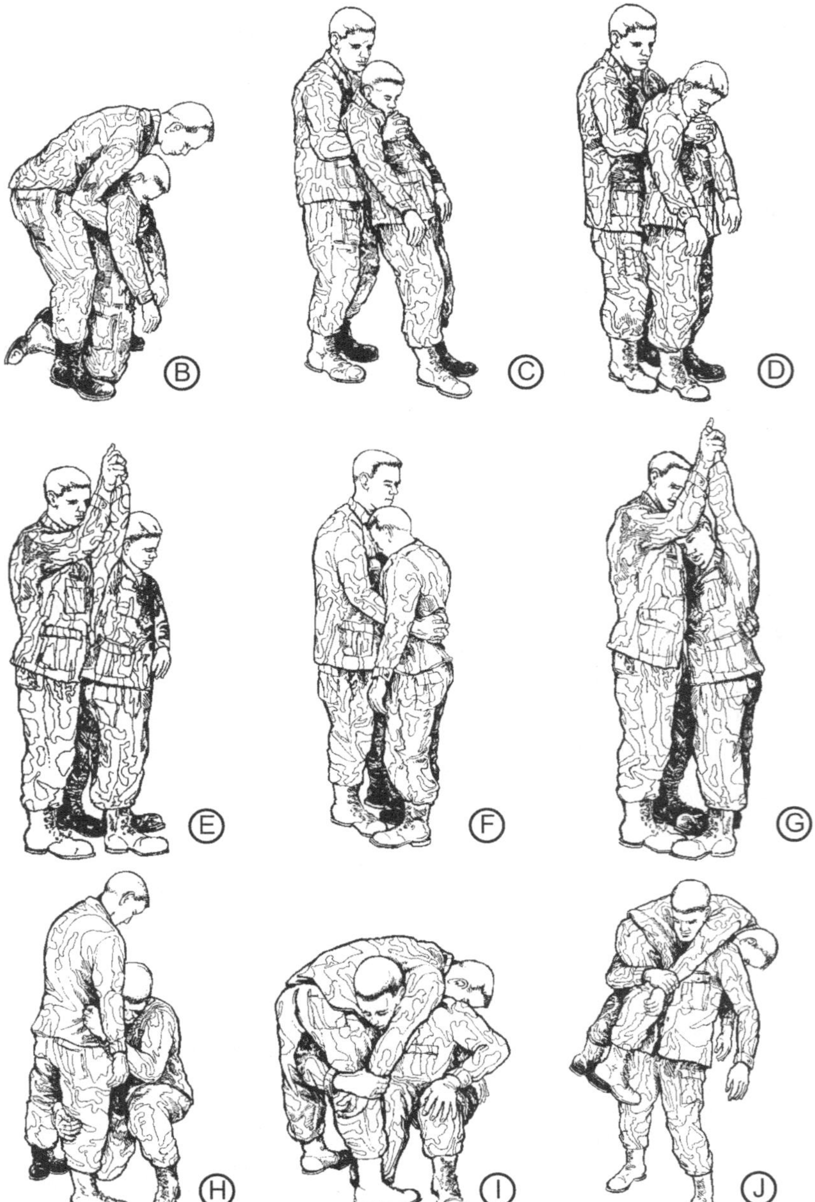

Figure B-3. Fireman's carry (Illustrated A—J) (Continued).

(2) The alternate method of the *fireman's carry* for raising a casualty from the ground is illustrated in Figure B-4; however, it should be used only when the bearer believes it to be safer for the casualty because of the location of his wounds. When the alternate method is used, care must be taken to prevent the casualty's head from snapping back and causing a neck injury. The steps for raising a casualty from the ground for the fireman's carry are also used in other one-man carries.

(*a*) Kneel on one knee at the casualty's head and face his feet. Extend your hands under his armpits, down his sides, and across his back.

(*b*) As you rise, lift the casualty to his knees. Then secure a lower hold and raise him to a standing position with his knees locked.

Figure B-4. Fireman's carry (alternate method) for lifting a casualty to a standing position (Illustrated A—B).

(3) In the *supporting carry* (Figure B-5), the casualty must be able to walk or at least hop on one leg, using the bearer as a crutch. This carry can be used to assist him as far as he is able to walk or hop.

(*a*) Raise the casualty from the ground to a standing position by using the fireman's carry.

(*b*) Grasp the casualty's wrist and draw his arm around your neck.

(*c*) Place your arm around his waist. The casualty is now able to walk or hop using you as a support.

Figure B-5. Supporting carry.

(4) The *arms carry* (Figure B-6) is useful in carrying a casualty for a short distance (up to 50 meters) and for placing him on a litter.

(*a*) Raise or lift the casualty from the ground to a standing position, as in the fireman's carry.

(*b*) Place one arm under the casualty's knees and your other arm around his back.

(*c*) Lift the casualty.

(*d*) Carry the casualty high to lessen fatigue.

Figure B-6. Arms carry.

(5) Only a conscious casualty can be transported by the *saddleback carry* (Figure B-7), because he must be able to hold onto the bearer's neck. To use this technique—

(*a*) Raise the casualty to an upright position, as in the fireman's carry.

(*b*) Support the casualty by placing an arm around his waist. Move to the casualty's side. Have the casualty put his arm around your neck and move in front of him with your back to support him.

(*c*) Have the casualty encircle his arms around your neck

(*d*) Stoop, raise him on your back and clasp your hands together beneath his thighs, if possible.

Figure B-7. Saddleback carry.

(6) In the *pack-strap carry* (Figure B-8), the casualty's weight rests high on the your back. This makes it easier for you to carry the casualty a moderate distance (50 to 300 meters). To eliminate the possibility of injury to the casualty's arms, you must hold his arms in a palms-down position.

(*a*) Lift the casualty from the ground to a standing position, as in the fireman's carry.

(*b*) Support the casualty with your arms around him and grasp his wrist closer to you.

(*c*) Place his arm over your head and across your shoulders.

(*d*) Move in front of him while still supporting his weight against your back.

(*e*) Grasp his other wrist and place this arm over your shoulder.

(*f*) Bend forward and raise or hoist the casualty as high on your back as possible so that his weight is resting on your back.

NOTE

Once the casualty is positioned on the bearer's back, the bearer remains as erect as possible to prevent straining or injuring his back.

Figure B-8. Pack-strap carry.

(7) The *pistol-belt carry* (Figure B-9) is the best one-man carry for a long distance (over 300 meters). The casualty is securely supported upon your shoulders by a belt. Both your hands and the casualty's (if conscious) are free for carrying a weapon or equipment, or climbing obstacles. With your hands free and the casualty secured in place, you are also able to creep through shrubs and under low-hanging branches.

(*a*) Link two pistol belts (or three, if necessary) together to form a sling. Place the sling under the casualty's thighs and lower back so that a loop extends from each side.

NOTE

If pistol belts are not available for use, other items such as a rifle sling, two cravat bandages, two litter straps, or any other suitable material, which will not cut or bind the casualty may be used.

(*b*) Lie face up between the casualty's outstretched legs. Thrust your arms through the loops and grasp his hands and trouser leg on his injured side.

(*c*) Roll toward the casualty's uninjured side onto your abdomen, bringing him onto your back. Adjust the sling, if necessary.

(*d*) Rise to a kneeling position. The belt will hold the casualty in place.

(*e*) Place one hand on your knee for support and rise to an upright position. (The casualty is supported on your shoulders.)

(*f*) Carry the casualty with your hands free for use in rifle firing, climbing, or surmounting obstacles.

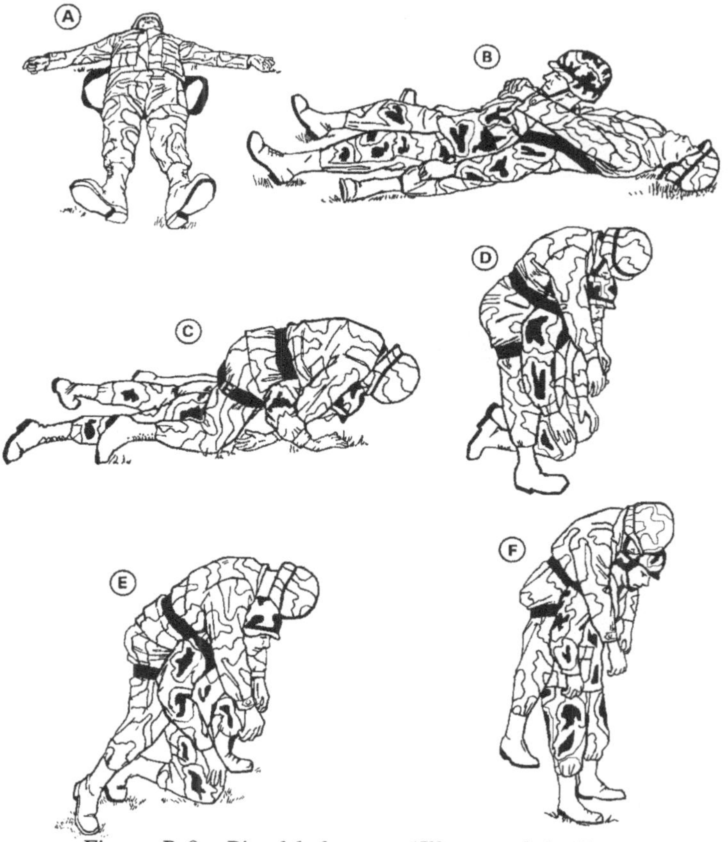

Figure B-9. Pistol-belt carry (Illustrated A—F).

(8) The *pistol-belt drag* (Figure B-10), as well as other drags, is generally used for short distances (up to 50 meters). This drag is useful in combat, since both the bearer and the casualty can remain closer to the ground than in any other drags.

(*a*) Extend two pistol belts or similar objects to their full length and join them together to make a continuous loop.

(*b*) Roll the casualty onto his back, as in the fireman's carry.

(*c*) Pass the loop over the casualty's head, and position it across his chest and under his armpits. Then cross the remaining portion of the loop, thus forming a figure eight. Keep tension on the belts so they do not come unhooked.

(*d*) Lie on your side facing the casualty.

(*e*) Slip the loop over your head and turn onto your abdomen. This enables you to drag the casualty as you crawl.

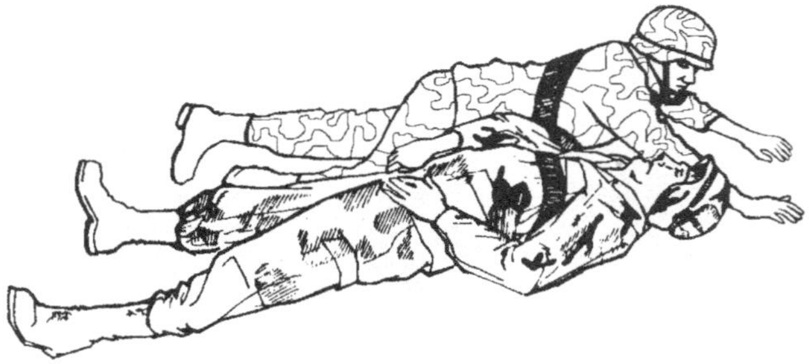

Figure B-10. Pistol-belt drag.

(9) The *neck drag* (Figure B-11) is useful in combat because the bearer can transport the casualty as he creeps behind a low wall or shrubbery, under a vehicle, or through a culvert. If the casualty is unconscious, his head must be protected from the ground. The neck drag cannot be used if the casualty has a broken arm.

NOTE

If the casualty is conscious, he may clasp his hands together around your neck.

(*a*) Tie the casualty's hands together at the wrists.

(*b*) Straddle the casualty in a kneeling face-to-face position.

(*c*) Loop the casualty's tied hands over and around your neck.

(*d*) Crawl forward dragging the casualty with you.

NOTE

If the casualty is unconscious, protect his head from the ground.

Figure B-11. Neck drag.

(10) The *cradle drop drag* (Figure B-12) is effective in moving a casualty up or down steps.

(*a*) Kneel at the casualty's head (with him lying on his back). Slide your hands, with palms up, under the casualty's shoulders and get a firm hold under his armpits.

(*b*) Rise (partially), supporting the casualty's head on one of your forearms. (You may bring your elbows together and let the casualty's head rest on both of your forearms.)

(*c*) Rise and drag the casualty backward. (The casualty is in a semisitting position.)

(*d*) Back down the steps, supporting the casualty's head and body and letting his hips and legs drop from step to step.

NOTE

If the casualty needs to be moved up the steps, you should back up the steps, using the same procedure.

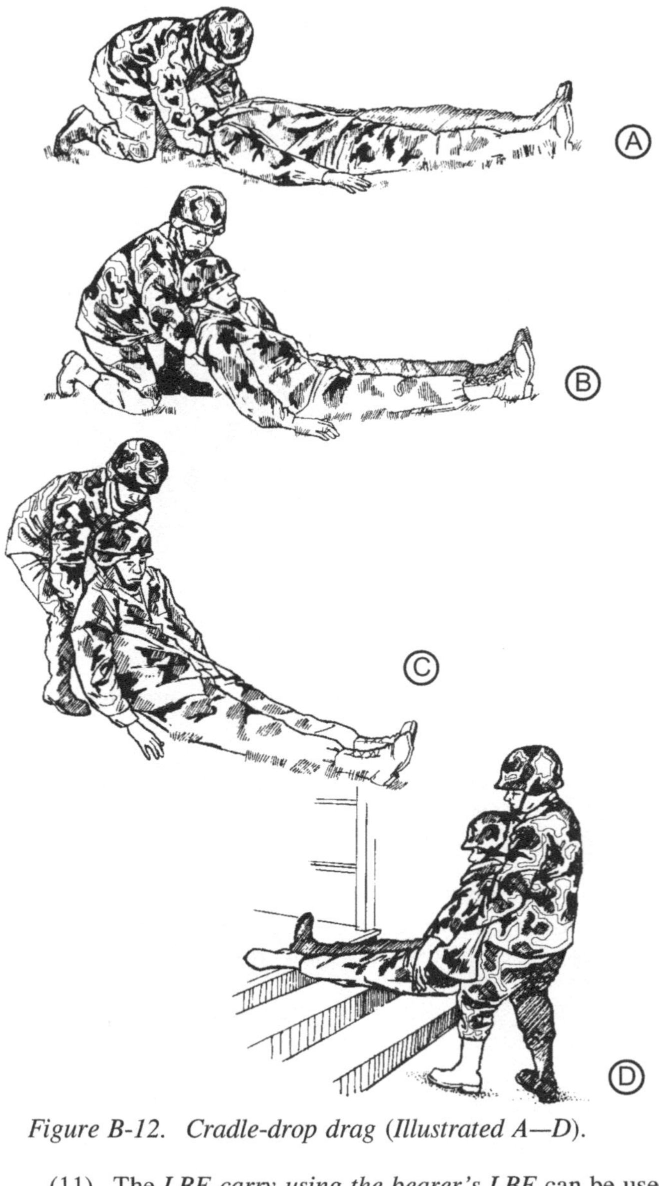

Figure B-12. Cradle-drop drag (Illustrated A—D).

(11) The *LBE carry using the bearer's LBE* can be used with a conscious casualty (Figure B-13).

(*a*) Loosen all suspenders on your LBE.

(*b*) Have the casualty place one leg into the loop formed by your suspenders and pistol belt.

(*c*) Squat in front of the standing casualty. Have him place his other leg into the loop, also.

(*d*) Have the casualty place his arms over your shoulders, lean forward onto your back, and lock his hands together.

(*e*) Stand up and lean forward into a comfortable position.

(*f*) Continue the mission.

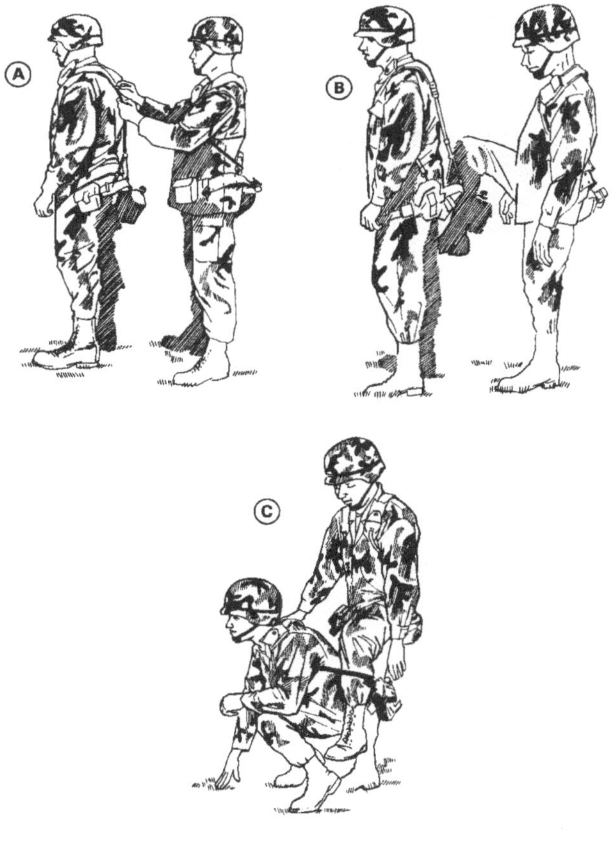

Figure B-13. Load bearing equipment carry using bearer's LBE (conscious casualty) (Illustrated A—F).

Figure B-13. Load bearing equipment carry using bearer's LBE (conscious casualty) (Illustrated A—F) (Continued).

(12) The *LBE carry using the bearer's LBE* can be used with an unconscious casualty or one who cannot stand (Figure B-14).

(*a*) Position the casualty on the flat of his back.

(*b*) Remove your LBE and loosen all suspender straps.

(*c*) Lift the casualty's leg and place it through the loop formed by your suspenders and pistol belt. Then place the other leg through the same loop. The LBE is moved up until the pistol belt is behind the casualty's thighs.

(*d*) Lay between the casualty's legs; work your arms through the LBE suspenders.

(*e*) Grasp the casualty's hand (on the injured side), and roll the casualty (on his uninjured side) onto your back.

(*f*) Rise to one knee and then push into a standing position.

(*g*) Bring the casualty's arms over your shoulders. Grasp his hands and secure them if the casualty is unconscious. If the casualty is conscious, have him lock his hands in front if he is able to do so.

(*h*) Lean forward into a comfortable position and continue the mission.

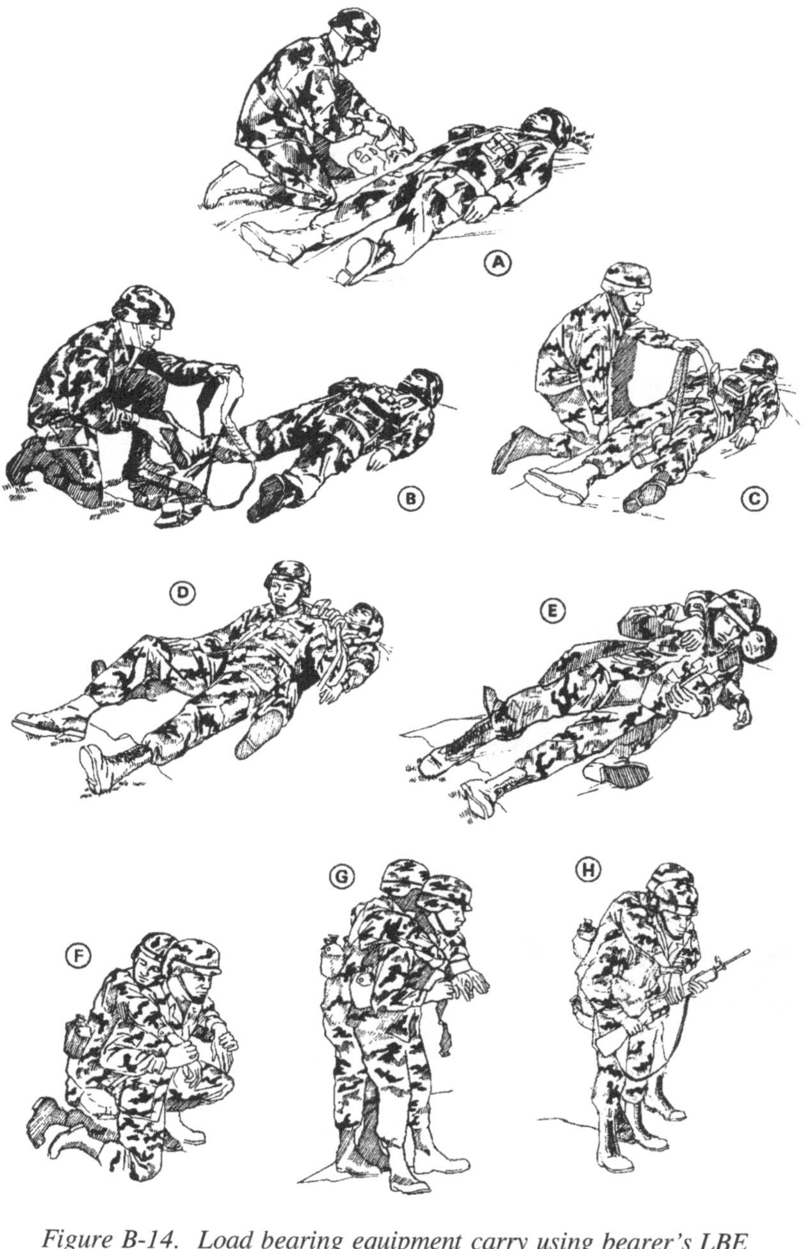

Figure B-14. Load bearing equipment carry using bearer's LBE (unconscious casualty or one that cannot stand) (Illustrated A—H).

(13) The *LBE carry using the casualty's LBE* (Figure B-15) can be used with a conscious or unconscious casualty.

(*a*) Position the casualty on his back with his LBE on.

(*b*) Loosen the casualty's two front suspenders.

(*c*) Position yourself between the casualty's legs, and slip your arms into the casualty's two front suspenders (up to your shoulders).

(*d*) Work his arms out of his LBE suspenders.

(*e*) Grasp the casualty's hand (on the injured side), and roll him (on his uninjured side) onto your back.

(*f*) Rise to one knee, then into a standing position.

(*g*) Grasp the casualty's hands and secure them, if the casualty is unconscious. Have the casualty lock his hands in front of you, if he is conscious.

(*h*) Lean forward into a comfortable position and continue the mission.

Figure B-15. Load bearing equipment carry using casualty's LBE (Illustrated A—G).

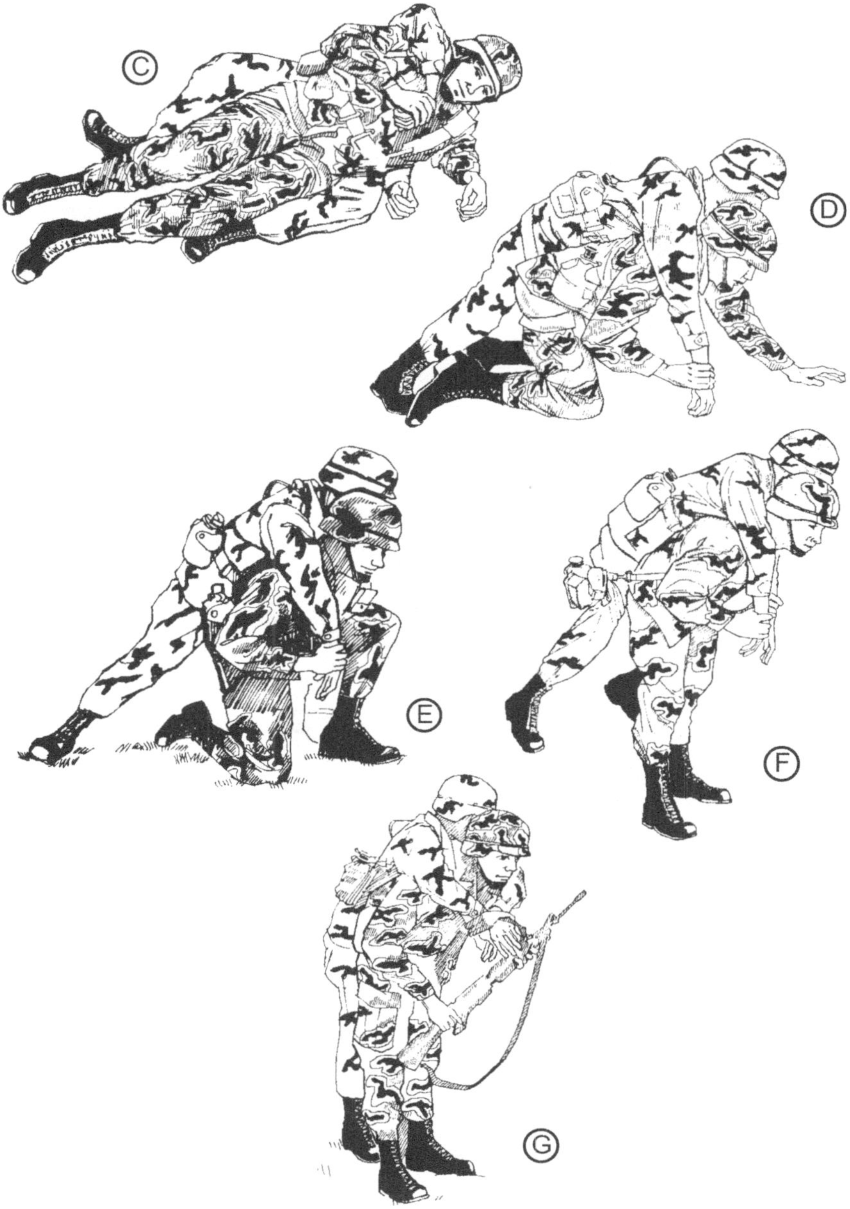

Figure B-15. Load bearing equipment carry using casualty's LBE (Illustrated A—G) (Continued).

b. *Two-man Carries*. These carries should be used whenever possible. They provide more casualty comfort, are less likely to aggravate injuries, and are less tiring for the bearers. Five different two-man carries can be used.

(1) The *two-man support carry* (Figure B-16) can be used in transporting either conscious or unconscious casualties. If the casualty is taller than the bearers, it may be necessary for the bearers to lift the casualty's legs and let them rest on their forearms. The bearers—

(*a*) Help the casualty to his feet and support him with their arms around his waist.

(*b*) Grasp the casualty's wrists and draw his arms around their necks.

Figure B-16. Two-man supporting carry.

(2) The *two-man arms carry* (Figure B-17) is useful in carrying a casualty for a moderate distance (50 to 300 meters) and placing him on a litter. To lessen fatigue, the bearers should carry the casualty high and as close to their chests as possible. In extreme emergencies when there is no time to obtain a spine board, this carry is the safest one for transporting a casualty with a back injury. If possible, two additional bearers should be used to keep the casualty's head and legs in alignment with his body. The bearers—

(*a*) Kneel at one side of the casualty; then they place their arms beneath the casualty's back, waist, hips, and knees.

(*b*) Lift the casualty while rising to their knees.

(*c*) Turn the casualty toward their chests, while rising to a standing position. Carry the casualty high to lessen fatigue.

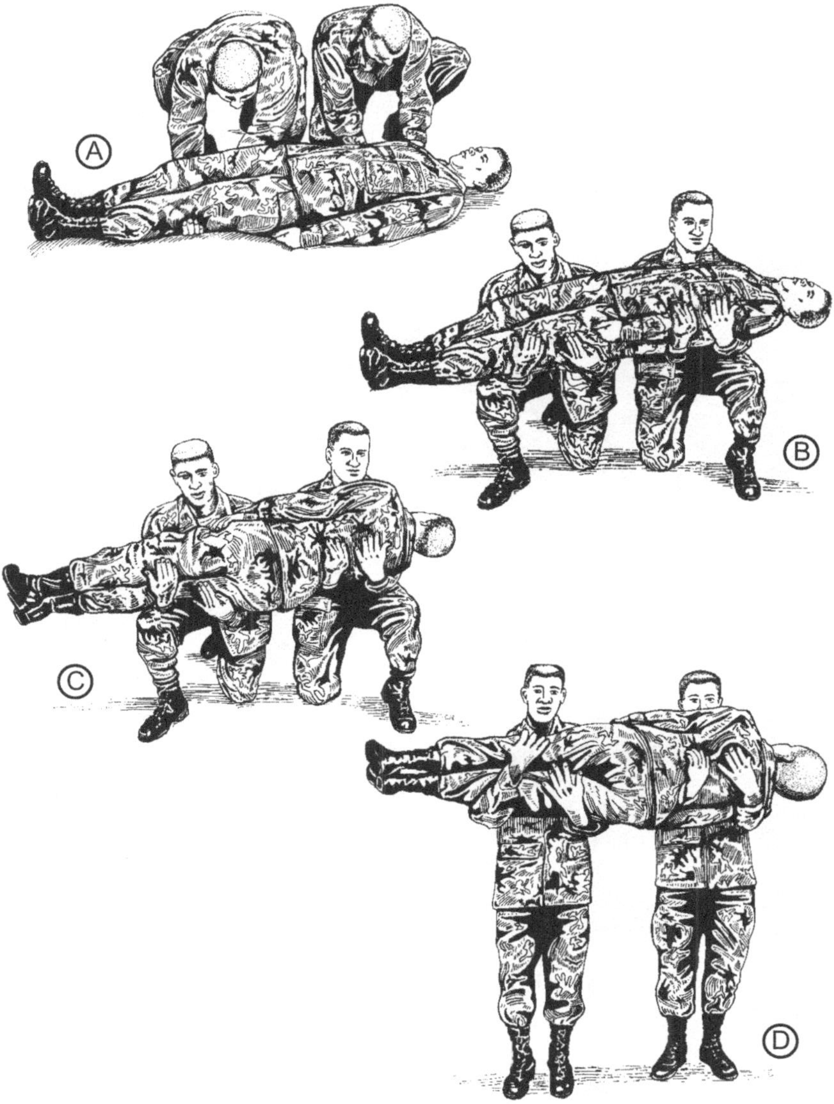

Figure B-17. Two-man arms carry (Illustrated A—D).

(3) The *two-man fore-and aft-carry* (Figure B-18) is a useful *two-man carry* for transporting a casualty for a long distance (over 300 meters). The taller of the two bearers should position himself at the casualty's head. By altering this carry so that both bearers face the casualty, it is useful for placing a casualty on a litter.

(*a*) The shorter bearer spreads the casualty's legs and kneels between them with his back to the casualty. He positions his hands behind the casualty's knees. The other bearer kneels at the casualty's head, slides his hands under the arms, across the chest, and locks his hands together.

(*b*) The two bearers rise together, lifting the casualty.

Figure B-18. Two-man fore-and-aft carry (Illustrated A—B).

(4) Only a conscious casualty can be transported with the *four-hand seat carry* (Figure B-19) because he must help support himself by placing his arms around the bearers' shoulders. This carry is especially useful in transporting a casualty with a head or foot injury for a moderate distance (50 to 300 meters). It is also useful for placing a casualty on a litter.

(*a*) Each bearer grasps one of his wrists and one of the other bearer's wrists, thus forming a packsaddle.

(*b*) The two bearers lower themselves sufficiently for the casualty to sit on the packsaddle; then, they have the casualty place his arms around their shoulders for support. The bearers then rise to an upright position.

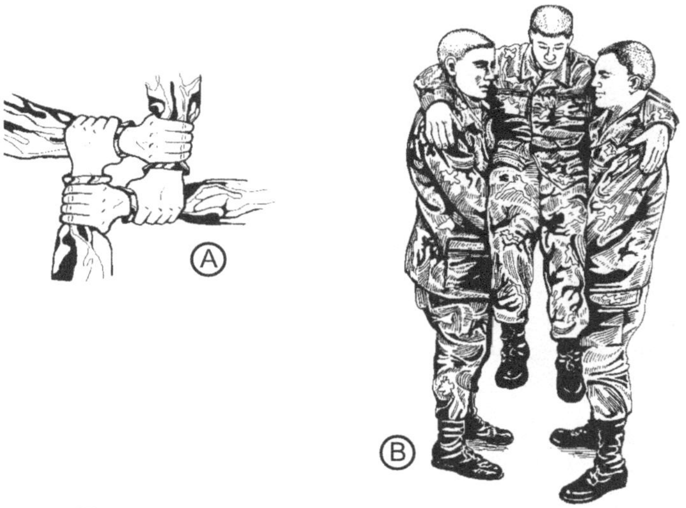

Figure B-19. Four-hand seat carry (Illustrated A—B).

(5) The *two-hand seat carry* (Figure B-20) is used when carrying a casualty for a short distance or for placing him on a litter. With the casualty lying on his back, a bearer kneels on each side of the casualty at his hips. Each bearer passes his arms under the casualty's thighs and back, and grasps the other bearer's wrists. The bearers rise lifting the casualty.

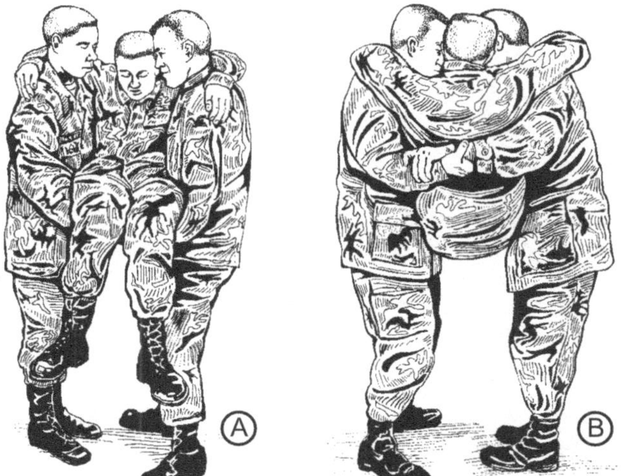

Figure B-20. Two-hand seat carry (Illustrated A—B).

B-9. Improvised Litters

Two men can support or carry a casualty without equipment for only short distances. By using available materials to improvise equipment, the casualty can be transported greater distances by two or more rescuers.

a. There are times when a casualty may have to be moved and a standard litter is not available. The distance may be too great for manual carries or the casualty may have an injury (such as a fractured neck, back, hip, or thigh) that would be aggravated by manual transportation. In these situations, litters can be improvised from materials at hand. Improvised litters must be as well constructed as possible to avoid risk of dropping or further injuring the casualty. Improvised litters are emergency measures and must be replaced by standard litters at the first opportunity.

b. Many different types of litters can be improvised, depending upon the materials available. A satisfactory litter can be made by securing poles inside such items as a blanket, poncho, shelter half, tarpaulin, mattress cover, jacket, shirt, or bed ticks, bags, and sacks (Figure B-18). Poles can be improvised from strong branches, tent supports, skis, lengths of pipe or other objects. If objects for improvising poles are not available, a blanket, poncho, or similar item can be rolled from both sides toward the center so the rolls can be gripped for carrying a patient. Most flat-surface objects of suitable size can be used as litters. Such objects include doors, boards, window shutters, benches, ladders, cots, and chairs. If possible, these objects should be padded for the casualty's comfort.

(1) To improvise a litter using a blanket and poles (Figure B-21), the following steps should be used.

Figure B-21. Litter made with blanket and poles.

(*a*) Open the blanket and lay one pole lengthwise across the center; then fold the blanket over the pole.

(*b*) Place the second pole across the center of the folded blanket.

(*c*) Fold the free edges of the blanket over the second pole and across the first pole.

(2) To improvise a litter using shirts or jackets (Figure B-22), button the shirt or jacket and turn it inside out, leaving the sleeves inside, (more than one shirt or jacket may be required), then pass the pole through the sleeves.

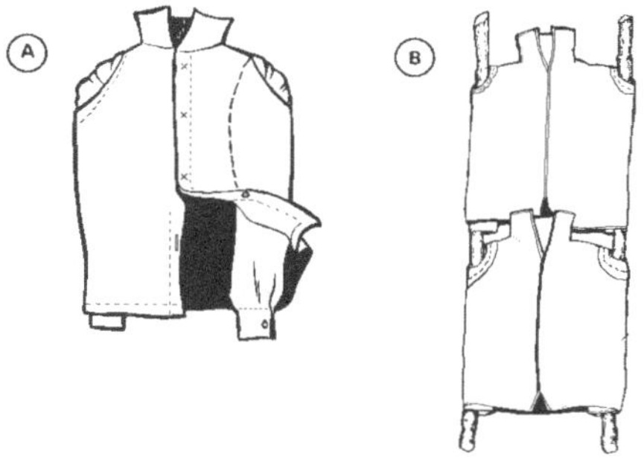

Figure B-22. Litter improvised from jackets and poles (Illustrated A—B).

(3) To improvise a litter from bed sacks and poles (Figure B-23), rip open the corners of bed ticks, bags, or sacks; then pass the poles through them.

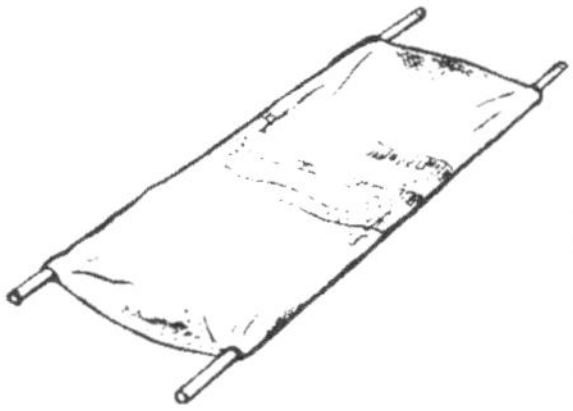

Figure B-23. Litter improvised from bed sacks and poles.

(4) If no poles are available, roll a blanket, shelter half, tarpaulin, or similar item from both sides toward the center (Figure B-24). Grip the rolls to carry the casualty.

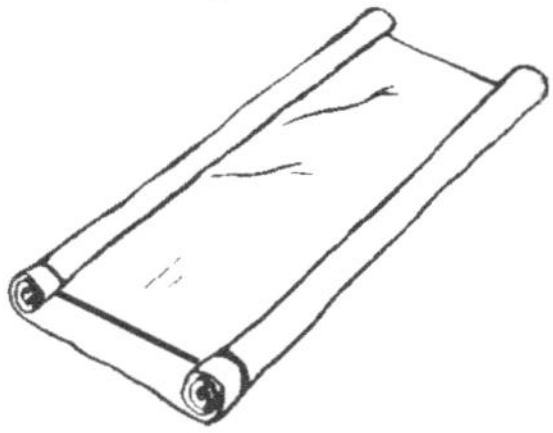

Figure B-24. Rolled blanket used as a litter.

c. Any of the appropriate carries may be used to place a casualty on a litter. These carries are:

- The one-man arms carry (Figure B-6).
- The two-man arms carry (Figure B-17).
- The two-man fore-and-aft carry (Figure B-18).
- The two-hand seat carry (Figure B-20).
- The four-hand seat carry (Figure B-19).

WARNING

Unless there is an immediate life-threatening situation (such as fire, explosion), DO NOT move a casualty with a suspected back or neck injury. Seek medical personnel for guidance on how to transport.

d. Either two or four service members (head/foot) may be used to lift a litter. To lift the litter, follow the procedure below.

(1) Raise the litter at the same time as the other carriers/ bearers.

(2) Keep the casualty as level as possible.

NOTE

Use caution when transporting on a sloping incline/hill.

GLOSSARY

ACRONYMS, ABBREVIATIONS, AND DEFINITIONS

AC hydrogen cyanide
AFMAN Air Force Manual
AOC area of concentration
AR Army regulation
ATM advanced trauma management
ATNAA Antidote Treatment, Nerve Agent, Autoinjector
attn attention

BDO battle dress overgarment
BDU battle dress uniform
BZ anticholinergic drugs

C Celsius
CANA Convulsant Antidote for Nerve Agent
CASEVAC casualty evacuation
cc cubic centimeter
CG phosgene
CHS combat health support
CK cyanogen chloride
Cl chlorine
CLS Combat Lifesaver
CNS central nervous system
CO_2 carbon dioxide
COSR combat and operational stress reactions
CSR combat stress reaction
CTA common table of allowance
CX phosgene oxime

DA Department of the Army
DD Department of Defense
DM diphenylaminochloroarsine (adamsite)
DNBI disease and nonbattle injury
DOD Department of Defense
DP diphosgene
DS direct support

EMT emergency medical treatment

F Fahrenheit
FM field manual

H mustard

HD mustard
HM Hospital Corpsman
HN nitrogen mustard
HSS health service support

IPE individual protective equipment
IV intravenous

JSLIST Joint Services Light Weight Integrated Suit Technology

L lewisite
lasers laser means Light Amplification by Stimulated Emission of Radiation and sources include range finders, weapons/guidance, communication systems, and weapons simulations such as MILES [Multiple Integrated Laser Engagement System].
LBE load bearing equipment
LX lewisite and mustard

MCRP Marine Corps Reference Publication
MILES Multiple Integrated Laser Engagement System
ml milliliter
MOPP mission-oriented protective posture
MOS military occupational specialty
MTF medical treatment facility

NAPP Nerve Agent Pyridostigmine Pretreatment
NATO North Atlantic Treaty Organization
NBC nuclear, biological, and chemical
NCO noncommissioned officer
NTRP Navy Tactical Reference Publication

occlusive dressing air tight transparent dressing used to seal and cover wounds
oz ounce

PAM pamphlet
PS chloropicrin
PTSD post-traumatic stress disorder

QSTAG Quadripartite Standardization Agreement

SOP standing operating procedure
STANAG standardization agreement
STP soldier training publication

2 PAM Cl pralidoxime chloride
TB MED technical bulletin medical
TM technical manual
TSOP tactical standing operating procedure

US United States

WP white phosphorus

REFERENCES

DOCUMENTS NEEDED

These documents must be available to the intended users of this publication.

NATO STANAGs

These agreements are available on request using DD Form 1425 from Standardization Document Order Desk, 700 Robin Avenue, Building 4, Section D, Philadelphia, Pennsylvania 19111-5094.

2122. *Medical Training in First Aid, Basic Hygiene and Emergency Care*. 10 December 1975.
2126. *First Aid Kits and Emergency Medical Care Kits*. 27 September 1983.
2358. *First Aid and Hygiene Training in NBC Operations*. 3 March 1989.
2871. *First Aid Material for Chemical Injuries*. 10 March 1986.

ABCA QSTAGs

These agreements are available on request using DD Form 1425 from Standardization Document Order Desk, 700 Robin Avenue, Building 4, Section D, Philadelphia, Pennsylvania 19111-5094.

535. *Medical Training in First Aid, Basic Hygiene and Emergency Care*. 12 November 1979.

Joint and Multiservice Publications

FM 21-10. *Field Hygiene and Sanitation*. MCRP 4-11.1D. 21 June 2000.

Army Publications

AR 350-41. *Training In Units*. 19 March 1993.
DA PAM 350-59. *Army Correspondence Course Program Catalog*. 26 October 2001.
FM 3-4. *NBC Protection*. FMFM 11-9. 29 May 1992 (Reprinted with basic including Change 1, 28 October 1992; Change 2, 26 February 1996.)
FM 3-5. *NBC Decontamination*. MCWP 3-37.3. 28 July 2000. (Change 1, 31 January 2002.)

FM 3-100. *Chemical Operations Principles and Fundamentals.* MCWP 3-3.7.1. 8 May 1996.
FM 4-02.33 (8-33). *Control of Communicable Diseases Manual* (17th Edition). 3 January 2000.
FM 8-10-6 (4-02.2). *Medical Evacuation in a Theater of Operations—Tactics, Techniques, and Procedures.* 14 April 2000.
FM 4-02.7 (8-10-7). *Health Service Support in a Nuclear, Biological, and Chemical Environment.* 1 October 2002.
FM 8-284 (4-02.284). *Treatment of Biological Warfare Agent Casualties.* NAVMED P-5042; AFMAN (I) 44-156; MCRP 4-11.1C. 17 July 2000. (Change 1, 8 July 2002.)
FM 8-285 (4-02.285). *Treatment of Chemical Agent Casualties and Conventional Military Chemical Injuries.* NAVMED P-5041; AFJMAN 44-149; FMFM 11-11. 22 December 1995.
FM 22-51 (4-02.22). *Leaders' Manual for Combat Stress Control.* 29 September 1994.
CTA 8-100. *Army Medical Department Expendable/Durable Items.* 31 August 1994.
CTA 50-900. *Clothing and Individual Equipment.* 1 September 1994.
STP 21-1-SMCT. *Soldier's Manual of Common Tasks Skill Level I.* 1 October 2001.

Department of Defense Forms

DD Form 1425. *Specifications and Standards Requisition.* March 1986.

READINGS RECOMMENDED

These readings contain relevant supplemental information.

Joint and Multiservice Publications

FM 8-9 (4-02.11). *NATO Handbook on the Medical Aspects of NBC Defensive Operations AMEDP-6 (B), Part I—Nuclear, Part II—Biological, Part III—Chemical.* NAVMED P-5059; AFJMAN 44-151V1V2V3. 1 February 1996.
TB MED 81. *Cold Injury.* NAVMED P-5052-29; AFP 161-11. 30 September 1976.
TB MED 507. *Occupational and Environmental Health Prevention, Treatment and Control of Heat Injury.* NAVMED P-5052-5; AFP 160-1. 25 July 1980.

Army Publications

AR 310-25. *Dictionary of United States Army Terms (Short Title: AD)*. 15 October 1983. (Reprinted with basic including Change 1, 21 May 1986.)

AR 310-50. *Authorized Abbreviations and Brevity Codes*. 15 November 1985.

TM 3-4230-216-10. *Operator's Manual for Decontaminating Kit, Skin: M258A1 (NSN 4230-01-101-3984) and Training Aid, Skin Decontaminating: M58A1 (6910-01-101-1768)*. 17 May 1985. (Change 1, 22 January 1997.)

INDEX

References are to paragraph numbers except where specified otherwise.

Made in United States
Cleveland, OH
05 November 2026

47672475R00317